Laboratory Manual
to accompany
Biology

Eleventh Edition

Sylvia S. Mader

Connect
Learn
Succeed™

Vice President, Editor-in-Chief: *Marty Lange*
Vice President, EDP: *Kimberly Meriweather David*
Senior Director of Development: *Kristine Tibbetts*
Publisher: *Michael S. Hackett*
Sponsoring Editor: *Eric Weber*
Senior Developmental Editor: *Rose M. Koos*
Director of Digital Content Development: *Tod Duncan, Ph.D.*
Senior Marketing Manager: *Tamara Maury*
Project Manager: *Mary Jane Lampe*
Senior Buyer: *Sandy Ludovissy*
Senior Designer: *Laurie B. Janssen*
Cover Designer: *Elise Lansdon*
Cover Image: *Corbis RF/Alamy*
Senior Photo Research Coordinator: *Lori Hancock*
Photo Research: *Evelyn Jo Johnson*
Compositor: *Lachina Publishing Services*
Typeface: *11/13 Slimbach*
Printer: *Quad/Graphics*

www.mhhe.com

Contents

Preface

To the Instructor

The laboratory exercises in this manual are coordinated with *Biology*, a general biology text that covers the entire field of biology from an evolutionary perspective.

Although each laboratory is referenced to the appropriate chapter(s) in *Biology*, this manual may also be used in coordination with other general biology texts. In addition, this laboratory manual can be adapted to a variety of course orientations and designs. There are a sufficient number of laboratories and exercises within each lab to tailor the laboratory experience as desired. Then, too, many exercises may be performed as demonstrations rather than as student activities, thereby shortening the time required to cover a particular concept.

The Exercises

All exercises have been tested for student interest, preparation time, time of completion, and feasibility. The following features are particularly appreciated by adopters:

Integrated opening: Each laboratory begins with a list of Learning Outcomes organized according to the major sections of the laboratory. The major sections of the laboratory are numbered on the opening page and in the laboratory text material. This organization will help students better understand the goals of each laboratory session.

Self-contained content: Each laboratory contains all the background information necessary to understand the concepts being studied and to answer the questions asked. This feature will reduce student frustration and increase learning.

Scientific process: All laboratories stress the scientific process, and many opportunities are given for students to gain an appreciation of the scientific method. The first laboratory of this edition explicitly explains the steps of the scientific method and gives students an opportunity to use them. I particularly recommend this laboratory because it utilizes the pillbug, a living subject.

Student activities: A color bar is used to designate each student activity. Some student exercises are Observations and some are Experimental Procedures. An icon appears whenever a procedure requires a period of time before results can be viewed. Sequentially numbered steps guide students as they perform each activity.

Live materials: Although students work with living material during some part of almost all laboratories, the exercises are designed to be completed within one laboratory session. This facilitates the use of the manual in multiple-section courses.

Virtual labs: New to this edition, virtual labs currently available on the *Biology* website **www.mhhe.com/maderbiology11** are announced and described whenever the announcement seems appropriate. Instructors can use the virtual labs as separate assignments or integrate them into the laboratory experience. In four instances (see Laboratories 5, 13, 14, and 19) I have revised the instructions for the virtual lab to create a lab that stresses the scientific method. These instructions have become part of the laboratory itself. Instructors are still free to use the virtual labs as they see fit.

Laboratory safety: Laboratory safety is of prime importance, and the listing on page viii will assist instructors in making the laboratory experience a safe one.

Improvements in This Edition

All laboratories have been revised to:

1. Improve the Introduction so that it becomes an integral part of the laboratory experience. For example, in Laboratory 1, pillbug anatomy is more clearly described in the Introduction.
2. Improve the Laboratory Review so that the questions better reflect the Learning Outcomes.

The following laboratories have been extensively revised:

Laboratory 1: Scientific Method
   The laboratory was rewritten to improve the flow and better emphasize the scientific method.

Laboratory 7: Cellular Respiration
   The laboratory now begins with an Introduction that includes the function of the mitochondrion, and better stresses the role of the mitochondrion in cellular respiration.

Laboratory 9: Mendelian Genetics
   The laboratory was rewritten to streamline the experiments and improve their clarity.

Laboratory 13: Natural Selection
   The laboratory was rewritten to stress gene pool changes during the process of natural selection. The difference between natural selection and genetic drift is clarified. A virtual lab experience illustrates the possible role of mutations in allowing natural selection to occur.

Laboratory 14: Bacteria and Protists
   The laboratory is improved by the addition of a virtual lab experience that gives the classification of bacteria an evolutionary emphasis and asks students to complete an evolutionary tree based on the data collected.

Laboratory 22: Introduction to Invertebrates
   The laboratory was rewritten to better stress a hands on examination of living hydras and planarians, vinegar eels, and rotifers.

Laboratory 24: The Vertebrates
   The portion of the laboratory dealing with vertebrate anatomy was rewritten to remove discrepancies and to better allow students to come to conclusions about the vertebrate comparative anatomy.

Laboratory 28: Chemical Aspects of Digestion
   The laboratory now provides a better correlation between the experiments and the digestive tract of humans.

Laboratory 29: Homeostasis
   The laboratory was rewritten to introduce more hands-on activities, including blood pressure and lung volume measurement and comparative urine analysis to diagnose particular illnesses.

Customized Editions

The 34 laboratories in this manual are now available as individual "lab separates," so instructors can custom-tailor the manual to their particular course needs.

Laboratory Resource Guide

The *Laboratory Resource Guide*, an essential aid for instructors and laboratory assistants, free to adopters of the *Biology Laboratory Manual*, is online at **www.mhhe.com/maderbiology11**. The answers to the Laboratory Review questions are in the Resource Guide.

To the Student

Special care has been taken in preparing the *Biology Laboratory Manual* to enable you to **enjoy** the laboratory experience as you **learn** from it. The instructions and discussions are written clearly so that you can understand the material while working through it. Student aids are designed to help you focus on important aspects of each exercise.

Student Learning Aids

Student learning aids are carefully integrated throughout this manual. The Learning Outcomes set the goals of each laboratory session and help you review the material for a laboratory practical or any other kind of exam. The major sections of each laboratory are numbered, and the Learning Outcomes are grouped according to these topics. This system allows you to study the chapter in terms of the outcomes presented.

The Introduction reviews much of the necessary background information required for comprehending upcoming experiments. Color bars bring attention to exercises that require your active participation by highlighting Observations and Experimental Procedures, and an icon indicates a timed experiment. Throughout, space is provided for recording answers to questions and the results of investigations and experiments. Each laboratory ends with a set of review questions covering the day's work.

Appendices at the end of the book provide useful information on preparing a laboratory report, the metric system, and a Tree of Life that shows how organisms are related through the process of evolution. When needed, you can find practical examination answer sheets on the *Biology* website.

Laboratory Preparation

Read each exercise before coming to the laboratory. ***Study*** the introductory material and the procedures. If necessary, to obtain a better understanding, read the corresponding chapter in your text. If your text is *Biology*, by Sylvia S. Mader and Michael Windelspecht, see the "text chapter reference" column in the table of contents at the beginning of the *Laboratory Manual*.

Explanations and Conclusions

Throughout a laboratory, you are often asked to formulate explanations or conclusions. To do so, you will need to synthesize information from a variety of sources, including the following:

1. Your experimental results and/or the results of other groups in the class. If your data are different from those of other groups in your class, do not erase your answer; add the other groups' answers in parentheses.
2. Your knowledge of underlying principles. Obtain this information from the laboratory Introduction or the appropriate section of the laboratory and from the corresponding chapter of your text.
3. Your understanding of how the experiment was conducted and/or the materials used. *Note:* Ingredients can be contaminated or procedures incorrectly followed, resulting in reactions that seem inappropriate. If this occurs, consult with other students and your instructor to see if you should repeat the experiment.

In the end, be sure you are truly writing an explanation or conclusion and not just restating the observations made.

Color Bar and Icon

Throughout each laboratory, a color bar designates either an Observation or an Experimental Procedure.

Observation: An activity in which models, slides, and preserved or live organisms are observed to achieve a learning outcome.

Experimental Procedure: An activity in which a series of steps uses laboratory equipment to gather data and come to a conclusion.

⏰ Time: An icon is used to designate when time is needed for an Experimental Procedure. Allow the designated amount of time for this activity. Start these activities at the beginning of the laboratory, proceed to other activities, and return to these when the designated time is up.

Laboratory Review

Each laboratory ends with a number of thought questions that will help you determine whether you have accomplished the learning outcomes for the laboratory.

Student Feedback

If you have any suggestions for how this laboratory manual could be improved, you can send your comments to:

The McGraw-Hill Companies
Product Development—General Biology
501 Bell St.
Dubuque, Iowa 52001

Acknowledgments

We gratefully acknowledge the following for their assistance in the development of this lab manual:

Robert E. Bailey
Central Michigan University

Isaac Barjis
New York City College of Technology

Kathryn Blair
Horry Georgetown Technical College

Angela Bruni
Mississippi Gulf Coast Community College

Kathryn Craven
Armstrong Atlantic State University

James Crowder
Brookdale Community College

Jill Crowder
Milwaukee Area Technical College

Tammy Dennis
Bishop State Community College

Gregory S. Farley
Chesapeake College

Eugene J. Fenster
MCC Longview

Christina B. Fieber
Horry Georgetown Technical College

Julie Fischer
Wallace Community College

Raul Galvan
South Texas College

Carla Gardner
Coastal Carolina Community College

Jared Gilmore
San Jacinto College Central

Melanie Glasscock
Wallace State Community College

Shashuna Gray
Germanna Community College

Chris Haynes
Shelton State Community College

Jane J. Henry
Baton Rouge Community College

Randy Howell
Southern Union State Community College

Allen Daniel Johnson
Wake Forest University

KJ Lodrigue, Jr.
Baton Rouge Community College

Sheryl Love
Temple University

Margaret Major
Georgia Perimeter College

Scott Matthews
Northern Virginia Community College

Christopher Milne
Pellissippi State Community College

Linda Smith-Staton
Pellissippi State Community College

Barbara Stegenga
University of North Carolina at Chapel Hill

Patricia Steinke
San Jacinto College Central

Laboratory Safety

The following is a list of practices required for safety purposes in the biology laboratory and in outdoor activities. Following rules of lab safety and using common sense throughout the course will enhance your learning experience by increasing your confidence in your ability to safely use chemicals and equipment. Pay particular attention to oral and written safety instructions given by the instructor. If you do not understand a procedure, ask the instructor, rather than a fellow student, for clarification. Be aware of your school's policy regarding accident liability and any medical care needed as a result of a laboratory or outdoor accident.

The following rules of laboratory safety should become a habit:

1. To prevent possible hazards to eyes or contact lenses, wear safety glasses or goggles during exercises in which glassware and solutions are heated, or when dangerous fumes may be present.
2. Assume that all reagents are poisonous and act accordingly. Read the labels on chemical bottles for safety precautions and know the nature of the chemical you are using. If chemicals come into contact with skin, wash immediately with water.
3. **DO NOT**
   a. ingest any reagents.
   b. eat, drink, or smoke in the laboratory. Toxic material may be present, and some chemicals are flammable.
   c. carry reagent bottles around the room.
   d. pipette anything by mouth.
   e. put chemicals in the sink or trash unless instructed to do so.
   f. pour chemicals back into containers unless instructed to do so.
   g. operate any equipment until you are instructed in its use.
4. **DO**
   a. note the location of emergency equipment such as a first aid kit, eyewash bottle, fire extinguisher, switch for ceiling showers, fire blanket(s), sand bucket, and telephone (911).
   b. become familiar with the experiments you will be doing before coming to the laboratory. This will increase your understanding, enjoyment, and safety during exercises. Confusion is dangerous. Completely follow the procedure set forth by the instructor.
   c. keep your work area neat, clean, and organized. Before beginning, remove everything from your work area except the lab manual, pen, and equipment used for the experiment. Wash hands and desk area, including desk top and edge, before and after each experiment. Use clean glassware at the beginning of each exercise, and wash glassware at the end of each exercise or before leaving the laboratory.
   d. wear clothing that, if damaged, would not be a serious loss, or use aprons or laboratory coats, since chemicals may damage fabrics.
   e. wear shoes as protection against broken glass or spillage that may not have been adequately cleaned up.
   f. handle hot glassware with a test tube clamp or tongs. Use caution when using heat, especially when heating chemicals. Do not leave a flame unattended; do not light a Bunsen burner near a gas tank or cylinder; do not move a lit Bunsen burner; do keep long hair and loose clothing well away from the flame; do make certain gas jets are off when the Bunsen burner is not in use. Use proper ventilation and hoods when instructed.
   g. read chemical bottle labels; be aware of the hazards of all chemicals used. Know the safety precautions for each.
   h. stopper all reagent bottles when not in use. Immediately wash reagents off yourself and your clothing if they spill on you, and immediately inform the instructor. If you accidentally get any reagent in your mouth, rinse the mouth thoroughly, and immediately inform your instructor.
   i. use extra care and wear disposable gloves when working with glass tubing and when using dissection equipment (scalpels, knives, or razor blades), whether cutting or assisting.
   j. administer first aid immediately to clean, sterilize, and cover any scrapes, cuts, and burns where the skin is broken and/or where there may be bleeding. Wear bandages over open skin wounds.
   k. report all accidents to the instructor immediately, and ask your instructor for assistance in cleaning up broken glassware and spills.
   l. report to the instructor any condition that appears unsafe or hazardous.
   m. use caution during any outdoor activities. Watch for snakes, poisonous insects or spiders, stinging insects, poison oak, poison ivy, and so on. Be careful near water.

I understand the safety rules as presented. I agree to follow them and all other instructions given by the instructor.

Name _____   Date _____

Laboratory Class and Time: _____

LABORATORY

1
Scientific Method

Learning Outcomes

Introduction
- In general, describe pillbug external anatomy and lifestyle. 1–2

1.1 Using the Scientific Method
- Outline the steps of the scientific method. 2–3
- Distinguish among observations, hypotheses, conclusions, and theories. 2–3

1.2 Observing a Pillbug
- Observe and describe the external anatomy of a pillbug, *Armadillidium vulgare*.* 4
- Observe and describe how a pillbug moves. 5

1.3 Formulating Hypotheses
- Formulate a hypothesis based on appropriate observations. 6

1.4 Performing the Experiment and Coming to a Conclusion
- Design an experiment that can be repeated by others. 6
- Reach a conclusion based on observation and experimentation. 6–7

Introduction

This laboratory will provide you with an opportunity to use the scientific method in the same manner as scientists. Today your subject will be the pillbug, *Armadillidium vulgare*,* a type of crustacean that lives on land.

Pillbugs have an exoskeleton consisting of overlapping "armored" plates that make them look like little armadillos. As pillbugs grow, they molt (shed the exoskeleton) four or five times during a lifetime. A pillbug can roll up into such a tight ball that its legs and head are no longer visible, earning it the nickname "roly-poly." They have three body parts: head, thorax, and abdomen. The head bears compound eyes and two pairs of antennae. The thorax bears pairs of walking legs; gills are located at the top of the first five pairs. The gills must be kept slightly moist, which explains why pillbugs are usually found in damp places. The final pair of appendages, the uropods, which are sensory and defensive in function, project from the abdomen of the animal.

Pillbugs are commonly found in damp leaf litter, under rocks, and in basements or crawl spaces under houses. Following an inactive winter, pillbugs mate in

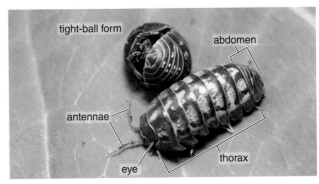

Pillbugs on leaf

*The garden snail, *Helix aspersa*, or the earthworm, *Lumbricus terrestris*, can be substituted as desired.

the spring. Several weeks later, the eggs hatch and remain for six weeks in a brood pouch on the underside of the female's body. Once they leave the pouch, they eat primarily dead organic matter, including decaying leaves. Therefore, they are easy to find and maintain in a moist terrarium with leaf litter, rocks, and wood chips. You are encouraged to collect some for your experiment. Since they live in the same locations as snakes, be careful when collecting them.

1.1 Using the Scientific Method

Some scientists work alone but often scientists belong to a community of scientists who are working together to study some aspect of the natural world. For example, many scientists from different institutions come together to study Arctic ecology (Fig. 1.1).

Figure 1.1 Scientists work together.
These scientists are studying the ecology of the Arctic.

> You will share your study of pillbugs with the other members of the class.

Even though the methodology can vary, scientists often use the **scientific method** (Fig. 1.2) when doing research. The scientific method involves:

Making observations. Observations help scientists begin their study of a particular topic.

> To learn about pillbugs you will visually observe one. You could also do a Google search of the Web or talk to someone who has worked with pillbugs for a long time.

Why does the scientific method begin with observations? _____

Formulating a hypothesis. Based on their observations, scientists come to a tentative decision, called a hypothesis, about their topic. Formulating hypotheses helps scientists decide how an experiment will be conducted.

> Based on your observations you might hypothesize that a pillbug will be attracted to fruit juice.

Now you know what you will actually do. What is the benefit of formulating a hypothesis? _____

_____

Testing the hypothesis. Scientists make further observations or perform experiments in order to test the hypothesis.

> **Virtual Lab Mealworm Behavior** A virtual lab called Mealworm Behavior is available on the *Biology* website **www.mhhe.com/maderbiology11**. This virtual lab demonstrates how investigators conducted an experiment similar to the one you will be doing.

> You could decide to expose a pillbug to fruit juice and observe its reaction.

A well-designed experiment must have a negative control—that is, a sample or event—that is not exposed to the testing procedure. If the negative control and the test produce the same results, either the procedure is flawed or the hypothesis is false.

> Water can substitute for fruit juice and be the control in your experiment.

Scientists call the results of their experiments the **data**. It is very important for scientists to keep accurate records of all their data.

You will record your data in a table that can be easily examined by another person.

When another person repeats the same experiment, and the data is the same, both experiments have merit. Why must a scientist keep complete records of an experiment? _____

Coming to a conclusion. Scientists come to a conclusion as to whether their data support or do not support the hypothesis.

If a pillbug is attracted to fruit juice, your hypothesis is supported. If the pillbug is not attracted to fruit juice, your hypothesis is not supported.

A scientist never says that a hypothesis has been proven true because, after all, some future knowledge might have a bearing on the experiment. What is the purpose of the conclusion? _____

_____

Developing a theory. A *theory* in science is an encompassing conclusion based on many individual conclusions in the same field. For example, the gene theory states that organisms inherit coded information that controls their anatomy, physiology, and behavior. It takes many years for scientists to develop theory and, therefore, we will not be developing any theories today. How is a scientific theory different from a conclusion? _____

_____

Figure 1.2 Flow diagram for the scientific method.
Often, scientists use this methodology to come to conclusions and develop theories about the natural world. The return arrow shows that scientists often choose to retest the same hypothesis or test a related hypothesis before arriving at a conclusion.

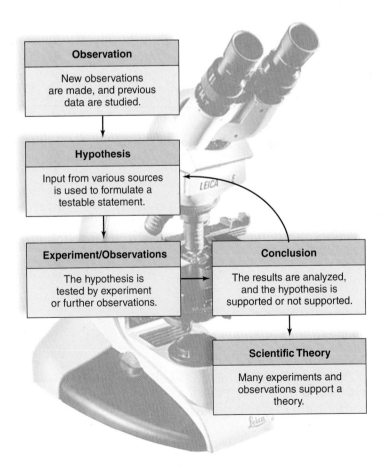

Observation
New observations are made, and previous data are studied.

Hypothesis
Input from various sources is used to formulate a testable statement.

Experiment/Observations
The hypothesis is tested by experiment or further observations.

Conclusion
The results are analyzed, and the hypothesis is supported or not supported.

Scientific Theory
Many experiments and observations support a theory.

1.2 Observing a Pillbug

Wash your hands before and after handling pillbugs. Please handle them carefully so they are not crushed. When touched, they roll up into a ball or "pill" shape as a defense mechanism. They will soon recover if left alone.

Observation: Pillbug's External Anatomy

Obtain a pillbug that has been numbered with white correction fluid or tape tags. Put the pillbug in a small glass or plastic dish to keep it contained.

1. Examine the exterior of the pillbug with the unaided eye and with a magnifying lens or dissecting microscope.

   - How can you recognize the head end of a pillbug? _____

   - How many segments and pairs of walking legs are in the thorax? _____

   - The abdomen ends in uropods, appendages with a sensory and defense function. (Females have leaflike growths at the base of some legs where developing eggs and embryos are held in pouches.)

2. In the following space, draw an outline of your pillbug (at least 7 cm long). Label the head, thorax, abdomen, antennae, eyes, uropods, and one of the seven pairs of legs.

3. Draw a pillbug rolled into a ball.

1. Watch a pillbug's underside as the pillbug moves up a transparent surface, such as the side of a graduated cylinder or beaker.

   a. Describe the action of the feet and any other motion you see. _____

   b. Allow a pillbug to crawl on your hand. Describe how it feels. _____

   c. Does a pillbug have the ability to move directly forward? _____

   d. Do you see evidence of mouthparts on the underside of the pillbug? _____

2. As you watch the pillbug, identify

   a. the anatomical parts that allow a pillbug to identify and take in food. _____

   b. behaviors that will help the pillbug acquire food. For example, is the ability of the pillbug to move directly forward a help in acquiring food? Explain. _____

      What other behaviors allow a pillbug to acquire food? _____

   c. a behavior that helps a pillbug avoid dangerous situations. _____

      If a pillbug rolls up into a ball, wait a few minutes and it may uncurl itself.

3. Measure the speed of three pillbugs.

   a. Place each pillbug on a metric ruler, and use a stopwatch to measure the number of seconds it takes for the pillbug to move several centimeters. Quickly record here the number of centimeters moved and the time in seconds.

      pillbug 1 _____

      pillbug 2 _____

      pillbug 3 _____

   b. Convert the number of centimeters traveled to millimeters and record this in the first column of Table 1.1. Record the total time taken in the second column.

   c. Use the space above to calculate the speed of each pillbug in mm/sec and record the speed for each pillbug in the last column of Table 1.1.

   d. Average the speed for your three pillbugs. (Since you have already calculated the mm/sec for each pillbug, it is only necessary to take an average of the millimeters moved.) Record the average speed of pillbug motion in Table 1.1.
      When you conduct the experiment in Section 1.4 you will have to be patient with your pillbug as it moves toward or away from a substance.

Table 1.1 Pillbug Speed			
Pillbug	Millimeters Traveled	Time (sec)	Speed (mm/sec)
1			
2			
3			
		Average speed:	

1.3 Formulating Hypotheses

You will be testing whether pillbugs are attracted to (move toward and eat), repelled by (move away from), or are unresponsive to (don't move away from and do not move toward and eat) the particular substances, which are potential foods. If a pillbug simply rolls into a ball, nothing can be concluded, and you may wish to choose another pillbug or wait a minute or two to check for further response.

1. Choose
   a. two dry substances, such as flour, cornstarch, coffee creamer, or baking soda. Fine sand will serve as a control for dry substances. Record your "dry" choices as 1, 2, and 3 in the first column of Table 1.2.
   b. two liquids, such as milk, orange juice, ketchup, applesauce, or carbonated beverage. Water will serve as a control for liquid substances. Record your "wet" choices as 4, 5, and 6 in the first column of Table 1.2.
2. In the second column of Table 1.2, hypothesize how you expect the pillbug to respond to each substance. Use a plus (+) sign if you hypothesize that the pillbug will move toward and eat the substance; a minus (−) sign if you hypothesize that the pillbug will be repelled by the substance; and a zero (0) if you expect the pillbug to show neither behavior.
3. In the third column of Table 1.2 offer a reason for your hypothesis based on your knowledge of pillbugs from the introduction and your examination of the animal.

Table 1.2 Hypotheses About Pillbug's Response to Potential Foods		
Substance	Hypothesis About Pillbug's Response	Reason for Hypothesis
1		
2		
3	(control)	
4		
5		
6	(control)	

1.4 Performing the Experiment and Coming to a Conclusion

A good experimental design would be to keep your pillbug in a petri dish to test its reaction to the chosen substances. During your experiment, no substance must be put directly on the pillbug, nor can the pillbug be placed directly onto the substances.

Experimental Procedure: Pillbug's Response to Potential Foods

1. Before testing the pillbug's reaction, fill in the first column of Table 1.3. It will look exactly like the first column of Table 1.2.
2. Since pillbugs tend to walk around the edge of a petri dish, you could put the wet or dry substance there; or for the wet substance you could put liquid-soaked cotton in the pillbug's path.
3. Rinse your pillbug between procedures by spritzing it with distilled water from a spray bottle. Then put it on a paper towel to dry it off.
4. Watch the pillbug's response to each substance, and record it in Table 1.3, using +, −, or 0 as before.

Table 1.3 Pillbug's Response to Potential Foods

Substance		Pillbug's Response	Hypothesis Supported?
1			
2			
3	(control)		
4			
5			
6	(control)		

Conclusion

5. Do your results support your hypotheses? Answer yes or no in the last column of Table 1.3.
6. Are there any hypotheses that were not supported by the experimental results (data)? Does this

   difference give you more insight into pillbug behavior? Explain. _____

   _____

   _____

7. **Class Results.** Compare your results with those of other students who tested the same substance. Calculate the proportional response to each potential food (% +, % −, % 0) and record your calculations in Table 1.4. As a group, your class can decide what proportion is needed to designate this response as typical. For example, if the pillbugs as a whole were attracted to a substance 70 % or more of the time, you can call that response the "typical response."

Table 1.4 Pillbug's Response to Potential Foods: Class Results

Substance		Pillbug's Response			Hypothesis Supported?
1		% +	% −	% 0	
2		% +	% −	% 0	
3	(control)	% +	% −	% 0	
4		% +	% −	% 0	
5		% +	% −	% 0	
6	(control)	% +	% −	% 0	

8. On the basis of the class data do you need to revise your conclusion for any particular pillbug

   response? _____ Scientists prefer to come to conclusions on the basis of many

   trials. Why is this the best methodology? _____

   _____

9. Did the pillbugs respond as expected to the controls, i.e., did not eat them? _____ If they

   did not, what can you conclude about your experimental results? _____

   _____

Laboratory 1   Scientific Method   **7**

1. What are the essential steps of the scientific method?

   _____

2. What is a hypothesis?

   _____

3. Is it sufficient to do a single experiment to test a hypothesis—why or why not?

   _____

4. What do you call a sample that goes through all the steps of an experiment but does not contain the factor being tested?

   _____

5. What part of a pillbug is for protection, and what does it do to protect itself?

   _____

6. Name one observation that you used to formulate your hypotheses regarding pillbug responses toward various substances.

   _____

7. Why is it important to test one substance at a time when doing an experiment?

   _____

Indicate whether statements 8 and 9 are hypotheses, conclusions, or theories.

8. The data show that vaccines protect people from disease. _____

9. All living things are made of cells. _____

10. How should an affirmative conclusion always be worded? _____

Biology **Website**

Enhance your study of the text and laboratory manual with study tools, practice tests, and virtual labs. Also ask your instructor about the resources available through ConnectPlus, including the media-rich eBook, interactive learning tools, and animations.

www.mhhe.com/maderbiology11

McGraw-Hill Access Science Website

An Encyclopedia of Science and Technology Online which provides more information including videos that can enhance the laboratory experience.

www.accessscience.com

2

Metric Measurement and Microscopy

Learning Outcomes

2.1 The Metric System
- Use metric units of measurement for length, weight, volume, and temperature.   10–13

2.2 Microscopy
- Describe similarities and differences between the stereomicroscope (dissecting microscope), compound light microscope, and the electron microscope.   14–15

2.3 Stereomicroscope (Dissecting Microscope)
- Identify the parts and tell how to focus the stereomicroscope.   16–17

2.4 Use of the Compound Light Microscope
- Identify and give the function of the basic parts of the compound light microscope.   18–19
- List, in proper order, the steps for bringing an object into focus with the compound light microscope.   19–20
- Describe how the image is inverted by the compound light microscope.   20
- Calculate the total magnification and the diameter of field for both low- and high-power lens systems.   21
- Explain how a slide of colored threads provides information on the depth of field.   22

2.5 Microscopic Observations
- Name and describe the kinds of cells studied in this exercise.   23–25
- State three differences between human epithelial cells and onion epidermal cells.   24
- Examine a wet mount of *Euglena* and pond water. Contrast the organisms observed in pond water.   24–25

Introduction

This laboratory introduces you to the metric system, which biologists use to indicate the sizes of cells and cell structures. This laboratory also examines the features, functions, and use of the stereomicroscope (dissecting microscope) and the compound light microscope. Transmission and scanning electron microscopes are explained, and micrographs produced using these microscopes appear throughout this lab manual. The stereomicroscope and the scanning electron microscope view the surface and/or the three-dimensional structure of an object. The compound light microscope and the transmission electron microscope can view only extremely thin sections of a specimen. If a subject was sectioned lengthwise for viewing, the interior of the projections at the top of the cell, called cilia, would appear in the micrograph. A lengthwise cut through any type of specimen is called a **longitudinal section (l.s.).** On the other hand, if the subject in Figure 2.1 was sectioned crosswise below the area of the cilia, you would see other portions of the interior of the subject. A crosswise cut through any type of specimen is called a **cross section (c.s.).**

Figure 2.1   Longitudinal and cross sections.
a. Transparent view of a cell. **b.** A longitudinal section would show the cilia at the top of the cell. **c.** A cross section shows only the interior where the cut is made.

a. Cell

b. Longitudinal
section

c. Cross section

2.1 The Metric System

The **metric system** is the standard system of measurement in the sciences, including biology, chemistry, and physics. It has tremendous advantages because all conversions, whether for volume, mass (weight), or length, are in units of ten. Refer to Appendix B, page A-5, for an in-depth look at the units of the metric system.

Length

Metric units of length measurement include the **meter (m), centimeter (cm), millimeter (mm), micrometer (µm)**, and **nanometer (nm)** (Table 2.1). The prefixes *milli-* (10^{-3}), *micro-* (10^{-6}), and *nano-* (10^{-9}) are used with length, weight, and volume.

Table 2.1	Metric Units of Length Measurement			
Unit	Meters	Centimeters	Millimeters	Relative Size
Meter (m)	1 m	100 cm	1,000 mm	Largest
Centimeter (cm)	0.01 (10^{-2}) m	1 cm	10 mm	
Millimeter (mm)	0.001 (10^{-3}) m	0.1 cm	1.0 mm	
Micrometer (µm)	0.000001 (10^{-6}) m	0.0001 (10^{-4}) cm	0.001 (10^{-3}) mm	
Nanometer (nm)	0.000000001 (10^{-9}) m	0.0000001 (10^{-7}) cm	0.000001 (10^{-6}) mm	Smallest

Experimental Procedure: Length

1. Obtain a small ruler marked in centimeters and millimeters. How many centimeters are represented? _____ One centimeter equals how many millimeters? _____ To express the size of small objects, such as cell contents, biologists use even smaller units of the metric system than those on the ruler. These units are the micrometer (µm) and the nanometer (nm). According to Table 2.1, 1 µm = _____ mm, and 1 nm = _____ mm. Therefore, 1 mm = _____ µm = _____ nm.

2. Measure the diameter of the circle shown below to the nearest millimeter. This circle is _____ mm = _____ µm = _____ nm.

For example, to convert mm to µm:

$$\text{_____ mm} \times \frac{1,000\,\mu m}{mm} = \text{_____ } \mu m$$

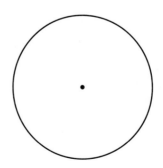

3. Obtain a meterstick. On one side, find the numbers 1 through 39, which denote inches. One meter equals 39.37 inches; therefore, 1 meter is roughly equivalent to 1 yard. Turn the meterstick over, and observe the metric subdivisions. How many centimeters are in a meter? _____ How many millimeters are in a meter? _____ The prefix *milli-* means _____.

4. Use the meterstick and the method shown in Figure 2.2 to measure the length of two long bones from a disarticulated human skeleton. Lay the meterstick flat on the lab table. Place a long bone next to the meterstick between two pieces of cardboard (each about 10 cm × 30 cm), held upright at right angles to the stick. The narrow end of each piece of cardboard should touch the meterstick. The length between the cards is the length of the bone in centimeters. For example, if the bone measures from the 22 cm mark to the 50 cm mark, the length of the bone is _____ cm. If the bone measures from the 22 cm mark to midway between the 50 cm and 51 cm marks, its length is _____ cm = _____ mm.

5. Record the length of two bones. First bone: _____ cm = _____ mm.
Second bone: _____ cm = _____ mm.

Figure 2.2 Measurement of a long bone.
How to measure a long bone using a meterstick.

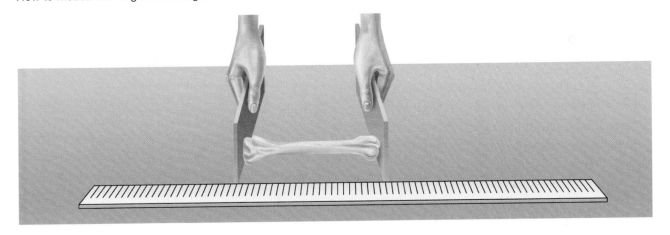

Weight

Two metric units of weight are the **gram (g)** and the **milligram (mg)**. A paper clip weighs about 1 g, which equals 1,000 mg. 2 g = _____ mg; 0.2 g = _____ mg; and 2 mg = _____ g.

Experimental Procedure: Weight

1. Use a balance scale to measure the weight of a wooden block small enough to hold in the palm of your hand.
2. Measure the weight of the block to the tenth of a gram. The weight of the wooden block is _____ g = _____ mg.
3. Measure the weight of an item small enough to fit inside the opening of a 50 mL graduated cylinder. The item, a(n) _____, is _____ g = _____ mg.

Volume

Two metric units of volume are the **liter (L)** and the **milliliter (mL)**. One liter = 1,000 mL.

Experimental Procedure: Volume

1. Volume measurements can be related to those of length. For example, use a millimeter ruler to measure the wooden block used in the previous Experimental Procedure to get its length, width, and depth.

   length = _____ cm; width = _____ cm; depth = _____ cm

   The volume, or space, occupied by the wooden block can be expressed in cubic centimeters (cc or cm^3) by multiplying: length × width × depth = _____ cm^3. For purposes of this Experimental Procedure, 1 cubic centimeter equals 1 milliliter; therefore, the wooden block has a volume of = _____ mL.

2. In the biology laboratory, liquid volume is usually measured directly in liters or milliliters with appropriate measuring devices. For example, use a 50 mL graduated cylinder to add 20 mL of water to a test tube. First, fill the graduated cylinder to the 20 mL mark. To do this properly, you have to make sure that the lowest margin of the water level, or the **meniscus** (Fig. 2.3), is at the 20 mL mark. Place your eye directly parallel to the level of the meniscus, and add water until the meniscus is at the 20 mL mark. (Having a dropper bottle filled with water on hand can help you do this.) A large, blank, white index card held behind the cylinder can also help you see the scale more clearly. Now pour the 20 mL of water into the test tube.

3. Hypothesize how you could find the total volume of the test tube. _____

   _____

   _____

   What is the test tube's total volume? _____

Figure 2.3   Meniscus.
The proper way to view the meniscus.

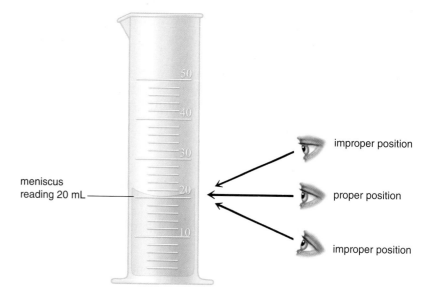

4. Fill a 50 mL graduated cylinder with water to about the 20 mL mark. Hypothesize how you could use this setup to calculate the volume of an object. _____

   _____

   Now perform the operation you suggested. The object, _____, has a volume of _____ mL.

5. Hypothesize how you could determine how many drops from the pipette of the dropper bottle equal 1 mL. _____

   Now perform the operation you suggested. How many drops from the pipette of the dropper bottle equal 1 mL? _____ Some pipettes are graduated and can be filled to a certain level as a way to measure volume directly. Your instructor will demonstrate this. Are pipettes customarily used to measure large or small volumes? _____

Temperature

There are two temperature scales: the **Fahrenheit (F)** and **Celsius (centigrade, C)** scales (Fig. 2.4). Scientists use the Celsius scale.

Experimental Procedure: Temperature

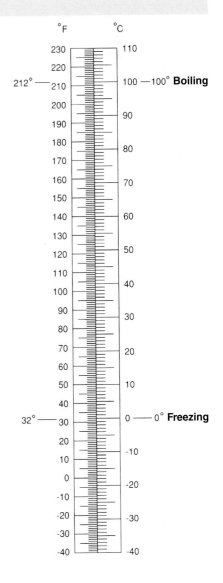

1. Study the two scales in Figure 2.4, and complete the following information:

   a. Water freezes at _____ °F = _____ °C.

   b. Water boils at _____ °F = _____ °C.

2. To convert from the Fahrenheit to the Celsius scale, use the following equation:

$$°C = (°F - 32°)/1.8$$

   or

$$°F = (1.8°C) + 32$$

   Human body temperature of 98°F is what temperature on the Celsius scale? _____

3. Record any two of the following temperatures in your lab environment. In each case, allow the end bulb of the Celsius thermometer to remain in or on the sample for one minute.

   Room temperature = _____ °C

   Surface of your skin = _____ °C

   Cold tap water in a 50 mL beaker = _____ °C

   Hot tap water in a 50 mL beaker = _____ °C

   Ice water = _____ °C

Figure 2.4   Temperature scales.
The Fahrenheit (°F) scale is on the left, and the Celsius (°C) scale is on the right.

2.2 Microscopy

Because biological objects can be very small, we often use a microscope to view them. Many kinds of instruments, ranging from the hand lens to the electron microscope, are effective magnifying devices. A short description of two kinds of light microscopes and two kinds of electron microscopes follows.

Light Microscopes

Light microscopes use light rays passing through lenses to magnify the object. The **stereomicroscope (dissecting microscope)** is designed to study entire objects in three dimensions at low magnification. The **compound light microscope** is used for examining small or thinly sliced sections of objects under higher magnification than that of the stereomicroscope. The term **compound** refers to the use of two sets of lenses: the ocular lenses located near the eyes and the objective lenses located near the object. Illumination is from below, and visible light passes through clear portions but does not pass through opaque portions. To improve contrast, the microscopist uses stains or dyes that bind to cellular structures and absorb light. Photomicrographs, also called light micrographs, are images produced by a compound light microscope (Fig. 2.5a).

Figure 2.5 Comparative micrographs.
Micrographs of a lymphocyte, a type of white blood cell. **a.** A photomicrograph (light micrograph) shows less detail than a **(b)** transmission electron micrograph (TEM). **c.** A scanning electron micrograph (SEM) shows the cell surface in three dimensions.

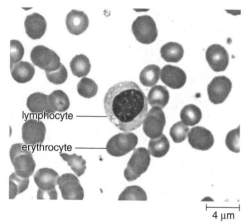

lymphocyte —————

erythrocyte —————

⊢————⊣
4 μm

a. Photomicrograph or light micrograph (LM)

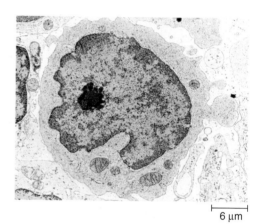

⊢————⊣
6 μm

b. Transmission electron micrograph (TEM)

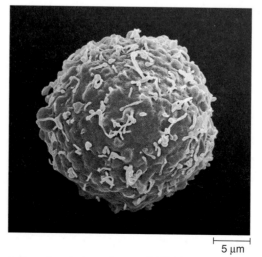

⊢————⊣
5 μm

c. Scanning electron micrograph (SEM)

Electron Microscopes

Electron microscopes use beams of electrons to magnify the object. The beams are focused on a photographic plate by means of electromagnets. The **transmission electron microscope** is analogous to the compound light microscope. The object is ultra-thinly sliced and treated with heavy metal salts to improve contrast. Figure 2.5*b* is a micrograph produced by this type of microscope. The **scanning electron microscope** is analogous to the dissecting light microscope. It gives an image of the surface and dimensions of an object, as is apparent from the scanning electron micrograph in Figure 2.5*c*.

The micrographs in Figure 2.5 demonstrate that an object is magnified more with an electron microscope than with a compound light microscope. The difference between these two types of microscopes, however, is not simply a matter of magnification; it is also the electron microscope's ability to show detail. The electron microscope has greater resolving power. **Resolution** is the minimum distance between two objects at which they can still be seen, or resolved, as two separate objects. The use of high-energy electrons rather than light gives electron microscopes a much greater resolving power since two objects that are much closer together can still be distinguished as separate points. Table 2.2 lists several other differences between the compound light microscope and the transmission electron microscope.

Table 2.2 Comparison of the Compound Light Microscope and the Transmission Electron Microscope	
Compound Light Microscope	**Transmission Electron Microscope**
1. Glass lenses	1. Electromagnetic lenses
2. Illumination by visible light	2. Illumination due to beam of electrons
3. Resolution $\cong$ 200 nm	3. Resolution $\cong$ 0.1 nm
4. Magnifies to 2,000$\times$	4. Magnifies to 1,000,000$\times$
5. Costs up to tens of thousands of dollars	5. Costs up to hundreds of thousands of dollars

Conclusions: Microscopy

- Which two types of microscopes view the surface of an object? _____
- Which two types of microscopes view objects that have been sliced and treated to improve contrast? _____
- Of the microscopes just mentioned, which one resolves the greater amount of detail?

    _____

2.3 Stereomicroscope (Dissecting Microscope)

The **stereomicroscope (dissecting microscope)** allows you to view objects in three dimensions at low magnifications. It is used to study entire small organisms, any object requiring lower magnification, and opaque objects that can be viewed only by reflected light. It is called a stereomicroscope because it produces a three-dimensional image.

Identifying the Parts

After your instructor has explained how to carry a microscope, obtain a stereomicroscope and a separate illuminator, if necessary, from the storage area. Place it securely on the table. Plug in the power cord, and turn on the illuminator. There is a wide variety of stereomicroscope styles, and your instructor will discuss the specific style(s) available to you. Regardless of style, you should be able to locate these parts and *label them in Figure 2.6*.

1. **Binocular head:** Holds two eyepiece lenses that move to accommodate for the various distances between different individuals' eyes.
2. **Eyepiece lenses**: The two lenses located on the binocular head. What is the magnification of your eyepieces? _____ Some models have one **independent focusing eyepiece** with a knurled knob to allow independent adjustment of each eye. The nonadjustable eyepiece is called the **fixed eyepiece**.
3. **Focusing knob:** A large, black or gray knob located on the arm; used for changing the focus of both eyepieces together.

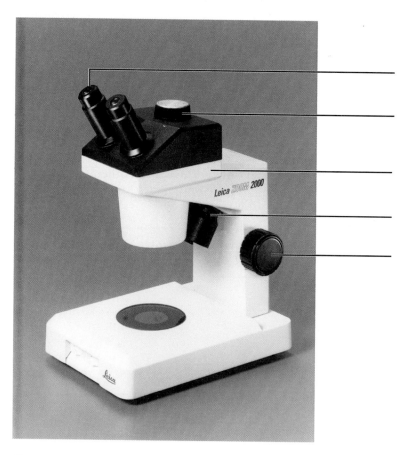

Figure 2.6 Stereomicroscope.
Label this stereomicroscope with the help of the text material.

4. **Magnification changing knob:** A knob, often built into the binocular head, used to change magnification in both eyepieces simultaneously. This may be a **zoom** mechanism or a **rotating lens** mechanism of different powers that clicks into place.
5. **Illuminator:** Used to illuminate an object from above; may be built into the microscope or may be separate.

Focusing the Stereomicroscope

1. In the center of the stage, place a plastomount that contains small organisms.
2. Adjust the distance between the eyepieces on the binocular head so that they comfortably fit the distance between your eyes. You should be able to see the object with both eyes as one three-dimensional image.
3. Use the focusing knob to bring the object into focus.
4. Does your microscope have an independent focusing eyepiece? _____ If so, use the focusing knob to bring the image in the fixed eyepiece into focus, while keeping the eye at the independent focusing eyepiece closed. Then adjust the independent focusing eyepiece so that the image is clear, while keeping the other eye closed. Is the image inverted? _____
5. Turn the magnification changing knob, and determine the kind of mechanism on your microscope. A zoom mechanism allows continuous viewing while changing the magnification. A rotating lens mechanism blocks the view of the object as the new lenses are rotated. Be sure to click each lens firmly into place. If you do not, the field will be only partially visible. What kind of mechanism is on your stereomicroscope? _____
6. Set the magnification changing knob on the lowest magnification. Sketch the object in the following circle as though this represents your entire field of view:

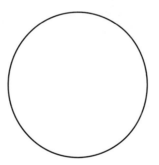

7. Rotate the magnification changing knob to the highest magnification. Draw another circle within the one provided to indicate the reduction of the field of view.
8. Experiment with various objects at various magnifications until you are comfortable with using the stereomicroscope.

Rules for Microscope Use

Observe the following rules for using a microscope:

1. The lowest power objective (scanning or low) should be in position, both at the beginning and end of microscope use.
2. Use only lens paper for cleaning lenses.
3. Do not tilt the microscope as the eyepieces could fall out, or wet mounts could be ruined.
4. Keep the stage clean and dry to prevent rust and corrosion.
5. Do not remove parts of the microscope.
6. Keep the microscope dust-free by covering it after use.
7. Report any malfunctions.

2.4 Use of the Compound Light Microscope

As mentioned, the name **compound light microscope** indicates that it uses two sets of lenses and light to view an object. The two sets of lenses are the ocular lenses located near the eyes and the objective lenses located near the object. Illumination is from below, and the light passes through clear portions but does not pass through opaque portions. This microscope is used to examine small or thinly sliced sections of objects under higher magnification than would be possible with the stereomicroscope.

Identifying the Parts

Obtain a compound light microscope from the storage area, and place it securely on the table. The following will help you identify the parts of your microscope. *Label them in Figure 2.7.*

1. **Eyepieces** (ocular lenses): Some compound light microscopes are **monocular** and only one eye is used to view the object; others are **binocular** and both eyes are used to view the object. What is the magnifying power of the ocular lenses on your microscope? _____

_____

2. **Viewing head:** Holds the ocular lenses.
3. **Arm:** Supports upper parts and provides carrying handle.
4. **Nosepiece:** Revolving device that holds objectives.
5. **Objectives** (objective lenses):
   a. **Scanning objective:** This is the shortest of the objective lenses and is used to scan the whole slide. The magnification is stamped on the housing of the lens. It is a number followed by an ×. What is the magnifying power of the scanning objective lens on your microscope? _____
   b. **Low-power objective:** This lens is longer than the scanning objective lens and is used to view objects in greater detail. What is the magnifying power of the low-power objective lens on your microscope? _____
   c. **High-power objective:** If your microscope has three objective lenses, this lens will be the longest. It is used to view an object in even greater detail. What is the magnifying power of the high-power objective lens on your microscope? _____
   d. **Oil immersion objective:** (on microscopes with four objective lenses): Holds a $95\times$ (to $100\times$) lens and is used in conjunction with immersion oil to view objects with the greatest magnification. Does your microscope have an oil immersion objective? _____ If this lens is available, your instructor will discuss its use when the lens is needed.

Figure 2.7 Compound light microscope.
Compound light microscope with binocular head and mechanical stage.

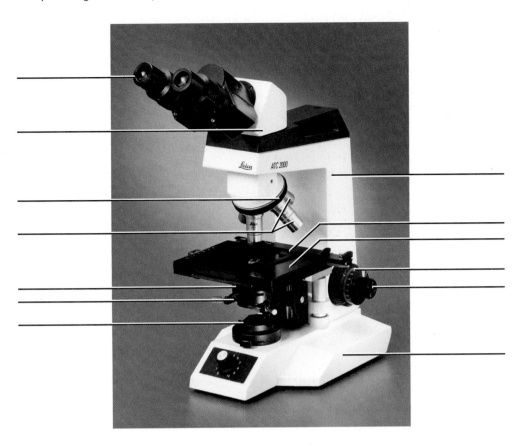

6. **Stage:** Holds and supports microscope slides. A mechanical stage is a movable portion that aids in the accurate positioning of the slide. Does your microscope have a mechanical stage? _____

   a. **Stage clips:** Holds a slide in place on the stage.
   b. **Mechanical stage control knobs:** Two knobs that control forward/reverse movement and right/left movement, respectively.

7. **Coarse-adjustment knob:** Knob used to bring object into approximate focus; used only with low-power objective.

8. **Fine-adjustment knob:** Knob used to bring object into final focus.

9. **Condenser:** Gathers light from the lamp and directs it toward the object being viewed.

   a. **Diaphragm or diaphragm control lever:** Controls the amount of light passing through the condenser.

10. **Light source:** An attached lamp that directs a beam of light up through the object.

11. **Base:** The flat surface of the microscope that rests on the table.

Focusing the Compound Light Microscope—Lowest Power

1. Turn the nosepiece so that the *lowest* power objective on your microscope is in straight alignment over the stage.

2. Always begin focusing with the *lowest* power objective on your microscope (4× [scanning] or 10× [low power]).

3. With the coarse-adjustment knob, lower the stage (or raise the objectives) until it stops.

4. Place a slide of the letter *e* on the stage, and stabilize it with the clips. (If your microscope has a mechanical stage, pinch the spring of the slide arms on the stage, and insert the slide.) Center the *e* as best you can on the stage or use the two control knobs located below the stage (if your microscope has a mechanical stage) to center the *e*.
5. Again, be sure that the lowest power objective is in place. Then, as you look from the side, decrease the distance between the stage and the tip of the objective lens until the lens comes to an automatic stop or is no closer than 3 mm above the slide.
6. While looking into the eyepiece, rotate the diaphragm (or diaphragm control lever) to give the maximum amount of light.
7. Using the coarse-adjustment knob, slowly increase the distance between the stage and the objective lens until the object—in this case, the letter *e*—comes into view, or focus.
8. Once the object is seen, you may need to adjust the amount of light. To increase or decrease the contrast, rotate the diaphragm slightly.
9. Use the fine-adjustment knob to sharpen the focus if necessary.
10. Practice having both eyes open when looking through the eyepiece, as this greatly reduces eyestrain.

Inversion

Inversion refers to the fact that a microscopic image is upside down and reversed.

Observation: Inversion

1. Draw the letter *e* as it appears on the slide (with the unaided eye, not looking through the eyepiece). _____

2. Draw the letter *e* as it appears when you look through the eyepiece. _____

3. What differences do you notice? _____

4. Move the slide to the right. Which way does the image appear to move? _____
   Explain. _____

Focusing the Compound Light Microscope—Higher Powers

Compound light microscopes are **parfocal;** that is, once the object is in focus with the lowest power, it should also be almost in focus with the higher power.

1. Bring the object into focus under the lowest power by following the instructions in the previous section.
2. Make sure that the letter *e* is centered in the field of the lowest objective.
3. Move to the next higher objective (low power [10×] or high power [40×]) by turning the nosepiece until you hear it click into place. Do not change the focus; **parfocal** microscopes stay in focus when you change the objective lenses. (If you are on low power [10×], proceed to high power [40×] before going on to step 4.)
4. If any adjustment is needed, use only the *fine*-adjustment knob. (*Note:* Always use only the fine-adjustment knob with high power, and do not use the coarse-adjustment knob.)
5. On a drawing of the letter *e* to the right, draw a circle around the portion of the letter that you are now seeing with high-power magnification. The letter *e* will not disappear because your microscope is parcentric (the focus remains near the center).
6. When you have finished your observations of this slide (or any slide), rotate the nosepiece until the lowest power objective clicks into place, and then remove the slide.

Total Magnification

Total magnification is calculated by multiplying the magnification of the ocular lens (eyepiece) by the magnification of the objective lens. The magnification of a lens is imprinted on the lens casing.

Observation: Total Magnification

Calculate total magnification figures for your microscope, and record your findings in Table 2.3.

Table 2.3 Total Magnification			
Objective	**Ocular Lens**	**Objective Lens**	**Total Magnification**
Scanning power (if present)			
Low power			
High power			
Oil immersion (if present)			

Field of View

A microscope's **field of view** is the circle visible through the lenses. The **diameter of field** is the length of the field from one edge to the other.

Observation: Field of View

Low-Power (10×) Diameter of Field

1. Place a clear, plastic ruler across the stage so that the edge of the ruler is visible as a horizontal line along the diameter of the low-power (not scanning) field. Be sure that you are looking at the millimeter side of the ruler.

2. Estimate the number of millimeters, to tenths, that you see along the field: _____ mm. (*Hint:* Start with one of the millimeter markers at the edge of the field.) Convert the figure to micrometers: _____ μm. This is the **low-power diameter of field (LPD)** for your microscope in micrometers.

High-Power (40×) Diameter of Field

1. To compute the **high-power diameter of field (HPD),** substitute these data into the formula given:
   a. LPD = low-power diameter of field (in micrometers) = _____
   b. LPM = low-power total magnification (from Table 2.3) = _____
   c. HPM = high-power total magnification (from Table 2.3) = _____

Example: If the diameter of field is about 2 mm, then the LPD is 2,000 μm. Using the LPM and HPM values from Table 2.3, the HPD would be 500 μm.

$$HPD = LPD \times \frac{LPM}{HPM}$$

$$HPD = (\quad) \times \frac{(\quad)}{(\quad)} = \underline{\quad}$$

Conclusions: Total Magnification and Field of View

- Does low power or high power have a larger field of view (one that allows you to see more of the object)? _____

- Which has a smaller field but magnifies to a greater extent? _____

- To locate small objects on a slide, first find them under _____ ; then place them in the center of the field before rotating to _____ .

Depth of Field

When viewing an object on a slide under high power, the **depth of field** (Fig. 2.8) is the area—from top to bottom—that comes into focus while slowly focusing up and down with the microscope's fine-adjustment knob.

Observation: Depth of Field

1. Obtain a prepared slide with three or four colored threads mounted together, or prepare a wet-mount slide with three or four crossing threads or hairs of different colors. (Directions for preparing a wet mount are given in Section 2.5.)

2. With low power, find a point where the threads or hairs cross. Slowly focus up and down. Notice that when one thread or hair is in focus, the others seem blurred. *Determine the order of the threads or hairs, and complete Table 2.4.* Remember, as the stage moves upward (or the objectives move downward), objects on top come into focus first.

3. Switch to high power, and notice that the depth of field is more shallow with high power than with low power. Constant use of the fine-adjustment knob when viewing a slide with high power will give you an idea of the specimen's three-dimensional form. For example, viewing a number of sections allows reconstruction of the three-dimensional structure, as demonstrated in Figure 2.8.

Figure 2.8   Depth of field.
A demonstration of how focusing at depths 1, 2, and 3 would produce three different images (views) that could be used to reconstruct the original three-dimensional structure of the object.

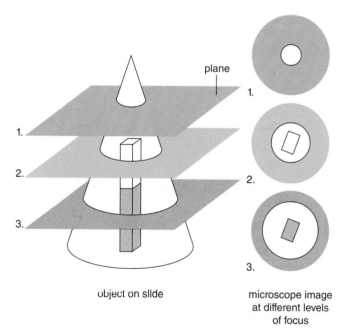

plane

1.

2.

3.

object on slide

microscope image at different levels of focus

Table 2.4 Order of Threads (or Hairs)	
Depth	**Thread (or Hair) Color**
Top	
Middle	
Bottom	

2.5 Microscopic Observations

When a specimen is prepared for observation, the object should always be viewed as a **wet mount.** A wet mount is prepared by placing a drop of liquid on a slide or, if the material is dry, by placing it directly on the slide and adding a drop of water or stain. The mount is then covered with a coverslip, as illustrated in Figure 2.9. Dry the bottom of your slide before placing it on the stage.

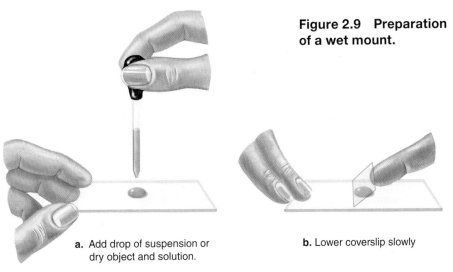

Figure 2.9 Preparation of a wet mount.

a. Add drop of suspension or dry object and solution.

b. Lower coverslip slowly

> ⚠ **Methylene blue** Avoid ingestion, inhalation, and contact with skin, eyes, and mucous membranes. Exercise care in using this chemical. If any should spill on your skin, wash the area with mild soap and water. Methylene blue will also stain clothing. Follow your instructor's directions for its disposal.

Observation: Onion Epidermal Cells

Epidermal cells cover the surfaces of plant organs, such as leaves. The bulb of an onion is made up of fleshy leaves.

> ⚠ **Scalpel** Exercise care when using a scalpel.

1. With a scalpel, strip a small, thin, transparent layer of cells from the inside of a fresh onion leaf.
2. Place it gently on a clean, dry slide, and add a drop of *iodine solution* (or *methylene blue*). Cover with a coverslip.
3. Observe under the microscope.
4. Locate the cell wall and the nucleus. *Label Figure 2.10.*
5. Count the number of onion cells that line up end to end in a single line across the diameter of the high-power (40×) field._____

   Based on what you learned in Section 2.4 about measuring diameter of field, what is your high-power diameter of field (HPD) in micrometers?

   _____ μm

   Calculate the length of each onion cell

   (HPD ÷ number of cells): _____ μm
6. Wash and reuse this slide for the next exercise (Human Epithelial Cells).

Figure 2.10 Onion epidermal cells.
Label the cell wall and the nucleus.

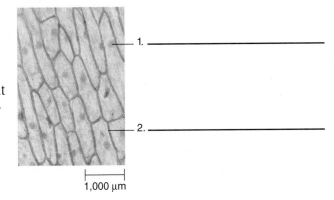

1. _____

2. _____

1,000 μm

Observation: Human Epithelial Cells

Epithelial cells cover the body's surface and line its cavities.

1. Obtain a prepared slide, or make your own as follows:
   a. Obtain a prepackaged flat toothpick (or sanitize one with alcohol or alcohol swabs).
   b. Gently scrape the inside of your cheek with the toothpick, and place the scrapings on a clean, dry slide. Discard used toothpicks in the biohazard waste container provided.
   c. Add a drop of very weak *methylene blue* or *iodine solution*, and cover with a coverslip.
2. Observe under the microscope.
3. Locate the nucleus (the central, round body), the cytoplasm, and the plasma membrane (outer cell boundary). *Label Figure 2.11.*
4. Because your epithelial slides are biohazardous, they must be disposed of as indicated by your instructor.
5. Note some obvious differences between the onion cells and the human cheek cells, and list them in Table 2.5.

Figure 2.11   Cheek epithelial cells.
Label the nucleus, the cytoplasm, and the plasma membrane.

1. _____
2. _____
3. _____

3 μm

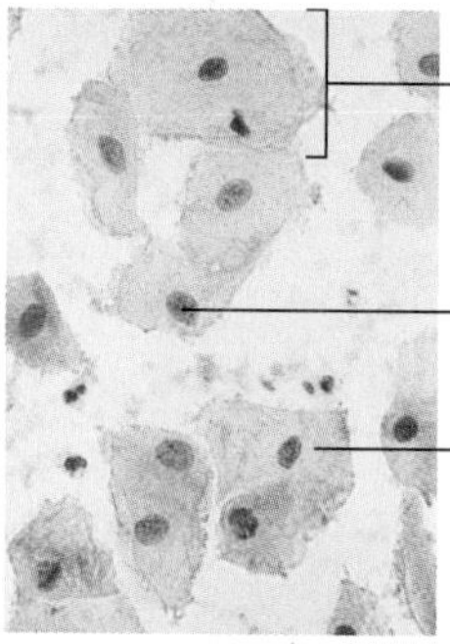

Table 2.5   Differences Between Onion Epidermal and Human Epithelial Cells

Differences	Onion Epidermal Cells	Human Epithelial Cells (Cheek)
Shape		
Orientation		
Boundary		

Observation: Euglena

Examination of *Euglena* (a unicellular organism with a flagellum to facilitate movement) will test your ability to utilize depth of field and to control illumination in order to heighten contrast.

1. Make a wet mount of *Euglena* by using a drop of a *Euglena* culture and adding a drop of Protoslo® (methyl cellulose solution) onto a slide. The Protoslo® slows the organism's swimming.
2. Mix thoroughly with a toothpick, and add a coverslip.
3. Scan the slide for *Euglena*: Start at the upper left-hand corner, and move the slide forward and back as you work across the slide from left to right. The *Euglena* may be at the edge of the slide because they show an aversion to Protoslo®.
4. Experiment by using scanning, low-power, and high-power objective lenses; by focusing up and down with the fine-adjustment knob; and by adjusting the light so that it is not too bright.
5. Compare your *Euglena* specimens with Figure 2.12 (top left). List the labeled features that you can actually see. _____

_____

Examination of pond water will also test your ability to observe objects with the microscope, to utilize depth of field, and to control illumination to heighten contrast. Use Figure 2.12 to help you identify the organisms you see in pond water.

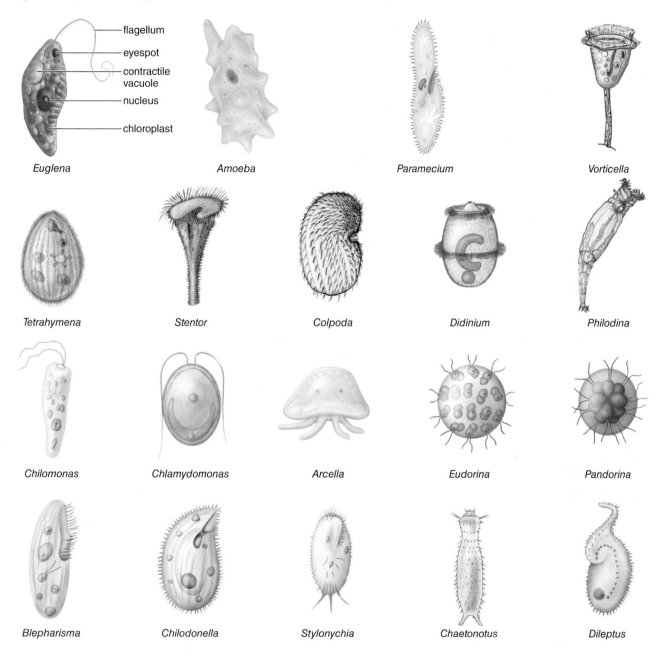

Figure 2.12 Microorganisms found in pond water.
Drawings are not the actual sizes of the organisms.

1. Make the following conversions:

   a. 1 mm = _____ μm = _____ cm

   b. 15 mm = _____ cm = _____ μm

   c. 50 mL = _____ liter

   d. 5 g = _____ mg

2. Explain the designation "compound light" microscope:

   a. compound _____

   b. light _____

3. What function is performed by the diaphragm of a microscope?

   _____

4. Why is it helpful for a microscope to be parfocal?

   _____

5. Why is locating an object more difficult if you start with the high-power objective than with the low-power objective?

   _____

6. How much larger than actual size does an object appear with a low-power objective? _____

   _____

7. A virus is 50 nm in size.

   a. Would you recommend using a stereomicroscope, a compound light microscope, or an electron

   microscope to see it? _____ Why? _____

   _____

   b. How many micrometers is the virus? _____ μm

8. If the diameter of a field is 1.6 mm, and you count 40 consecutive cells from one end of the field to the

   other, how wide is each cell in micrometers? _____ μm

9. What type of microscope, aside from the compound light microscope, might you use to observe the

   organisms found in pond water? _____

10. Briefly describe the necessary steps for observing a slide at low power under the compound
    light microscope.

    _____

    _____

3

Chemical Composition of Cells

Learning Outcomes

Introduction
- Describe how cells build and break down polymers.   27
- Distinguish between positive and negative controls and between positive and negative results.   27–28

3.1 Carbohydrates
- State the monomer for starch, and distinguish sugars from starch in terms of monomer number.   28
- Describe a test for the detection of starch and another for the detection of sugars.   29–30
- Explain the varied results of the Benedict's test.   31

3.2 Proteins
- State the function of various types of proteins.   32
- State the monomer for a polypeptide, and explain how this monomer joins with others to form a polypeptide.   32–33
- Describe a test for the detection of a protein versus a peptide.   33

3.3 Lipids
- Name several types of lipids and state the components of fat.   34
- Describe a simple test for the detection of fat and tell how an emulsifier works.   35–36
- Tell where fat (adipose) tissue is located in the body and describe its appearance.   37

3.4 Testing Foods and Unknowns
- Explain a procedure for testing the same food for all three components—carbohydrates, protein, and fat.   37–38

Introduction

All living things consist of basic units of matter called **atoms.** Molecules form when atoms bond with one another. Inorganic molecules are often associated with nonliving things, and organic molecules are associated with living organisms. In this laboratory, you will be studying the organic molecules of cells: **carbohydrates** (monosaccharides, disaccharides, polysaccharides), **proteins,** and **lipids** (i.e., fat).

> **Planning Ahead** To save time, your instructor may have you start the boiling water bath needed for the experiments on page 31 and page 38 at the beginning of the laboratory session.

Large organic molecules form during *dehydration reactions* when smaller molecules bond as water is given off. During *hydrolysis reactions,* bonds are broken as water is added. A fat contains one glycerol and three fatty acids. Proteins and some carbohydrates (called polysaccharides) are **polymers** because they are made up of smaller molecules called **monomers.** Proteins contain a large number of amino acids (the monomer) joined together by a peptide bond. A polysaccharide, such as starch, contains a large number of glucose molecules (the monomer) joined together.

Various chemicals will be used in this laboratory to test for the presence of cellular organic molecules. If a color change is observed, a particular molecule is present; if the color change is not observed, the molecule is not present. Therefore, each testing procedure will contain a positive control and a negative control as described in the box on page 28.

What Is a Control?

The experiments in today's laboratory have both a positive control and a negative control, *which should be saved for comparison purposes until the experiment is complete.* The **positive control** goes through all the steps of the experiment and does contain the substance being tested. Therefore, positive results are expected. The **negative control** goes through all the steps of the experiment except it does not contain the substance being tested. Therefore, negative results are expected.

For example, if a test tube contains glucose (the substance being tested) and Benedict's reagent (blue) is added, a red color develops upon heating. This test tube is the positive control; it tests positive for glucose. If a test tube does not contain glucose and Benedict's reagent is added, Benedict's is expected to remain blue. This test tube is the negative control; it tests negative for glucose.

What benefit is a positive control? Positive controls give you a standard by which to tell if the substance being tested is present (or acting properly) in an unknown sample. Negative controls ensure that the experiment is giving reliable results; after all, if a negative control should happen to give a positive result, then the entire experiment may be faulty and unreliable.

3.1 Carbohydrates

Carbohydrates include sugars and molecules that are chains of sugars. **Glucose,** which has only one sugar unit, is a monosaccharide; **maltose,** which has two sugar units, is a disaccharide (Fig. 3.1). Glycogen, starch, and cellulose are polysaccharides, made up of chains of many glucose units (Fig. 3.2).

Glucose is used by all organisms as an energy source. Energy is released when glucose is broken down to carbon dioxide and water. This energy is used by the organism to do work. Animals store glucose as glycogen and plants store glucose as starch. Plant cell walls are composed of cellulose.

Figure 3.1 Formation of a disaccharide.
During a dehydration reaction, a disaccharide, such as maltose, forms when a glucose joins with a glucose as a water molecule is removed. During a hydrolysis reaction, water is added and the bond is broken.

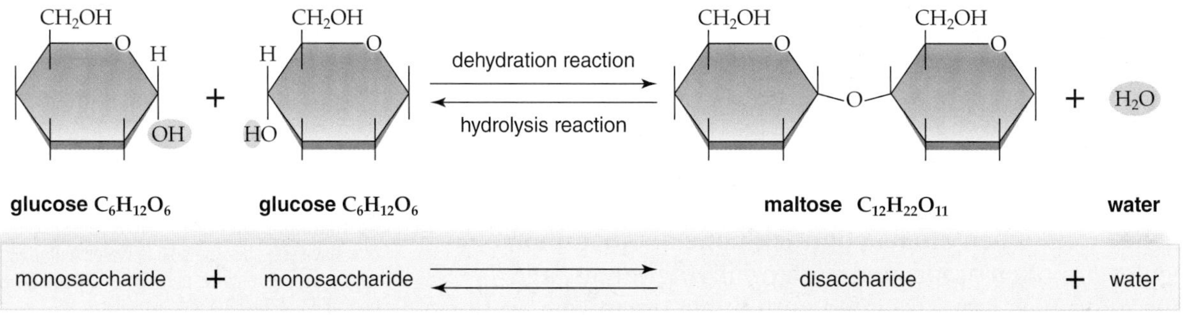

Test for Starch

In the presence of starch, iodine solution (yellowish-brown) reacts chemically with starch to form a blue-black color as noted in Table 3.1. This positive result is explained thus: Iodine molecules lodge in the coiled structure of starch. This causes a change in the way that iodine molecules are able to reflect light and in the color we observe.

Table 3.1   Iodine Test for Starch

	Starch
Iodine solution	Black

Figure 3.2   Starch.
Starch is a polysaccharide composed of many glucose units. **a.** Photomicrograph of starch granules in plant cells such as those in a potato. **b.** Structure of starch. Starch consists of amylose that is nonbranched and amylopectin that is branched.

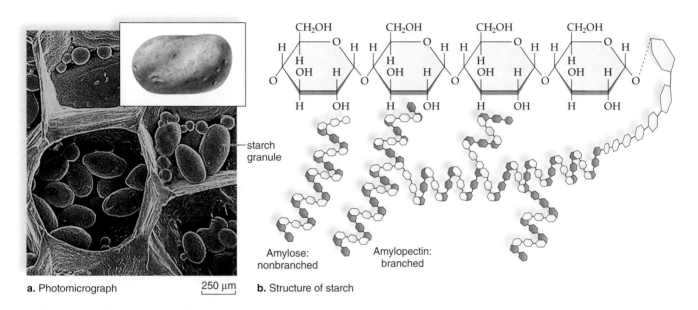

starch granule

Amylose: nonbranched

Amylopectin: branched

a. Photomicrograph   250 μm   **b.** Structure of starch

Experimental Procedure: Test for Starch

With a wax pencil, label and mark five clean test tubes at the 1 cm level.

*Tube 1
1. Fill to the 1 cm mark with *water*, and add five drops of *iodine solution.*
2. Note the final color change, and record your results in Table 3.2.

Tube 2
1. It is very important to shake the *starch suspension* well before taking your sample. After shaking, fill this tube to the 1 cm mark with the *starch suspension.* Add five drops of *iodine solution.* Label and save this tube for Section 3.4.
2. Note the final color change, and record your results in Table 3.2.

Tube 3
1. Add a few drops of *onion juice* to the test tube. (Obtain the juice by adding water and crushing a small piece of onion with a mortar and pestle. Clean mortar and pestle after using.) Add five drops of *iodine solution.*
2. Note the final color change, and record your results in Table 3.2.

Tube 4
1. Add a few drops of *potato juice* to the test tube. (Obtain the juice by adding water and crushing a small piece of potato with a mortar and pestle. Clean mortar and pestle after using.) Add five drops of *iodine solution.*
2. Note the final color change, and record your results in Table 3.2.

Tube 5
1. Fill to the 1 cm mark with *glucose solution,* and add five drops of *iodine solution.*
2. Note the final color change, and record your results in Table 3.2.

*To test a sample for starch, use this procedure. Instead of only water, use a liquefied sample. If starch is present, a blue-black color appears.

Table 3.2 Iodine Test for Starch

Tube	Contents	Color	Conclusions
1	Water		
2	Starch suspension		
3	Onion juice		
4	Potato juice		
5	Glucose solution		

Conclusions: Starch

- From your test results, draw conclusions about what organic compound is present in each tube. Write these conclusions in Table 3.2.

- Does the potato or the onion store glucose as starch? _____ How do you know? _____
  _____

- If your results are not as expected, offer an explanation. Then inform your instructor, who will advise you how to proceed.

Experimental Procedure: Microscopic Study of Potato and Onion

Potato

1. With a sharp razor blade, slice a very thin piece of potato. Place it on a microscope slide, add a drop of *water* and a coverslip, and observe under low power with your compound light microscope. Compare your slide with the photomicrograph of starch granules (see Fig. 3.2a). Find the cell wall (large, geometric compartments) and the starch grains (numerous clear, oval-shaped objects).
2. Without removing the coverslip, place two drops of *iodine solution* onto the microscope slide so that the iodine touches the coverslip. Draw the iodine under the coverslip by placing a small piece of paper towel in contact with the water on the **opposite** side of the coverslip.
3. Microscopically examine the potato again on the side closest to where the iodine solution was applied.

   What is the color of the small, oval bodies? _____ What is the chemical composition of

   these oval bodies? _____ What is the name of these oval bodies? _____

Onion

1. Peel a single layer of onion from the bulb. On the inside surface, you will find a thin, transparent layer of onion skin. Peel off a small section of this layer for use on your slide.
2. Add a large drop of *iodine solution.*
3. Does onion contain starch? _____

4. Are these results consistent with those you recorded for onions in Table 3.2? _____

Test for Sugars

Monosaccharides and some disaccharides will react with **Benedict's reagent** (blue color) after being heated in a boiling water bath. In this reaction, copper ion (Cu^{2+}) in the Benedict's reagent reacts with part of the sugar molecule and is reduced, causing a distinctive color change. The color change can range from green to red, and increasing concentrations of sugar will give a continuum of colored products (Table 3.3).

> ⚠️ **Benedict's reagent** Benedict's reagent is highly corrosive. Exercise care in using this chemical. If any should spill on your skin, wash the area with mild soap and water. Follow your instructor's directions for disposal of this chemical.

Table 3.3 Benedict's Test for Sugars (Some Typical Results)

Chemical	Chemical Category	Benedict's Reagent (After Heating)
Water	Inorganic	Blue (no change)
Glucose	Monosaccharide (carbohydrate)	Varies with concentration: very low—green / low—yellow / moderate—yellow-orange / high—orange / very high—orange-red
Maltose	Disaccharide (carbohydrate)	Varies with concentration—see "Glucose"
Starch	Polysaccharide (carbohydrate)	Blue (no change)

Experimental Procedure: Test for Sugars

With a wax pencil, label and mark five clean test tubes at the 1 cm level. If possible, place all five test tubes in a boiling water bath at the same time.

*Tube 1
1. Fill to the 1 cm mark with *water,* then add about five drops of *Benedict's reagent.*
2. Heat in a boiling water bath for 5 to 10 minutes, note any color change, and record in Table 3.4.

Tube 2
1. Fill to the 1 cm mark with *glucose solution,* then add about five drops of *Benedict's reagent.*
2. Heat in a boiling water bath for 5 to 10 minutes, note any color change, and record in Table 3.4. Label and save this tube for Section 3.4.

Tube 3
1. Fill to the 1 cm mark with *starch suspension,* then add about five drops of *Benedict's reagent.*
2. Heat in a boiling water bath for 5 to 10 minutes, note any color change, and record in Table 3.4.

Tube 4
1. Place a few drops of *onion juice* in the test tube. (Obtain the juice by adding water and crushing a small piece of onion with a mortar and pestle. Clean mortar and pestle after using.)
2. Fill to the 1 cm mark with *water,* then add about five drops of *Benedict's reagent.*
3. Heat in a boiling water bath for 5 to 10 minutes, note any color change, and record in Table 3.4.

Tube 5
1. Place a few drops of *potato juice* in the test tube. (Obtain the juice by adding water and crushing a small piece of potato with a mortar and pestle.)
2. Fill to the 1 cm mark with *water,* then add about five drops of *Benedict's reagent.*
3. Heat in a boiling water bath for 5 to 10 minutes, note any color change, and record in Table 3.4.

*To test a sample for sugars, use this procedure. Instead of only water, use a liquefied sample. If sugar is present, a green to orange-red color appears.

Table 3.4 Benedict's Test for Sugars

Tube	Contents	Color (After Heating)	Conclusions
1	Water		
2	Glucose solution		
3	Starch suspension		
4	Onion juice		
5	Potato juice		

Conclusions: Sugars

- From your test results, conclude what kind of chemical is present. Enter your conclusions in Table 3.4.

- Indicate which tube served as a negative and which a positive control. _____

- Compare Table 3.2 with Table 3.4. In what form do potatoes store carbohydrate (glucose or starch)? _____ In what form do onions store carbohydrate (glucose or starch)? _____ Is this consistent with your results in Table 3.4? _____ Explain. _____

3.2 Proteins

Proteins have numerous functions in cells. Antibodies are proteins that combine with pathogens so that the pathogens are destroyed by the body. Transport proteins combine with and move substances from place to place. Hemoglobin transports oxygen throughout the body. Albumin is another transport protein in our blood. Regulatory proteins control cellular metabolism in some way. For example, the hormone insulin regulates the amount of glucose in blood so that cells have a ready supply. Structural proteins include keratin, found in hair, and myosin, found in muscle. **Enzymes** are proteins that speed chemical reactions. A reaction that could take days or weeks to complete can happen within an instant if the correct enzyme is present. Amylase is an enzyme that speeds the breakdown of starch in the mouth and small intestine.

Proteins are made up of **amino acids** joined together. About 20 different common amino acids are found in cells. All amino acids have an acidic group (—COOH) and an amino group (H_2N—). They differ by the **R group** (remainder group) attached to a carbon atom, as shown in Figure 3.3. The R groups have varying sizes, shapes, and chemical activities.

A chain of two or more amino acids is called a **peptide,** and the bond between the amino acids is called a **peptide bond.** A **polypeptide** is a very long chain of amino acids. A protein can contain one or more polypeptide chains. Insulin contains a single chain, while hemoglobin contains four polypeptides. A protein has a particular shape, which is important to its function. The shape comes about because the R groups of the polypeptide chain(s) can interact with one another in various ways.

Figure 3.3 Formation of a dipeptide.

During a dehydration reaction, a dipeptide forms when an amino acid joins with an amino acid as a water molecule is removed. The bond between amino acids is called a peptide bond. During a hydrolysis reaction, water is added and the peptide bond is broken.

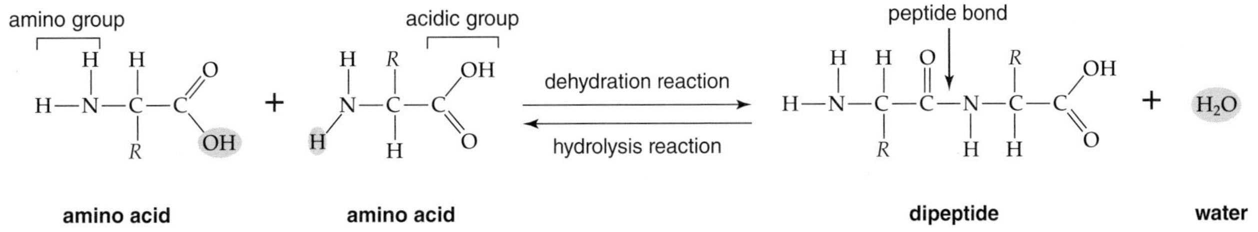

amino acid   amino acid   dipeptide   water

Test for Proteins

Biuret reagent (blue color) contains a strong solution of sodium or potassium hydroxide (NaOH or KOH) and a small amount of dilute copper sulfate ($CuSO_4$) solution. This reagent changes color in the presence of proteins or peptides because the peptide bonds of the protein or peptide chemically combine with the copper ions in biuret reagent (Table 3.5).

Table 3.5 Biuret Test for Protein and Peptides

	Protein	Peptides
Biuret reagent (blue)	Purple	Pinkish-purple

Experimental Procedure: Test for Proteins

With a millimeter ruler and a wax pencil, label and mark four clean test tubes at the 1 cm level. After filling a tube, cover it with Parafilm®, and swirl well to mix. (Do not turn upside down.) The reaction is almost immediate.

> ⚠️ **Biuret reagent** Biuret reagent is highly corrosive. Exercise care in using this chemical. If any should spill on your skin, wash the area with mild soap and water. Follow your instructor's directions for its disposal.

*Tube 1
  1. Fill to the mark with *distilled water,* and add about five drops of *biuret reagent.*
  2. Record the final color in Table 3.6.

Tube 2
  1. Fill to the mark with *albumin solution,* and add about five drops of *biuret reagent.*
  2. Record the final color in Table 3.6. Label and save this tube for Section 3.4.

Tube 3
  1. Fill to the mark with *pepsin solution,* and add about five drops of *biuret reagent.*
  2. Record the final color in Table 3.6.

Tube 4
  1. Fill to the mark with *starch solution,* and add about five drops of *biuret reagent.*
  2. Record the final color in Table 3.6.

Table 3.6 Biuret Test for Protein

Tube	Contents	Final Color	Conclusions
1	Distilled water		
2	Albumin		
3	Pepsin		
4	Starch		

*To test a sample for protein, use this procedure. Instead of only water, use a liquefied sample. If protein is present, a pinkish-purple color appears.

- From your test results, conclude if a protein is present or absent and explain. Enter your conclusions in Table 3.6.

- Pepsin is an enzyme. Your results indicate that enzymes are what type of organic molecule? _____

- According to your results, is starch a protein? _____

- Which of the four tubes is the negative control? _____

  _____

- Why do experimental procedures include controls? _____

  _____

- If your results are not as expected, inform your instructor, who will advise you how to proceed.

  _____

3.3 Lipids

Lipids are compounds that are insoluble in water and soluble in solvents, such as alcohol and ether. Lipids include fats, oils, phospholipids, steroids, and cholesterol. Typically, fat, such as in the adipose tissue of animals, and oils, such as the vegetable oils from plants, are composed of three molecules of fatty acids bonded to one molecule of glycerol (Fig. 3.4). Phospholipids have the same structure as fats, except in place of the third fatty acid there is a phosphate group (a grouping that contains phosphate). Steroids are derived from cholesterol and, like this molecule, have skeletons of four fused rings of carbon atoms, but they differ by functional groups (attached side chains). Fat, as we know, is long-term stored energy in the human body. Phospholipids are found in the plasma membrane of cells. In recent years, cholesterol, a molecule transported in the blood, has been implicated in causing cardiovascular disease. Regardless, steroids are very important compounds in the body; for example, the sex hormones are steroids.

Figure 3.4 Formation of a fat.
During a dehydration reaction, a fat molecule forms when glycerol joins with three fatty acids as three water molecules are removed. During a hydrolysis reaction, water is added, and the bonds are broken between glycerol and the three fatty acids.

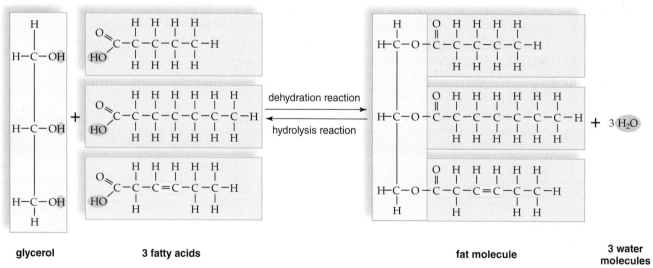

glycerol   3 fatty acids   fat molecule   3 water molecules

Test for Fat

Fats and oils do not evaporate from paper such as brown or loose-leaf paper; instead, they leave an oily spot.

Experimental Procedure: Test for Fat

*1. Place a small drop of *water* on a square of brown or loose-leaf paper. Describe the immediate effect.

_____

2. Place a small drop of *vegetable oil* on a square of the paper. Describe the immediate effect. _____

_____

3. Wait at least 15 minutes for the paper to dry. Evaluate which substance penetrates the paper and which is subject to evaporation. Record your observations and conclusions in Table 3.7. Save the paper for comparison when testing for fat in Section 3.4.

Table 3.7	Paper Test for Fat	
Sample	**Observations**	**Conclusions**
Water spot		
Oil spot		

Emulsification of Oil

Some molecules are **polar,** meaning that they have charged groups or atoms, and some are **nonpolar,** meaning that they have no charged groups or atoms. A water molecule is polar, and therefore, water is a good solvent for other polar molecules. When the charged ends of water molecules interact with the charged groups of polar molecules, these polar molecules disperse in water.

Water is not a good solvent for nonpolar molecules, such as fats. A fat has no polar groups to interact with water molecules. An **emulsifier,** however, can cause a fat to disperse in water. An emulsifier contains molecules with both polar and nonpolar ends. When the nonpolar ends interact with the fat and the polar ends interact with the water molecules, the fat disperses in water, and an *emulsion* results (Fig. 3.5).

Bile salts (emulsifiers found in bile produced by the liver) are used in the digestive tract. Commercially produced emulsifiers include detergents and the wetting agent Tween®. Today milk, such as 1% milk, has been homogenized so that fat droplets do not congregate and rise to the top of the container. Homogenization requires the addition of natural emulsifiers such as phospholipids—the phosphate part of the molecule is polar and the lipid portion is nonpolar.

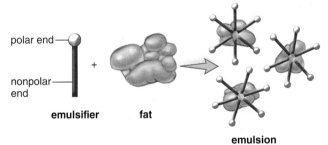

polar end
+
nonpolar end
emulsifier   **fat**
emulsion

Figure 3.5 Emulsification.
An emulsifier contains molecules with both a polar and a nonpolar end. The nonpolar ends are attracted to the nonpolar fat, and the polar ends are attracted to the water. This causes droplets of fat molecules to disperse.

*To test an unknown for fat, use this procedure. Do not liquefy the sample; place it as is on a piece of brown or loose-leaf paper over the sink, in order to facilitate cleanup. If lipids are present, an oily spot appears.

With a wax pencil, label three clean test tubes 1, 2, and 3. Mark tube 1 at the 3 cm and 4 cm levels. Mark tube 2 at the 2 cm, 3 cm, and 4 cm levels. Mark tube 3 at the 1 cm and 2 cm levels.

Tube 1
1. Fill to the 3 cm mark with *water* and to the 4 cm mark with *vegetable oil.* Shake.
2. Observe for the initial dispersal of oil, followed by rapid separation into two layers. Is vegetable oil soluble in water? _____
3. Let the tube settle for 5 minutes. Label a microscope slide as 1.
4. Use a dropper to remove a sample of the solution that is just below the layer of oil. Place the drop on the slide, add a coverslip, and examine with the low power of your compound light microscope.
5. Record your observations in Table 3.8.

Tube 2
1. Fill to the 2 cm mark with *water,* to the 3 cm mark with *vegetable oil,* and to the 4 cm mark with the available emulsifier (*Tween*® or *bile salts*). Shake well.
2. Describe how the distribution of oil in tube 2 compares with the distribution in tube 1.

_____

_____

3. Let the tube settle for 5 minutes. Label a microscope slide as 2.
4. Use a different dropper to remove a sample of the solution that is just below a thin layer of oil. Place the drop on the slide, add a coverslip, and examine with the low power of your compound light microscope.
5. Record your observations in Table 3.8.

Tube 3
1. Fill to the 1 cm mark with milk and to the 2 cm mark with *water.* Shake well.
2. Use a different dropper to remove a sample of the solution. Place a drop on a slide, add a coverslip, and examine with the low power of your compound light microspcope.
3. Record your observations in Table 3.8.

Table 3.8 Emulsification of Oil

Tube	Contents	Observations	Conclusions
1	Oil Water		
2	Oil Water Emulsifier		
3	Milk Water		

Conclusions: Emulsification of Oil

- From your observations, conclude why the contents of the tubes appear as they do under the microscope. Record your conclusions in Table 3.8.
- Explain the correlation between your macroscopic observations (how the tubes look to your unaided eye) and your microscopic observations. _____

_____

Adipose Tissue

Adipose tissue in animals such as humans stores droplets of fat. Adipose tissue is found beneath the skin, where it helps insulate and keep the body warm. It also forms a protective cushion around various internal organs.

Observation: Adipose Tissue

1. Obtain a slide of adipose tissue, and view it under the microscope at high power. Refer to Figure 3.6 for help in identifying the structures.
2. Notice how the fat droplets push the cytoplasm to the edges of the cells. The cytoplasm of cells is largely water. Explain why fat does not disperse in cytoplasm.

_____

_____

_____

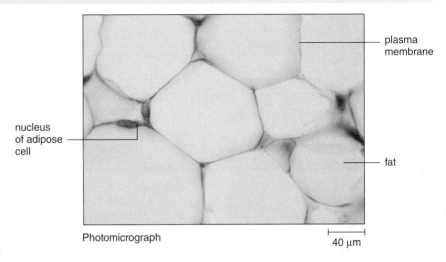

Photomicrograph   40 μm

Figure 3.6   Adipose tissue.
The cells are so full of fat that the nucleus is pushed to one side.

3.4   Testing Foods and Unknowns

It is common for us to associate the term *organic* with the foods we eat, including carbohydrate foods (Fig. 3.7), protein foods (Fig. 3.8), and lipid foods (Fig. 3.9). Though we may recognize foods as being organic, often we are not aware of what specific types of compounds are found in what we eat. In the following Experimental Procedure, you will use the same tests you used previously to determine the composition of everyday foods and unknowns.

Figure 3.7   Carbohydrate foods.   **Figure 3.8   Protein foods.**   **Figure 3.9   Lipid foods.**

Experimental Procedure: Testing Foods and Unknowns

Your instructor will provide you with several everyday foods including unknowns, and your task is to:

1. Develop instructions for a procedure that will allow you to test substances for carbohydrates (pages 29 and 31), protein (page 33), and fat (page 35) using tests from this laboratory manual.

_____

_____

_____

_____

2. Have your instructor okay your procedure, and then conduct the necessary tests.
3. Record your results as positive (+) or negative (-) in Table 3.9.

Table 3.9 Testing Foods and Unknowns

Sample Name	Starch (Iodine)	Sugar (Benedict's)	Protein (Bluret)	Fat (Brown or loose-leaf paper)
Unknown A				
Unknown B				

Conclusions: Testing Foods and Unknowns

• What foods tested positive for only one of the organic compounds? _____

   Explain. _____

• What foods tested positive for more than one of the organic compounds? _____

• What type of carbohydrate would give a positive iodine test but a negative Benedict's test? _____

_____

1. What three major classes of organic molecules studied today are present in cells? _____

   _____

2. You have been assigned the task of constructing a protein. What type of building block would you use?

   _____

3. A hydrolytic digestive enzyme breaks down starch to maltose molecules. Maltose is a _____

   _____ .

4. Why is it necessary to shake an oil and vinegar salad dressing before adding it to a salad? _____

   _____

5. How would you test an unknown solution for each of the following:

   a. Sugar _____

   b. Fat _____

   c. Starch _____

   d. Protein _____

6. Assume that you have tested an unknown sample with both biuret reagent and Benedict's reagent and
   that both tests result in a blue color. What have you learned? _____

   _____

7. What purpose is served when a test is done using water instead of a sample substance? _____

   _____

8. A test tube contains starch, a digestive enzyme for starch, and water. The biuret test is negative. After
   30 minutes, the Benedict's test is positive. What substance is present? _____

9. A test tube that contains either albumin or pepsin tests positive for protein. Explain. _____

   _____

10. A student adds iodine solution to egg white and waits for a color change. How long will the student
    have to wait? _____

4

Cell Structure and Function

Learning Outcomes

4.1 Prokaryotic Versus Eukaryotic Cells
- Distinguish between prokaryotic and eukaryotic cells by description and examples.   42

4.2 Animal Cell and Plant Cell Structure
- Label an animal cell diagram, and state a function for the structures labeled.   43–44
- Label a plant cell diagram, and state a function for the unique structures labeled.   45
- Use microscopic techniques to observe plant cell structure.   46

4.3 Diffusion
- Define diffusion and describe the process of diffusion as affected by the medium.   47–48
- Predict and observe which substances will or will not diffuse across a plasma membrane.   48–49

4.4 Osmosis: Diffusion of Water Across Plasma Membrane
- Define osmosis and explain the movement of water across a membrane.   49–50
- Define isotonic, hypertonic, and hypotonic solutions, and give examples in terms of NaCl concentrations.   50–51
- Predict the effect of different tonicities on animal (e.g., red blood) cells and on plant (e.g., *Elodea*) cells.   51–53

4.5 pH and Cells
- Predict the change in pH after the addition of acid to nonbuffered and buffered solutions.   54–55
- Explain the pH scale and predict a method by which it is possible to test the effectiveness of antacid medications.   55

Introduction

The molecules we studied in the last laboratory are not alive—the basic units of life are cells. The **cell theory** states that all living things are composed of cells and that cells come only from other cells. While we are accustomed to considering the heart, the liver, or the intestines as enabling the human body to function, it is actually cells that do the work of these organs.

Figure 2.11 shows human cheek epithelial cells as viewed by an ordinary compound light microscope available in general biology laboratories. It shows that the content of a cell, called the **cytoplasm,** is bounded by a **plasma membrane.** The plasma membrane regulates the movement of molecules into and out of the cytoplasm. In this lab, we will study how the passage of water into a cell depends on the difference in concentration of solutes (particles) between the cytoplasm and the surrounding medium or solution. The well-being of cells also depends upon the pH of the solution surrounding them. We will see how a buffer can maintain the pH within a narrow range and how buffers within cells can protect them against damaging pH changes.

> **Planning Ahead** To save time, your instructor may have you start a boiling water bath (page 49) and the potato strip experiment (page 53) at the beginning of the laboratory.

Because a photomicrograph shows only a minimal amount of detail, it is necessary to turn to the electron microscope to study the contents of a cell in greater depth. The models of plant and animal cells available in the laboratory today are based on electron micrographs.

4.1 Prokaryotic Versus Eukaryotic Cells

All living cells are classified as either prokaryotic or eukaryotic. One of the basic differences between the two types is that **prokaryotic cells** do not contain nuclei (*pro* means "before"; *karyote* means "nucleus"), while eukaryotic cells do contain nuclei (*eu* means "true"; *karyote* means "nucleus"). Only bacteria (including cyanobacteria) and archaea are prokaryotes; all other organisms are eukaryotes.

Prokaryotes also don't have all the organelles found in **eukaryotic cells**. **Organelles** are small, usually membranous bodies, each with a specific structure and function. Prokaryotes do have **cytoplasm,** the material bounded by a plasma membrane and cell wall. The cytoplasm contains ribosomes, small granules that coordinate the synthesis of proteins; thylakoids (only in cyanobacteria) that participate in photosynthesis; and innumerable enzymes. Prokaryotes also have a nucleoid, a region in the cell interior in which the DNA is physically organized but not enclosed by a membrane (Fig. 4.1).

Figure 4.1   Prokaryotic cell.
Prokaryotic cells lack membrane-bounded organelles, as well as a nucleus. Their DNA is in a nucleoid region.

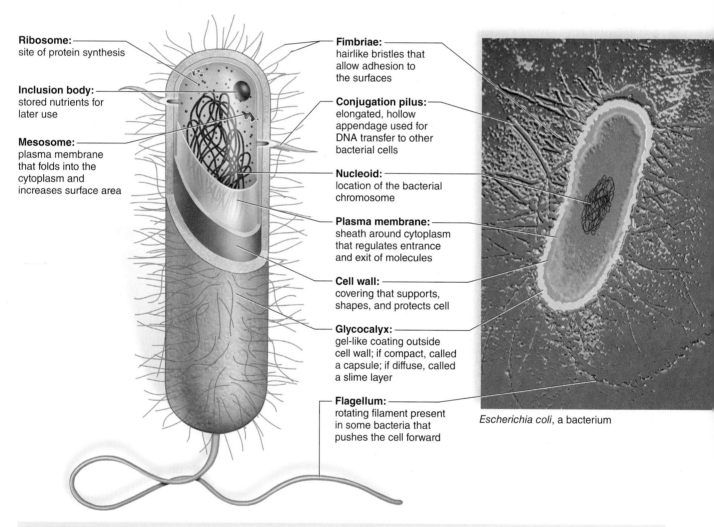

Ribosome:
site of protein synthesis

Inclusion body:
stored nutrients for later use

Mesosome:
plasma membrane that folds into the cytoplasm and increases surface area

Fimbriae:
hairlike bristles that allow adhesion to the surfaces

Conjugation pilus:
elongated, hollow appendage used for DNA transfer to other bacterial cells

Nucleoid:
location of the bacterial chromosome

Plasma membrane:
sheath around cytoplasm that regulates entrance and exit of molecules

Cell wall:
covering that supports, shapes, and protects cell

Glycocalyx:
gel-like coating outside cell wall; if compact, called a capsule; if diffuse, called a slime layer

Flagellum:
rotating filament present in some bacteria that pushes the cell forward

Escherichia coli, a bacterium

Observation: Prokaryotic/Eukaryotic Cells

Two microscope slides on display will show you the main difference between prokaryotic and eukaryotic cells.

1. Examine a prepared slide of a prokaryote. Do you agree that there are no nuclei in these cells? _____

2. Examine a prepared slide of cuboidal cells from a human kidney (see page 357). Do you agree that there are nuclei in these cells? _____

4.2 Animal Cell and Plant Cell Structure

Table 4.1 lists the structures found in animal and plant cells. The **nucleus** in a eukaryotic cell is bounded by a **nuclear envelope** and contains **nucleoplasm.** The **cytoplasm,** found between the plasma membrane and the nucleus, consists of a background fluid and the organelles. Many **organelles** are membranous, such as the nucleolus, endoplasmic reticulum, Golgi apparatus, vacuoles and vesicles, lysosomes, peroxisome, mitochondrion, and chloroplast.

Table 4.1 Eukaryotic Structures in Animal Cells and Plant Cells

Name	Composition	Function
Cell wall*	Contains cellulose fibrils	Provides support and protection
Plasma membrane	Phospholipid bilayer with embedded proteins	Outer cell surface that regulates entrance and exit of molecules
Nucleus	Enclosed by nuclear envelope; contains chromatin (threads of DNA and protein)	Stores genetic information; synthesizes DNA and RNA
Nucleolus	Concentrated area of chromatin	Produces subunits of ribosomes
Ribosome	Protein and RNA in two subunits	Coordinates protein synthesis
Endoplasmic reticulum (ER)	Membranous, flattened channels and tubular canals; rough ER and smooth ER	Synthesizes and/or modifies proteins and other substances; transport by vesicle formation
Rough ER	Studded with ribosomes	Protein synthesis
Smooth ER	Lacks ribosomes	Synthesizes lipid molecules
Golgi apparatus	Stack of membranous saccules	Processes, packages, and distributes proteins and lipids
Vesicle/vacuole	Membrane-bounded sac	Stores and transports substances
Lysosome	Vesicle containing hydrolytic enzymes	Digests macromolecules and cell parts
Peroxisome	Vesicle containing specific enzymes	Breaks down fatty acids and converts resulting hydrogen peroxide to water; various other functions
Mitochondrion	Membranous cristae bounded by double membrane	Carries out cellular respiration, producing ATP molecules
Chloroplast*	Membranous thylakoids bounded by double membrane	Carries out photosynthesis, producing sugars
Cytoskeleton	Microtubules, intermediate filaments, actin filaments	Maintains cell shape and assists movement of cell parts
Cilia and flagella	9 + 2 pattern of microtubules	Movement of cell, substances
Centrioles in centrosome**	9 + 0 pattern of microtubules	Organizes microtubules in cilia and flagella

*Plant cells only
**Animal cells only

Study Table 4.1 to determine structures that are unique to plant cells and unique to animal cells, and write them below the examples given.

	Plant Cells	**Animal Cells**
Unique structures:	1. Large central vacuole	1. Small vacuoles
	2. _____	2. _____
	3. _____	

Animal Cell Structure

With the help of Table 4.1, give a function for each of these structures, and label Figure 4.2.

Structure	Function
Plasma membrane	_____
Nucleus	_____
Nucleolus	_____
Ribosome	_____
Endoplasmic reticulum	_____
Rough ER	_____
Smooth ER	_____
Golgi apparatus	_____
Vesicles	_____
Lysosome	_____
Mitochondrion	_____
Centrioles in centrosome	_____
Cytoskeleton	_____

Figure 4.2   Animal cell structure.

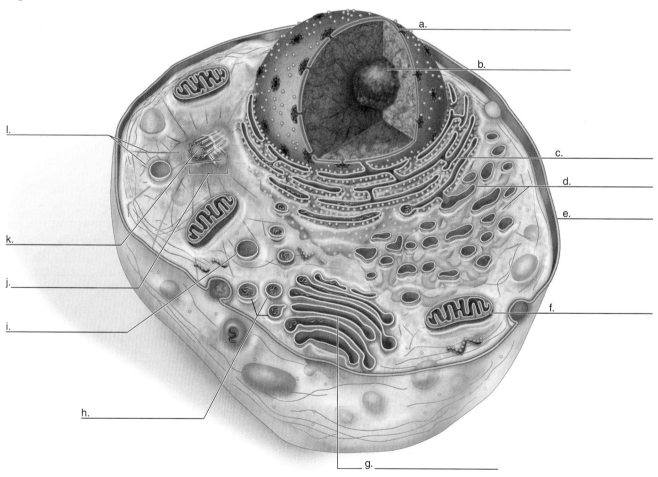

Plant Cell Structure

With the help of Table 4.1, give a function for these structures unique to plant cells, and label Figure 4.3.

Structure   **Function**

Cell wall _____

Central vacuole, large _____

Chloroplast _____

Figure 4.3   Plant cell structure.

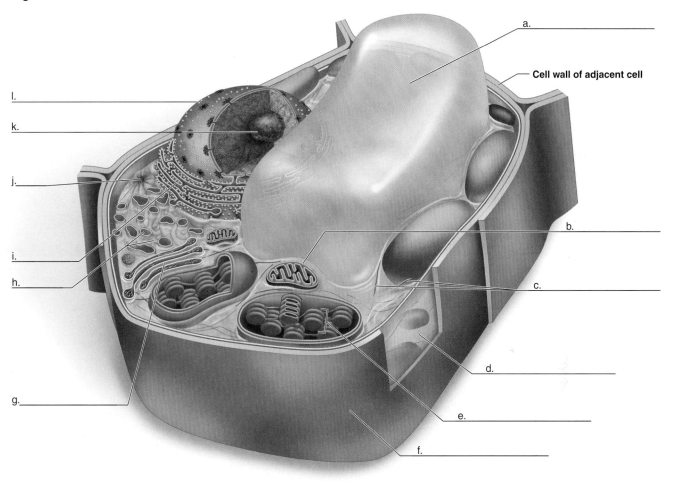

Cell wall of adjacent cell

Virtual Lab Cellular Pursuit   A Virtual Laboratory called Cellular Pursuit is available on the *Biology* website **www.mhhe.com/maderbiology11**. After opening this virtual lab, be sure to:

- Click on the continue button at the bottom of screen **only after** reading all the instructions by dragging down the side bar.
- Use the **tab** key to enter the name of up to four players.
- Click the spinner to move on; usually two question categories are highlighted, click on one category and the category title appears in the box below the wheel.
- Ignore the number next to the spinner and take turns in designated order.
- Understand hidden term scoring. Type in your answer within 15 seconds. A correct response receives the maximum number of points. If incorrect, **wait** because another clue will appear. If unsuccessful, the correct hidden term will eventually be revealed.
- **At the start** of your turn, purchase an organelle by clicking on the organelle in the center of the wheel. The organelle will show up in your cell.

1. Prepare a wet mount of a small piece of young *Elodea* leaf in fresh water. *Elodea* is a multicellular, eukaryotic plant found in freshwater ponds and lakes.
2. Have the drop of *water* ready on your slide so that the leaf does not dry out, even for a few seconds. Take care that the leaf is mounted with its top side up.
3. Examine the slide using low power, focusing sharply on the leaf surface.
4. Select a cell with numerous chloroplasts for further study, and switch to high power.
5. Carefully focus on the side and end walls of the cell. The chloroplasts appear to be only along the sides of the cell because the large, fluid-filled, membrane-bounded central vacuole pushes the cytoplasm against the cell walls (Fig. 4.4*a*). Then focus on the surface and notice an even distribution of chloroplasts (Fig. 4.4*b*).

Figure 4.4 *Elodea* cell structure.

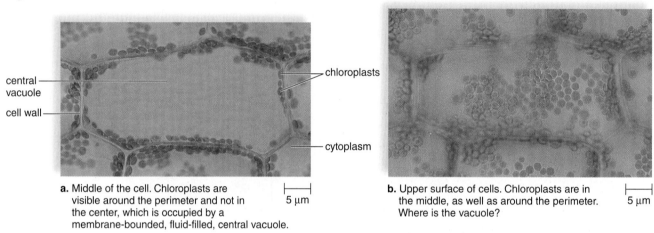

a. Middle of the cell. Chloroplasts are visible around the perimeter and not in the center, which is occupied by a membrane-bounded, fluid-filled, central vacuole.　5 µm

b. Upper surface of cells. Chloroplasts are in the middle, as well as around the perimeter. Where is the vacuole?　5 µm

6. Can you locate the cell nucleus? _____ It may be hidden by the chloroplasts, but when visible, it appears as a faint, grey lump on one side of the cell.
7. Why can't you see the other organelles featured in Figure 4.3?_____
_____
8. Can you detect movement of chloroplasts in this cell or any other cell? _____ The chloroplasts are not moving under their own power but are being carried by a streaming of the nearly invisible cytoplasm.
9. Save your slide for use later in this laboratory.

4.3 Diffusion

Diffusion is the movement of molecules from a higher to a lower concentration until equilibrium is achieved and the molecules are distributed equally (Fig. 4.5). At this point, molecules may still be moving back and forth, but there is no net movement in any one direction.

Figure 4.5 Process of diffusion.
Diffusion is apparent when dye molecules have equally dispersed.

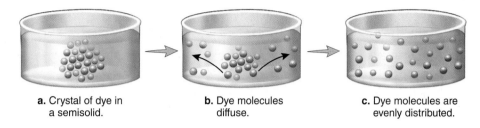

a. Crystal of dye in a semisolid. **b.** Dye molecules diffuse. **c.** Dye molecules are evenly distributed.

Diffusion is a general phenomenon in the environment. The speed of diffusion is dependent on such factors as the temperature, the size of the molecule, and the type of medium.

Experimental Procedure: Speed of Diffusion

Solute Diffusion Through a Semisolid

1. Observe a petri dish containing 1.5% gelatin (or agar) to which potassium permanganate ($KMnO_4$) was added in the center depression at the beginning of the lab.

   > ⚠ **Potassium permanganate/$KMnO_4$** is highly poisonous and is a strong oxidizer. Avoid contact with skin and eyes and any combustible materials. If spillage occurs, wash all surfaces thoroughly. $KMnO_4$ will also stain clothing.

2. Obtain *time zero* from your instructor, and record *time zero* and the *final time* (now) in Table 4.2. Calculate the length of time in hours and minutes. Convert the time to hours: _____ hr.

3. Using a ruler placed over the petri dish, measure (in mm) the movement of color from the center of the depression outward in one direction: _____ mm.

4. Calculate the speed of diffusion: _____ mm/hr.

5. Record all data in Table 4.2.

Solute Diffusion Through a Liquid

1. Add enough water to cover the bottom of a glass petri dish.
2. Place the petri dish over a thin, flat ruler.
3. With tweezers, add a crystal of potassium permanganate ($KMnO_4$) directly over a millimeter measurement line. Note the *time zero* in Table 4.2.
4. After 10 minutes, note the distance the color has moved. Record the *final time, length of time,* and *distance moved* in Table 4.2.
5. Multiply the length of time and the distance moved by 6 to calculate the *speed of diffusion:* _____ mm/hr. Record in Table 4.2.

Diffusion Through Air

1. Measure the distance from a spot designated by your instructor to your laboratory work area today. Record this distance in the fifth column of Table 4.2.
2. Record *time zero* in Table 4.2 when a perfume or similar substance is released into the air.
3. Note the time when you can smell the perfume. Record this as the *final time* in Table 4.2. Calculate the *length of time* since the perfume was released, and record it in Table 4.2.
4. Calculate the speed of diffusion: _____ mm/hr. Record in Table 4.2.

Table 4.2 Speed of Diffusion					
Medium	Time Zero	Final Time	Length of Time (hr)	Distance Moved (mm)	Speed of Diffusion (mm/hr)
Semisolid					
Liquid					
Air					

Conclusions: Speed of Diffusion

- In which experiment was diffusion the fastest? _____
- What accounts for the difference in speed? _____

Solute Diffusion Across the Plasma Membrane

Some molecules can diffuse across a plasma membrane, and some cannot. In general, small, noncharged molecules can cross a membrane by simple diffusion, but large molecules cannot diffuse across a membrane. The dialysis tube membrane in the experimental procedure simulates a plasma membrane.

Experimental Procedure: Solute Diffusion Across Plasma Membrane

At the start of the experiment,

1. Cut a piece of dialysis tubing approximately 40 cm (approximately 16 in) long. Soak the tubing in water until it is soft and pliable.
2. Close one end of the dialysis tubing with two knots.
3. Fill the bag halfway with *glucose solution*.
4. Add 4 full droppers of *starch solution* to the bag.
5. Hold the open end while you mix the contents of the dialysis bag. Rinse off the outside of the bag with *distilled water*.
6. Fill a beaker 2/3 full with *distilled water*.
7. Add droppers of *iodine solution* to the water in the beaker until an amber (tealike) color is apparent.
8. Record the color of the solution in the bag and the beaker in Table 4.3.
9. Place the bag in the beaker with the open end hanging over the edge. Secure the open end of the bag to the beaker with a rubber band as shown (Fig. 4.6). Make sure the contents do not spill into the beaker.

Figure 4.6   Placement of dialysis bag in water containing iodine.

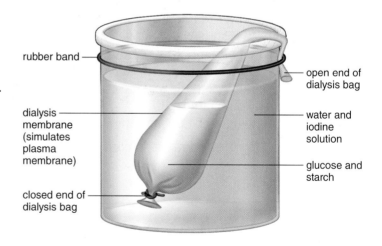

rubber band

dialysis membrane (simulates plasma membrane)

closed end of dialysis bag

open end of dialysis bag

water and iodine solution

glucose and starch

After about 5 minutes, at the end of the experiment:

10. You will note a color change. Record the color of the solution in the bag and the beaker in Table 4.3.
11. Mark off a test tube at 1 cm and 3 cm.
12. Draw solution from near the bag and at the bottom of the beaker for testing with Benedict's reagent. Fill the test tube to the first mark with this solution. Add *Benedict's reagent* to the 3 cm mark. Heat in a boiling water bath for 5 to 10 minutes, observe any color change, and record your results as positive or negative in Table 4.3. (Optional use of glucose test strip: Dip glucose test strip into beaker. Compare stick with chart provided by instructor.)
13. Remove the dialysis bag from the beaker. Dispose of it and the used Benedict's reagent solution in the manner directed by your instructor.

> ⚠️ **Benedict's reagent** is highly corrosive. Exercise care in using this chemical. If any should spill on your skin, wash the area with mild soap and water. Follow your instructor's directions for disposal of this chemical.

Table 4.3 Solute Diffusion Across Plasma Membrane

| | At Start of Experiment | | At End of Experiment | | |
	Contents	Color	Color	Benedict's Test Results	Conclusion
Bag	Glucose Starch			———	
Beaker	Water Iodine				

Conclusions: Solute Diffusion Across Plasma Membrane

- Based on the color change noted in the bag, conclude what solute diffused across the dialysis membrane from the beaker to the bag, and record your conclusion in Table 4.3.
- From the results of the Benedict's test on the beaker contents, conclude what solute diffused across the dialysis membrane from the bag to the beaker, and record your conclusion in Table 4.3.

- Which solute did not diffuse across the dialysis membrane from the bag to the beaker? _____

  How do you know? _____

4.4 Osmosis: Diffusion of Water Across Plasma Membrane

Osmosis is the diffusion of water across the plasma membrane of a cell. Just like any other molecule, water follows its concentration gradient and moves from the area of higher concentration to the area of lower concentration.

Experimental Procedure: Speed of Osmosis

To demonstrate osmosis, a thistle tube is covered with a differentially permeable membrane at its lower opening and partially filled with 50% starch solution. The whole apparatus is placed in a beaker containing distilled water, as described in the legend for Figure 4.7. Therefore, the water concentration in the beaker is 100%. Water molecules can move freely between the thistle tube and the beaker.

Figure 4.7 Osmosis demonstration.

a. A thistle tube, covered at the broad end by a differentially permeable membrane, contains a starch solution. The beaker contains distilled water. **b.** The solute (starch) is unable to pass through the membrane (red), but the water (arrows) passes through in both directions. There is a net movement of water toward the inside of the thistle tube, where there is a higher solute (and lower water) concentration. **c.** Due to the incoming water molecules, the level of the solution rises in the thistle tube.

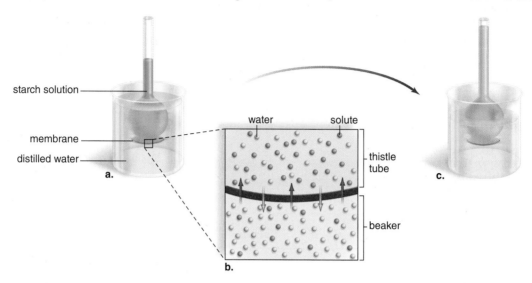

1. First note the level of liquid in the thistle tube, and then measure how far it travels up the thistle tube in 10 minutes: _____ mm.

2. Calculate the speed of osmosis under these conditions: _____ mm/hr.

Conclusions: Speed of Osmosis

- In which direction was there a net movement of water? _____
  Explain what is meant by "net movement" after examining the arrows in Figure 4.7b.

  _____

- If the starch molecules moved from the thistle tube to the beaker, would there have been a net movement of water into the thistle tube? _____ Why wouldn't large starch molecules be able to move across the membrane from the thistle tube to the beaker?

  _____

- Explain why the water level in the thistle tube rose: In terms of solvent concentration, water moved from the area of _____ water concentration to the area of _____ water concentration across a differentially permeable membrane.

Tonicity

Tonicity is the relative concentration of solute (particles), and therefore also of solvent (water), outside the cell compared with inside the cell.

- An **isotonic solution** has the same concentration of solute (and therefore of water) as the cell. When a cell is placed in an isotonic solution, there is no net movement of water. For example, a solution of 0.9% NaCl is isotonic to red blood cells. In such a solution, red blood cells maintain their normal appearance (Fig. 4.8a).

- A **hypertonic solution** has a higher solute (therefore, lower water) concentration than the cell. When cells are placed in a hypertonic solution, water moves out of the cell into the solution. For example, a solution greater than 0.9% NaCl is hypertonic to red blood cells. In such a solution, the cells shrivel up, a process called **crenation** (Fig. 4.8*b*).
- A **hypotonic solution** has a lower solute (therefore, higher water) concentration than the cell. When cells are placed in a hypotonic solution, water moves from the solution into the cell. For example, a solution of less than 0.9% NaCl is hypotonic to red blood cells. In such a solution, the cells swell to bursting, a process called **hemolysis** (Fig. 4.8*c*).

Figure 4.8 Tonicity and red blood cells.

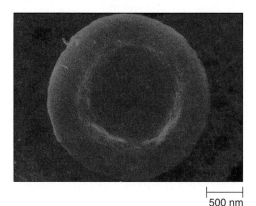 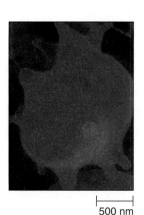 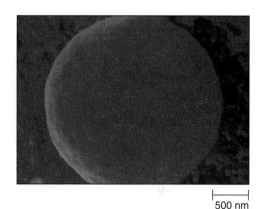

| 500 nm | 500 nm | 500 nm |

a. Isotonic solution. Red blood cell has normal appearance due to no net gain or loss of water.

b. Hypertonic solution. Red blood cell shrivels due to loss of water.

c. Hypotonic solution. Red blood cell fills to bursting due to gain of water.

Experimental Procedure: Effect of Tonicity on Red Blood Cells

Three stoppered test tubes on display have the following contents:
Tube 1: 0.9% NaCl plus a few drops of whole sheep blood
Tube 2: 10% NaCl plus a few drops of whole sheep blood
Tube 3: 0.9% NaCl *plus distilled water* and a few drops of whole sheep blood

> ⚠ Do not remove the stoppers of test tubes during this procedure.

1. In the second column of Table 4.4, record the tonicity of each tube in relation to red blood cells.
2. Hold each tube in front of one of the pages of your lab manual. Determine whether you can see the print on the page through the tube. Record your findings in the third column of Table 4.4.
3. Explain in the fourth column of Table 4.4 why you can or cannot see the print.

Table 4.4	Effect of Tonicity on Red Blood Cells		
Tube	**Tonicity**	**Print Visibility**	**Explanation**
1			
2			
3			

Laboratory 4   Cell Structure and Function   **51**

When plant cells are in a hypotonic solution, the large central vacuole gains water and exerts pressure, called **turgor pressure.** As depicted in Figure 4.9*a*, where do you find the chloroplasts? _____

When plant cells are in a hypertonic solution, the central vacuole loses water, and the cytoplasm pulls away from the cell wall. This is called **plasmolysis**. As depicted in Figure 4.9*b*, where do you find the chloroplasts?_____

Hypotonic Solution

1. If possible, use the *Elodea* slide you prepared earlier in this laboratory. If not, prepare a new wet mount of a small *Elodea* leaf using fresh water.
2. After several minutes, focus on the surface of the cells, and compare your slide with Figure 4.9*a*.

_____

3. Complete the portion of Table 4.5 that pertains to a hypotonic solution.

Hypertonic Solution

1. Prepare a new wet mount of a small *Elodea* leaf using a 10% NaCl solution.
2. After several minutes, focus on the surface of the cells, and compare your slide with Figure 4.9*b*.

_____

3. Complete the portion of Table 4.5 that pertains to a hypertonic solution.

Figure 4.9   *Elodea* cells.
a. Surface view of cells in a hypotonic solution *(above)* and longitudinal section diagram *(below)*. The large central vacuole, filled with water, pushes the cytoplasm, including the chloroplasts, right up against the cell wall. **b.** Surface view of cells in a hypertonic solution *(above)* and longitudinal section diagram *(below)*. When the central vacuole loses water, cytoplasm, including the chloroplasts, piles up in the center of the cell because the cytoplasm has pulled away from the cell wall. (*a:* Magnification 400×)

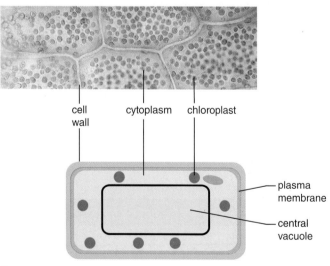

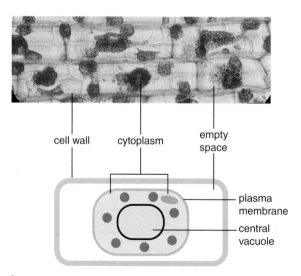

a.   b.

Table 4.5 Effect of Tonicity on *Elodea* Cells

Tonicity	Appearance of Cells	Due to (Scientific Term)
Hypotonic		
Hypertonic		

Conclusions: Hypotonic and Hypertonic Solutions

- In a hypotonic solution, the large central vacuole of plant cells exerts _____ pressure, and the chloroplasts are seen _____ the cell wall.

- In a hypertonic solution, the central vacuole loses water, and the cytoplasm including the chloroplasts have _____ the cell wall. This is called _____.

Experimental Procedure: Potato Strips

(This experimental procedure runs for one hour. Prior setup can maximize time efficiency.)

1. Cut two strips of potato, each about 7 cm long and 1.5 cm wide.
2. Label two test tubes 1 and 2. Place one *potato strip* in each tube.
3. Fill tube 1 with *water* to cover the potato strip.
4. Fill tube 2 with 10% *sodium chloride* (NaCl) to cover the potato strip.
5. After one hour, remove the potato strips from the test tubes and place them on a paper towel. Observe each strip for limpness (water loss) or stiffness (water gain). Which tube has the limp potato strip?

   _____ Use tonicity to explain why water diffused out of the potato strip in this tube? _____
   _____

   Which tube has the stiff potato strip? _____ Use tonicity to explain why water diffused into the potato strip in this tube? _____

6. Use this space to create a table to display your results. Give your table a title and columns for tube number and contents, tonicity, results, and explanation.

4.5 pH and Cells

The pH of a solution indicates its hydrogen ion concentration [H⁺]. The **pH scale** ranges from 0 to 14. A pH of 7 is neutral (Fig. 4.10). A pH lower than 7 indicates that the solution is acidic (has more hydrogen ions than hydroxide ions), whereas a pH greater than 7 indicates that the solution is basic (has more hydroxide ions than hydrogen ions).

The concept of pH is important in biology because living organisms are very sensitive to hydrogen ion concentrations. For example, in humans the pH of the blood must be maintained at about 7.4 or we become ill. All living things need mechanisms to maintain the hydrogen ion concentration, or pH, at a constant level. A **buffer** is a system of chemicals that takes up excess hydrogen ions or hydroxide ions, as appropriate.

Why are cells and organisms buffered? _____

Figure 4.10   The pH scale.
The proportionate amount of hydrogen ions (H⁺) to hydroxide ions (OH⁻) is indicated by the diagonal line.

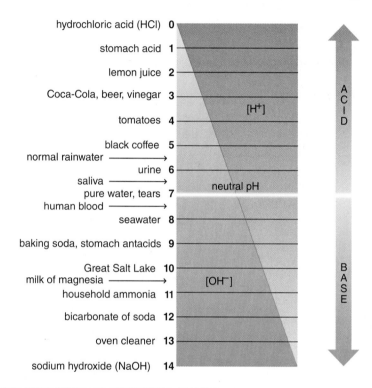

Experimental Procedure: pH and Cells

1. Label three test tubes, and fill them to the halfway mark as follows: tube 1: *water;* tube 2: *buffer* (a buffered inorganic solution); and tube 3: *cytoplasm* (a buffered protein solution).

2. Use pH paper to determine the pH of each tube. Dip the end of a stirring rod into the solution, and then touch the stirring rod to a 5 cm strip of pH paper. Read the current pH by matching the color observed with the color code on the pH paper package. Record your results in the "pH Before Acid" column in Table 4.6.

> ⚠ **Hydrochloric acid (HCl)** used to produce an acid pH is a strong, caustic acid. Exercise care in using this chemical. If any HCl spills on your skin, rinse immediately with clear water. Follow your instructor's directions for disposal of tubes that contain HCl.

3. Add 0.1 N hydrochloric acid (HCl) dropwise to each tube until you have added 5 drops—shake or swirl after each drop. Use pH paper as in step 2 to determine the new pH of each solution. Record your results in the "pH After Acid" column in Table 4.6.

Table 4.6 pH and Cells

Tube	Contents	pH Before Acid	pH After Acid	Explanation
1	Water			
2	Buffer			
3	Cytoplasm			

Conclusions: pH and Cells

- Enter your explanations in the last column of Table 4.6.
- Why would you expect cytoplasm to be as effective as the buffer in maintaining pH?_____

_____

Experimental Procedure: Effectiveness of Antacids

Perform this procedure to test the ability of commercial products such as Alka-Seltzer, Rolaids, Tums, or antacid tablets to absorb excess H^+.

1. Use a mortar and pestle to grind up the amount of *antacid* that is listed as one dose.
2. For each antacid tested, use a 100 mL of *phenol red solution* diluted to a faint pink to wash the antacid into a 250 mL beaker. Phenol red solution is a pH indicator that turns yellow in an acid and red in a base. Use a stirring rod to get the powder to dissolve.
3. Add and count the number of *0.1 N HCl drops* it takes for the solution to turn light yellow.
4. Record your results in Table 4.7.

Table 4.7 Effectiveness of Antacids

Antacid	Drops of Acid Needed to Reach End Point	Evaluation
1		
2		
3		

Conclusions: Effectiveness of Antacids

- Participate with others in concluding which of the antacids tested neutralizes the most acid.

_____

- Did a difference in dosage (convert to mg) have any effect on the results?_____

_____

- Which of the substances on the label could be a buffer?_____

1. Contrast the location of DNA in a eukaryotic cell with that of a prokaryotic cell. _____

2. What characteristics do all eukaryotic cells have in common? _____
   _____

3. Why would you predict that an animal cell, but not a plant cell, might burst when placed in a hypotonic solution? _____
   _____

4. Which of the cellular organelles would be included in a category called:
   a. Membranous canals and vacuoles? _____
   _____
   b. Energy-related organelles? _____
   _____

5. How do you distinguish between rough endoplasmic reticulum and smooth endoplasmic reticulum?
   a. Structure _____
   b. Function _____

6. If a dialysis bag filled with water is placed in a molasses solution, what do you predict will happen to the weight of the bag over time? _____
   Why? _____
   _____

7. How does plant cell structure aid the ability of plants to stand upright? _____
   _____
   _____

8. The police are trying to determine if some nondescript material removed from a crime scene was plant matter. What would you suggest they microscopically look for? _____
   _____

9. A test tube contains red blood cells and a salt solution. When the tube is held up to a page, you can see the print. With reference to a concentration of 0.9% sodium chloride (NaCl), how concentrated is the salt solution? _____

10. Predict the microscopic appearance of cells in the leaf tissue of a wilted plant. _____
    _____

Biology Website

Enhance your study of the text and laboratory manual with study tools, practice tests, and virtual labs. Also ask your instructor about the resources available through ConnectPlus, including the media-rich eBook, interactive learning tools, and animations.

www.mhhe.com/maderbiology11

McGraw-Hill Access Science Website

An Encyclopedia of Science and Technology Online which provides more information including videos that can enhance the laboratory experience.

www.accessscience.com

5

How Enzymes Function

Introduction

The cell carries out many chemical reactions. A possible chemical reaction can be indicated like this:

$$A + B \longrightarrow C + D$$
$$\text{reactants} \qquad \text{products}$$

In all chemical reactions, the **reactants** are molecules that undergo a change, which results in the **products.** The arrow stands for the change that produced the product(s). All the reactions that occur in a cell have an enzyme. **Enzymes** are organic catalysts that speed reactions. Because enzymes are **specific** and speed only one type of reaction, they are given names. In today's laboratory, you will be studying the action of the enzyme **catalase.** The reactants in an enzymatic chemical reaction are called the enzyme's **substrate(s)** (Fig. 5.1).

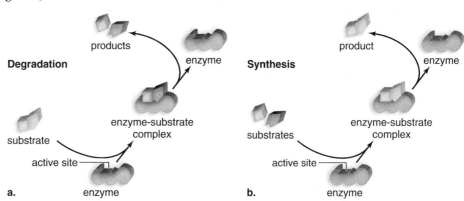

Figure 5.1   Enzymatic action.
The reaction occurs on the surface of the enzyme at the active site. The enzyme is reusable. **a.** Degradation: substrate is broken down. **b.** Synthesis: substrates are combined.

Enzymes are specific because they have a shape that accommodates the shape of their substrates. Enzymatic reactions can be indicated like this:

$$E + S \longrightarrow ES \longrightarrow E + P$$

In this reaction, E = enzyme, ES = enzyme-substrate complex, and P = product.

Two types of enzymatic reactions in cells are shown in Figure 5.1. During degradation reactions, the substrate is broken down to the product(s), and during synthesis reactions, the substrates are joined to form a product. A number of other types of reactions also occur in cells. The location where the enzyme and substrate form an enzyme-substrate complex is called the **active site** because the reaction occurs here. At the end of the reaction, the product is released, and the enzyme can then combine with its substrate again. A cell needs only a small amount of an enzyme because enzymes are used over and over. Some enzymes have turnover rates well in excess of a million product molecules per minute.

5.1 Catalase Activity

Catalase is involved in a degradation reaction: Catalase speeds the breakdown of hydrogen peroxide (H_2O_2) in nearly all organisms including bacteria, plants, and animals. A cellular organelle called a peroxisome, which contains catalase, is present in every plant and animal organ. This means that we could use any plant or animal organ as our source of catalase today. Commonly, school laboratories use the potato as a source of catalase because potatoes are easily obtained and cut up.

Catalase performs a useful function in organisms because hydrogen peroxide is harmful to cells. Hydrogen peroxide is a powerful oxidizer that can attack and denature cellular molecules like DNA! Knowing its harmful nature, humans use hydrogen peroxide as a commercial antiseptic to kill germs (Fig. 5.2). In reduced concentration, hydrogen peroxide is a whitening agent used to bleach hair and teeth. Skillful technicians use it to provide oxygen to aquatic plants and fish, but it is also used industrially to clean most anything from tubs to sewage. It's even put in glow sticks where it reacts with a dye that then emits light.

When catalase speeds the breakdown of hydrogen peroxide, water and oxygen are released.

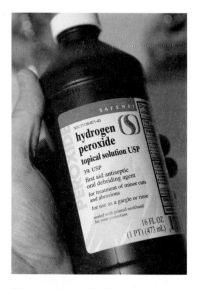

Figure 5.2 Hydrogen peroxide.
Bubbling occurs when you apply hydrogen peroxide to a cut because oxygen is being released when catalase, an enzyme present in the body's cells, degrades hydrogen peroxide.

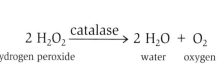

$$2\ H_2O_2 \xrightarrow{\text{catalase}} 2\ H_2O + O_2$$
hydrogen peroxide   water   oxygen

What is the reactant in this reaction? _____ What is the substrate for catalase? _____

What are the products in this reaction? _____ and _____ Bubbling occurs as the reaction proceeds. Why? _____

In the experimental procedure that follows, you will use bubble height to indicate the amount of product per unit time and therefore enzyme activity. Examine Table 5.1 and hypothesize which tube (1, 2, or 3) will have a greater bubble height. Explain your answer. _____

_____

With a wax pencil, label and mark three clean test tubes at the 1 cm and 5 cm levels.

Tube 1
  1. Fill to the first mark with *catalase* buffered at pH 7.0, the optimum pH for catalase.
  2. Fill to the second mark with *hydrogen peroxide*. Swirl well to mix, and wait at least 20 seconds for bubbling to develop.
  3. Measure the height of the bubble column (in millimeters), and record your results in Table 5.1.

Tube 2
  1. Fill to the first mark with *water*.
  2. Fill to the second mark with *hydrogen peroxide*. Swirl well to mix, and wait at least 20 seconds.
  3. Measure the height of the bubble column (in millimeters), and record your results in Table 5.1.

Tube 3
  1. Fill to the first mark with *catalase*.
  2. Fill to the second mark with *sucrose solution*. Swirl well to mix; wait 20 seconds.
  3. Measure the height of the bubble column, and record your results in Table 5.1.

Table 5.1	Catalase Activity		
Tube	**Contents**	**Bubble Column Height**	**Explanation**
1	Catalase Hydrogen peroxide		
2	Water Hydrogen peroxide		
3	Catalase Sucrose solution		

Conclusions: Catalase Activity

- Which tube showed the amount of bubbling you expected? _____ Record your explanation in Table 5.1.

- Which tube is a negative control? _____ If this tube showed bubbling, what could you conclude about your procedure? _____
Record your explanation in Table 5.1.

- Enzymes are specific; they speed only a reaction that contains their substrate. Which tube exemplifies this characteristic of an enzyme? _____ Record your explanation in Table 5.1.

5.2 Effect of Temperature on Enzyme Activity

The active sites of enzymes increase the likelihood that substrate molecules will find each other and interact. Therefore, enzymes lower the energy of activation (the temperature needed for a reaction to occur). Still, increasing the temperature is expected to increase the likelihood that active sites will be occupied because molecules move about more rapidly as the temperature rises. In this way, a warm temperature increases enzyme activity.

The shape of an enzyme and its active site must be maintained or else they will no longer be functional. A very high temperature, such as the one that causes water to boil, is likely to cause weak bonds of a protein to break; and if this occurs, the enzyme **denatures**—it loses its original shape and the active site will no longer function to bring reactants together. Now enzyme activity plummets.

With this information in mind, examine Table 5.2 below and hypothesize which tube (1, 2, or 3) will have more product per unit time as judged by bubble height. _____ Explain your answer. _____

Experimental Procedure: Effect of Temperature

With a wax pencil, label and mark three clean test tubes at the 1 cm and 5 cm levels.

1. Fill each tube to the first mark with *catalase* buffered at pH 7.0, the optimum pH for catalase.
2. Place tube 1 in a refrigerator or cold water bath, tube 2 in an incubator or warm water bath, and tube 3 in a boiling water bath. Complete the second column in Table 5.2. Wait 15 minutes.
3. As soon as you remove the tubes one at a time from the refrigerator, incubator, and boiling water, fill to the second mark with *hydrogen peroxide*.
4. Swirl well to mix, and wait 20 seconds.
5. Measure the height of the bubble column (in millimeters) in each tube, and record your results in Table 5.2. Plot your results in Figure 5.3.

Table 5.2 Effect of Temperature

Tube	Temperature °C	Bubble Column Height (mm)	Explanation
1 Refrigerator			
2 Incubator			
3 Boiling water			

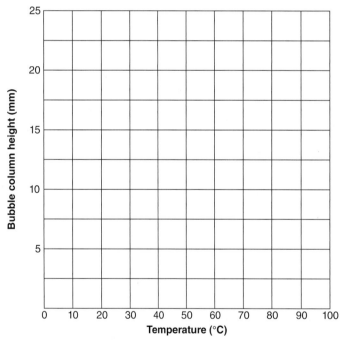

Figure 5.3 Effect of temperature on enzyme activity.

Conclusions: Effect of Temperature

- The bubble height indicates the degree of enzyme activity. Was your hypothesis supported? _____ Explain in Table 5.2 the degree of enzyme activity per tube.

- What is your conclusion concerning the effect of temperature on enzyme activity? _____

5.3 Effect of Concentration on Enzyme Activity

Consider that if you increased the number of caretakers per number of children, it is more likely that each child will have quality time with a caretaker. So it is if you increase the amount of enzyme per amount of substrate, it is more likely that the active sites of enzymes will be occupied and a reaction will take place. With this in mind, examine Table 5.3 and hypothesize which tube (1, 2, or 3) will have more product per unit time as judged by bubble height. _____ Explain your answer. _____

Experimental Procedure: Effect of Enzyme Concentration

With a wax pencil, label three clean test tubes.

Tube 1
1. Mark this tube at the 1 cm and 5 cm levels.
2. Fill to the first mark with *water* and to the second mark with *hydrogen peroxide*.
3. Swirl well to mix, and wait 20 seconds.
4. Measure the height of the bubble column (in millimeters), and record your results in Table 5.3.

Tube 2
1. Mark this tube at the 1 cm and 5 cm levels.
2. Fill to the first mark with buffered *catalase* and to the second mark with *hydrogen peroxide*.
3. Swirl well to mix, and wait 20 seconds.
4. Measure the height of the bubble column (in millimeters), and record your results in Table 5.3.

Tube 3
1. Mark this tube at the 3 cm and 7 cm levels.
2. Fill to the first mark with buffered *catalase* and to the second mark with *hydrogen peroxide*.
3. Swirl well to mix, and wait 20 seconds.
4. Measure the height of the bubble column (in millimeters), and record your results in Table 5.3.

Table 5.3 Effect of Enzyme Concentration

Tube	Amount of Enzyme	Bubble Column Height (mm)	Explanation
1	none		
2	1 cm		
3	3 cm		

Conclusions : Effect of Concentration

- The bubble indicates the degree of enzyme activity. Was your hypothesis supported? _____ Explain in Table 5.3 the degree of enzyme activity per tube.

- If unlimited time was allotted, would the results be the same in all tubes? _____ Explain why or why not. _____

- Would you expect similar results if the substrate concentration were varied in the same manner as the enzyme concentration? _____ Why or why not? _____

- What is your conclusion concerning the effect of concentration on enzyme activity? _____

5.4   Effect of pH on Enzyme Activity

Each enzyme has a pH at which the speed of the reaction is optimum (occurs best). Any higher or lower pH affects hydrogen bonding and the structure of the enzyme, leading to reduced activity.

> ⚠ **Hydrochloric acid (HCl)** used to produce an acid pH is a strong, caustic acid, and sodium hydroxide (NaOH) used to produce a basic pH is a strong, caustic base. Exercise care in using these chemicals, and follow your instructor's directions for disposal of tubes that contain these chemicals. If any acidic or basic solutions spill on your skin, rinse immediately with clear water.

Catalase is an enzyme found in cells where the pH is near 7 (called neutral pH). Other enzymes prefer different pHs. The pancreas secretes a slightly basic (below pH 7) juice into the digestive tract and the stomach wall releases a very acidic digestive juice which can be as low as pH 2. With this information about catalase in mind, examine Table 5.4 and hypothesize which tube (1, 2, or 3) will have more product per unit time as judged by the bubble height. _____ Explain your answer. _____

Experimental Procedure: Effect of pH

With a wax pencil, label and mark three clean test tubes at the 1 cm, 3 cm, and 7 cm levels. Fill each tube to the 1 cm level with nonbuffered *catalase*.

Tube 1
1. Fill to the second mark with *water* adjusted to pH 3 by the addition of *HCl*. Wait one minute.
2. Fill to the third mark with *hydrogen peroxide*.
3. Swirl to mix, and wait 20 seconds.
4. Measure the height of the bubble column (in millimeters), and record your results in Table 5.4.

Tube 2
1. Fill to the second mark with *water* adjusted to pH 7. Wait one minute.
2. Fill to the third mark with *hydrogen peroxide*.
3. Swirl to mix, and wait 20 seconds.
4. Measure the height of the bubble column (in millimeters), and record your results in Table 5.4.

Tube 3
1. Fill to the second mark with *water* adjusted to pH 11 by the addition of *NaOH*. Wait one minute.
2. Fill to the third mark with *hydrogen peroxide*.
3. Swirl to mix, and wait 20 seconds.
4. Measure the height of the bubble column (in millimeters), and record your results in Table 5.4.

Table 5.4	Effect of pH		
Tube	**pH**	**Bubble Column Height (mm)**	**Explanation**
1	3		
2	7		
3	11		

Plot your results from Table 5.4 here.

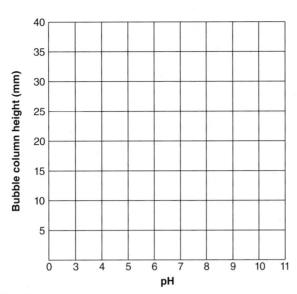

Figure 5.4 Effect of pH on enzyme activity.

Conclusions : Effect of pH

- The amount of bubble height indicates the degree of enzyme activity. Was your hypothesis supported? _____ Explain in Table 5.4 the degree of enzyme activity per tube.
- What is your conclusion concerning the effect of pH on enzyme activity? _____

5.5   Comparison of Factors That Affect Enzymatic Activity

In Table 5.5, summarize what you have learned about factors that affect the speed of an enzymatic reaction. For example, in general, what type of temperature promotes enzyme activity, and what type inhibits enzyme activity? Answer similarly for enzyme or substrate concentration and pH.

Table 5.5 Factors That Affect Enzyme Activity		
Factors	**Promote Enzyme Activity**	**Inhibit Enzyme Activity**
Enzyme specificity		
Temperature		
Enzyme or substrate concentration		
pH		

Conclusions: Factors That Affect Enzyme Activity

- Why does enzyme specificity promote enzyme activity? _____
  _____
- Why does a warm temperature promote enzyme activity? _____
  _____
- Why does increasing enzyme concentration promote enzyme activity? _____
  _____
- Why does optimum pH promote enzyme activity? _____
  _____

Virtual Lab **Enzyme-Controlled Reactions** A Virtual Laboratory called Enzyme-Controlled Reactions is available on the *Biology* website **www.mhhe.com/maderbiology11**. After opening, this virtual lab, click on the TV screen to watch a video about enzymes.

Experiment 1 Determining an Enzyme's Optimum pH

Hypothesis

Considering that the pH of the medium affects the shape of protein and the shape of an enzyme is necessary to its enzymatic function, formulate a general hypothesis about the relationship between pH and enzyme function.

Hypothesis: _____ .

Steps of the Experiment

In this experiment, it is possible to vary both the amount of substrate and the pH. Which of these factors should you hold constant during this experiment for determining an enzyme's optimum pH? _____ Knowing how enzymes work, why might you decide to use the maximum concentration of substrate available? _____

1. Sequentially change the pH to pH 3, pH 5, pH 7, pH 9, and pH 11 by using the down or up arrows beneath the test tubes.

2. Click and drag 8 g of substrate (far right only) to each test tube.

Results

3. Click the computer monitor to see the number of product molecules per minute formed for each test tube. In Table 1 below sequentially enter your results per pH when the substrate is held constant at 8 g.

Table 1 Product/min/pH at substrate concentration of 8 g					
pH	pH 3	pH 5	pH 7	pH 9	pH 11
Product/min					

Did the enzyme perform best at a particular pH? _____ What pH? _____

Use this graph to show your results:

Conclusion

Results of this experiment (support or do not support) the hypothesis that an enzyme performs best at a particular pH. _____ This particular pH is called the **optimum pH.**

Experiment 2 Determining the Optimum Substrate Concentration

Hypothesis
Knowing that enzymes perform best when their active site is always filled with substrate, hypothesize how increasing the substrate concentration could affect the product/min:

Hypothesis: _____

Steps of the Experiment
In this experiment what experimental factor will you hold constant? _____ Which pH would you use and why? _____

1. Click the reset button and this time hold the pH constant. Note that all test tubes already indicate pH 7.

2. Click and drag substrate concentrations from 0.5 to 8.0 g to the test tubes.

Results
3. Click the computer monitor to see the number of product molecules formed per minute formed for each test tube. In Table 2 below sequentially enter your results per amount of substrate when the pH is held constant at pH 7.

Table 2 Product/min at pH 7					
Amount of Substrate	**0.5 g**	**1.0 g**	**2.0 g**	**4.0 g**	**8.0 g**
Product/min					

Use this graph to show your results:

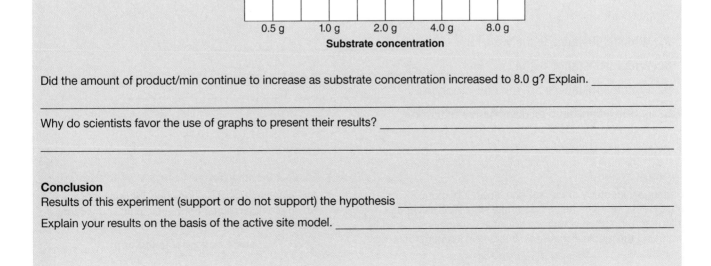

Did the amount of product/min continue to increase as substrate concentration increased to 8.0 g? Explain. _____

_____

Why do scientists favor the use of graphs to present their results? _____

_____

Conclusion
Results of this experiment (support or do not support) the hypothesis _____

Explain your results on the basis of the active site model. _____

1. What happens at the active site of an enzyme? _____
   _____

2. On the basis of the active site, explain why the following conditions speed a chemical reaction:

   a. More enzyme _____

   b. More substrate _____

3. Name three other conditions (other than the ones mentioned in question 2) that maximize enzymatic reactions.

   a. _____   **b.** _____   **c.** _____

4. Explain the necessity for each of the three conditions you listed in question 3.

   a. _____

   b. _____

   c. _____

5. Lipase is a digestive enzyme that digests fat droplets when the pH is basic due to the presence of $NaHCO_3$. Circle which of the following test tubes would show digestion following incubation at 37°C? Explain why this one would and the the others would not show digestion.

   Tube 1: Water, fat droplets _____

   Tube 2: Water, fat droplets, lipase _____

   Tube 3: Water, fat droplets, lipase, $NaHCO_3$ _____

   Tube 4: Water, lipase, $NaHCO_3$ _____

6. Fats are digested to fatty acids and glycerol. As the reaction described in question 5 proceeds, the

   solution will become what type pH? _____ Why? _____
   _____

Given the following reaction:

$$2\ H_2O_2 \xrightarrow{\text{catalase}} 2\ H_2O + O_2$$

hydrogen peroxide   water   oxygen

7. Which substance is the substrate? _____

8. Which substance is the enzyme? _____

9. Which substances are the end products? _____

10. Is this a synthetic or degradative reaction? _____ .

6
Photosynthesis

Introduction

The simplified overall equation for **photosynthesis** is

$$CO_2 + H_2O \xrightarrow{\text{solar energy}} (CH_2O) + O_2$$

In this equation, (CH_2O) represents any general carbohydrate. Sometimes, this equation is multiplied by 6 so that glucose $(C_6H_{12}O_6)$ appears as an end product of photosynthesis.

Photosynthesis takes place in chloroplasts (Fig. 6.1). Here membranous thylakoids are stacked in grana surrounded by the stroma. During the so-called light reactions, pigments within the *thylakoid* membranes absorb solar energy, water is split, and oxygen is released. The Calvin cycle reactions occur within the *stroma.* During these reactions, carbon dioxide (CO_2) is reduced to a carbohydrate (CH_2O).

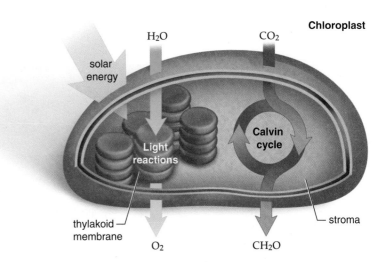

Figure 6.1 Overview of photosynthesis.
Photosynthesis includes the light reactions when energy is collected and O_2 is released and the Calvin cycle reactions when CO_2 is reduced and carbohydrate (CH_2O) is formed.

6.1 Photosynthetic Pigments

Later in this laboratory, we will learn that visible (white) light is composed of different colors of light (see Fig. 6.3). We can hypothesize that leaves contain various pigments that absorb the solar energy of these different colors. Restate this hypothesis here:

Hypothesis: _____

To test this hypothesis, we will use a technique called chromatography to separate the pigments located in leaves. Chromatography separates molecules from each other on the basis of their solubility in particular solvents. The solvents used in the following Experimental Procedure are petroleum ether and acetone, which have no charged groups and are therefore nonpolar. As a nonpolar solvent moves up the chromatography paper, the pigment moves along with it. The more nonpolar a pigment, the more soluble it is in a nonpolar solvent and the faster and farther it proceeds up the chromatography paper.

Experimental Procedure: Photosynthetic Pigments

1. Assemble a **chromatography apparatus** (Fig. 6.2a):
    *   Obtain a large, dry test tube and a cork with a hook.
    *   Attach a strip of precut **chromatography paper** (hold from the top) to the hook and test for fit. The paper should hang straight down and barely touch the bottom of the test tube; trim if necessary.
    *   Measure 2 cm from the bottom of the paper; place here a small dot with a pencil (not a pen).
    *   With the stopper in place, mark the test tube with a wax pencil 1 cm below where the dot is on the paper.
    *   Set the chromatography apparatus in a test tube rack.

Figure 6.2   Paper chromatography.
a. For a chromatography apparatus, the paper must be cut to size and arranged to hang down without touching the sides of a dry tube. **b.** The pigment (chlorophyll) solution is applied to a designated spot. **c.** The chromatogram, which develops after the spotted paper is suspended in the chromatography solution, will show these pigments.

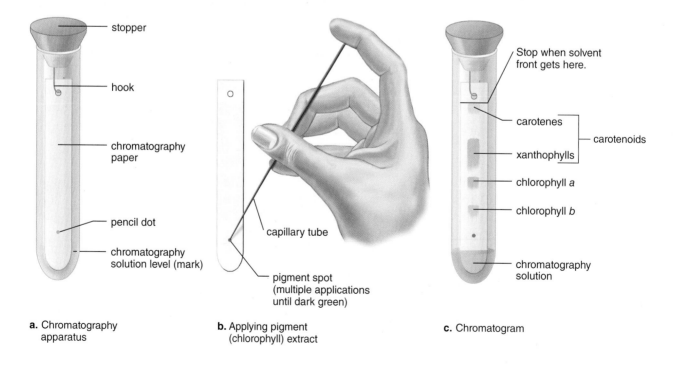

a. Chromatography apparatus

b. Applying pigment (chlorophyll) extract

c. Chromatogram

2. Prepare the chromatography paper (Fig. 6.2*b*):
   - Remove the chromatography paper from the test tube; place on a paper towel and apply **plant pigments** to the dot on the paper as directed by your instructor. Figure 6.2*b* shows how to apply pigment extract using a capillary tube.
   - In the **fume hood**, add **chromatography solution** to the mark you made on the test tube. Place the chromatography paper attached to the hook back in the test tube. The pigment spot should remain above the chromatography solution, but close the chromatography apparatus tightly.

> ⚠ The chromatography solution is toxic and extremely flammable. Do not breathe the fumes, and do not place the chromatography solution near any source of heat. A fume (ventilation) hood is recommended.

3. Develop the chromatogram (Fig. 6.2*c*):
   - Place the reassembled chromatography apparatus in the test tube rack and allow ten minutes for the chromatogram to develop, but check frequently so that the solution does not reach the top of the paper.
   - When the solvent has moved to within 1 cm of the upper edge of the paper, remove the paper. Close the empty apparatus tightly. With a pencil, lightly mark the location of the solvent front (where the solvent stopped on the paper) and allow the chromatogram to dry in the fume hood.

4. Read the chromatogram:
   - Compare your chromatogram to that shown in Figure 6.2*c*. Measure the distance in mm from the dark green pigment spot to the top of each individual pigment band, and record these values in Table 6.1.
   - Measure the distance the solvent moved from the dark green pigment spot to the solvent front and add this value to Table 6.1.
   - Use this formula to calculate the R_f (ratio-factor) values for each pigment, and record these values in Table 6.1:

   $$R_f = \frac{\text{distance moved by pigment}}{\text{distance moved by solvent}}$$

Table 6.1 R_f (Ratio-Factor) Values for Each Pigment

Pigments	Distance Moved (mm)	R_f Values
Carotenes		
Xanthophylls		
Chlorophyll *a*		
Chlorophyll *b*		
Solvent		————

Conclusions: Photosynthetic Pigments

- Do your results support the hypothesis that plant leaves contain various photosynthetic pigments?

_____

Explain. _____

6.2 Solar Energy

During light reactions of photosynthesis solar energy is absorbed by the photosynthetic pigments and is transformed into the chemical energy of a carbohydrate (CH_2O). Without solar energy, photosynthesis would be impossible.

Verify that photosynthesis releases oxygen by reviewing the overall equation for photosynthesis on page 67. Release of oxygen from a plant indicates that the light reactions of photosynthesis are occurring. The oxygen released during photosynthesis is taken up by a plant when cellular respiration occurs. This must be taken into account when the rate of photosynthesis is calculated.

Role of White Light

White (sun) light contains different colors of light, as is demonstrated when white light passes through a prism (Fig. 6.3). White light is the best for photosynthesis because it contains all the colors of light.

Figure 6.3 White light.
White light is made up of various colors, as can be seen when white light passes through a prism.

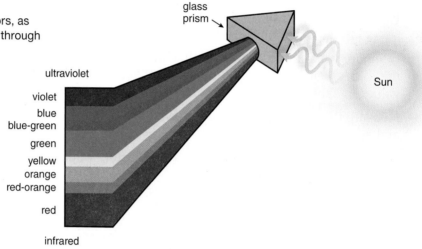

glass
prism

Sun

ultraviolet
violet
blue
blue-green
green
yellow
orange
red-orange
red
infrared

Experimental Procedure: White Light

1. Place a generous quantity of *Elodea* with the cut end up (make sure the cuts are fresh) in a test tube with a rubber stopper containing a piece of glass tubing, as illustrated in Figure 6.4. When assembled, this is your volumeter for studying the need for light in photosynthesis. (Do not hold the volumeter in your hand, as body heat will also drive the reaction forward.) Your instructor will show you how to fix the volumeter in an upright position.

2. Before stoppering the test tube, add sufficient 3% *sodium bicarbonate* ($NaHCO_3$) solution so that, when the rubber stopper is inserted into the tube, the solution comes to rest at about 1/4 the length of the upright glass tubing. Mark this location on the glass tubing with a wax pencil.

3. Place a beaker of plain water next to the *Elodea* tube to serve as a heat absorber. Place a lamp (150 watt) next to the beaker. The tube, beaker, and lamp should be as close to one another as possible.

4. Turn on the lamp. As soon as the edge of the solution in the tubing begins to move, time the reaction for ten minutes. Be careful not to bump the tubing or to readjust the stopper, or your readings will be altered. After ten minutes, mark the edge of the solution, and measure

   in millimeters the distance the edge moved upward: _____ mm/10 min. This is **net**

   photosynthesis, a measurement that does not take into account the oxygen that was used up for

   cellular respiration. Record your results in Table 6.2. Why did the edge move upward? _____

Figure 6.4 Volumeter.
A volumeter apparatus is used
to study the role of light in
photosynthesis.

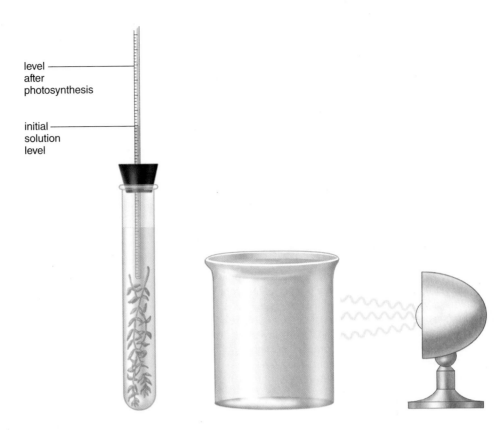

level
after
photosynthesis

initial
solution
level

5. Carefully wrap the tube containing *Elodea* in aluminum foil, and record here the length of time
it takes for the edge of the solution in the tubing to move downward 1 mm: _____. Convert
your measurement to _____ mm/10 min, and record this value for **cellular respiration** in
Table 6.2. (Do not use a minus sign, even though the edge moved downward.) Why does cellular
respiration, which occurs in a plant all the time, cause the edge to move downward? _____

_____

6. If the *Elodea* had not been respiring in step 4, how far would the edge have moved upward?
_____ mm/10 min. This is **gross photosynthesis** (net photosynthesis + cellular respiration).
Record this number in Table 6.2.

7. Calculate the **rate of photosynthesis** (mm/hr) by multiplying gross photosynthesis (mm/10 min)
by 6 (that is, 10 min × 6 = 60 min = 1 hr): _____ mm/hr. Record this value in Table 6.2.

Table 6.2 Rate of Photosynthesis (White Light)		
	Movement of Edge (mm/10 min)	**Rate of Photosynthesis (mm/hr)**
Net photosynthesis (white light)		_____
Cellular respiration (no light)		_____
Gross photosynthesis (net + cellular respiration)		_____

Role of Green Light

Green light is only one part of white light (see Fig. 6.3). The photosynthetic pigments absorb certain colors of light better than other colors (Fig. 6.5). According to Figure 6.5, what color light do the chlorophylls absorb best? _____ Least? _____

What color light do the carotenoids (carotenes and xanthophylls) absorb best? _____ Least? _____

Hypothesize which color light is minimally utilized for photosynthesis. _____ The following Experimental Procedure will test your hypothesis.

Figure 6.5 Action spectrum for photosynthesis.
The action spectrum for photosynthesis is the sum of the absorption spectrums for the pigments chlorophyll *a*, chlorophyll *b*, and carotenoids. The peaks in this diagram represent wavelengths of sunlight absorbed by photosynthetic pigments. The chlorophylls absorb predominantly violet-blue and orange-red light and reflect green light. The carotenoids (carotenes and xanthophylls) absorb mostly blue-green light and reflect yellow-red light.

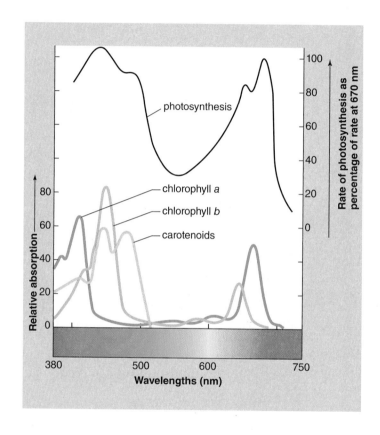

Experimental Procedure: Green Light

1. Add three drops of green dye (or use a green cellophane wrapper) to the beaker of water used in the previous Experimental Procedure until there is a distinctive green color. Remove all previous wax pencil marks from the glass tubing.

2. Record in Table 6.3 your data for gross photosynthesis (mm/10 min) and for rate of photosynthesis for white light (mm/hr) from Table 6.2.

3. Turn on the lamp. Mark the location of the edge of the solution on the glass tubing. As soon as the edge begins to move, time the reaction for ten minutes. After ten minutes, mark the edge of the solution, and measure in millimeters the distance the edge moved. Net photosynthesis for green light = _____ mm/10 min. As before, net photosynthesis does not take into account that oxygen was used for cellular respiration.

4. Carefully wrap the tube containing *Elodea* in aluminum foil, and record here the length of time it takes for the edge of the solution in the tubing to recede 1 mm: _____ . Convert your measurement to _____ mm/10 min. As before, this reading shows how much oxygen was used for cellular respiration.

5. Calculate gross photosynthesis for green light (mm/10 min) as you did for white light, and record your data below in Table 6.3. (As before, add your data for net photosynthesis to that for cellular respiration.)

6. Calculate rate of photosynthesis for green light (mm/hr) as you did for white light, and record your data below in Table 6.3. (As before, convert your data mm/10 min to mm/hr.)

7. Collect the white and green light rates of photosynthesis (mm/hr) from each group in your lab; then average all the rates including your own. In the last column of Table 6.3 record the average for white and green light rates of photosynthesis (mm/hr).

Table 6.3 Rate of Photosynthesis (White and Green Light)

Gross photosynthesis	Your Data	Class Data
White (from Table 6.2)	mm/10 min	_____
Green	mm/10 min	_____
Rate of photosynthesis		
White (from Table 6.2)	mm/hr	mm/hr
Green	mm/hr	mm/hr

8. Calculate the rate of photosynthesis (green light) as a percentage of the rate of photosynthesis (white light) using this equation.

$$\text{Percentage} = \frac{\text{rate of photosynthesis (green light)}}{\text{rate of photosynthesis (white light)}} \times 100$$

This percentage, based on your data recorded in Table 6.3, = _____ . This percentage, based on class data recorded in Table 6.3, = _____ .

Conclusions: Rate of Photosynthesis

- Do your results support the hypothesis that green light is minimally used by a land plant for photosynthesis? _____ Explain, with reference to Figure 6.5. _____

_____

- How does the percentage based on your data differ from that based on class data?

_____

6.3 Carbon Dioxide Uptake

During the Calvin cycle reactions of photosynthesis, the plant takes up carbon dioxide (CO_2) and reduces it to a carbohydrate, such as glucose ($C_6H_{12}O_6$). Therefore, the carbon dioxide in the solution surrounding *Elodea* should disappear as photosynthesis takes place.

Experimental Procedure: Carbon Dioxide Uptake

1. Temporarily remove the *Elodea* from the test tube. Empty the sodium bicarbonate ($NaHCO_3$) solution from the test tube, rinse the test tube thoroughly, and fill with a phenol red solution diluted to a faint pink. (Add more water if the solution is too dark.) Phenol red is a pH indicator that turns yellow in an acid and red in a base.

> ⚠️ **Phenol red** Avoid ingestion, inhalation, and contact with skin, eyes, and mucous membranes. Follow your instructor's directions for disposal of this chemical. Use protective eyewear when performing this experiment.

2. Blow *lightly* on the surface of the solution. Stop blowing as soon as the surface color changes to yellow. Then shake the test tube until the rest of the solution turns yellow.

   Blowing onto the solution adds what gas to the test tube? _____ When carbon dioxide combines with water, it forms carbonic acid; therefore, the solution appears yellow.

3. Thoroughly rinse the *Elodea* with distilled water, return it to the test tube which now contains a yellow solution, and assemble your volumeter as before.

4. The water in the beaker used to absorb heat should be clear.

5. Turn on the lamp, and wait until the edge of the solution just begins to move upward. Note the time. Observe until you note a change in color. How long did the color change take? _____

6. Why did the solution eventually turn red? _____

   _____

7. The carbon cycle includes all the many ways that organisms exchange carbon dioxide with the atmosphere. Figure 6.6 notes the relationship between cellular respiration and photosynthesis. Animals produce carbon dioxide used by plants to carry out photosynthesis. Plants produce the food (and oxygen) that they and animals require to carry out cellular respiration. Therefore, the same carbon atoms pass between animals and plants and between plants and animals (Fig. 6.6).

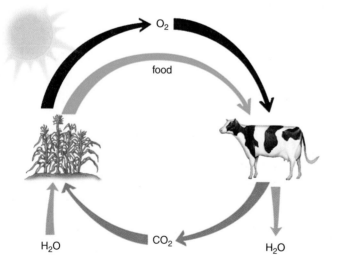

Figure 6.6 Photosynthesis and cellular respiration.
Animals are dependent on plants for a supply of oxygen, and plants are dependent on animals for a supply of carbon dioxide.

6.4 The Light Reactions and the Calvin Cycle Reactions

Review the introduction to this laboratory. Note the overall equation for photosynthesis and that photosynthesis consists of the light reactions and the Calvin cycle reactions. Solar energy is absorbed by photosynthetic pigments during the light reactions and energy is used during the Calvin cycle reactions to reduce carbon dioxide to a carbohydrate.

Light Reactions

1. Photosynthetic pigments

    a. What is the function of the photosynthetic pigments in photosynthesis? _____

    b. How does it benefit a plant to have a variety of photosynthetic pigments? (See Fig. 6.5.) _____

    _____

    c. Green light minimally promotes photosynthesis. Account for this observation with reference to

    the photosynthetic pigments. (See Fig. 6.5.) _____

    d. Does this explain why leaves are green? _____ How so? _____

2. Water and oxygen

    a. What happens to water during the light reactions? _____

    b. What happens to the released oxygen? _____

3. Location of the light reactions

    Fill in the blank: The light reactions take place in the _____ membranes.

4. In your own words, summarize the light reactions based on this laboratory.

Calvin Cycle Reactions

1. Carbon dioxide and carbohydrate

    What happens to carbon dioxide after it is taken up during the Calvin cycle reactions? _____

2. Location of the Calvin cycle reaction

    Fill in the blank: The Calvin cycle reactions take place in the _____.

3. In your own words, summarize the Calvin cycle reactions based on this laboratory.

Light Reactions and the Calvin Cycle Reactions

1. Examine the overall equation for photosynthesis and show that there is a relationship between the light reactions and the Calvin cycle reactions by drawing an arrow between the hydrogen atoms in water and the hydrogen atoms in the carbohydrate.

$$CO_2 + H_2O \longrightarrow (CH_2O) + O_2$$

2. Only because solar energy splits water can hydrogen atoms be used to reduce carbon dioxide. In this sense, solar energy is now stored in the carbohydrate. This energy sustains all the organisms in the biosphere.

Laboratory Review 6

1. How are plant pigments involved in photosynthesis? _____
   _____

2. Why is it beneficial to have several different plant pigments involved in photosynthesis? _____
   _____

3. On what basis does chromatography separate substances? _____
   _____

4. Some types of red algae carry on photosynthesis 70 m beneath the ocean surface. What color light do
   you predict does not penetrate to this depth? _____

Consider the following reaction:

$$CO_2 + H_2O \longrightarrow \underset{\text{carbonic acid}}{H_2CO_3} \longrightarrow H^+ + HCO_3^-$$

5. Phenol red, a pH indicator, turns yellow (indicating acid) when you breathe into a solution. How does
   the reaction explain why the solution turned acidic? _____
   _____

6. When light is available and a plant is added, the solution returns to its original red color. Why does this
   occur? _____
   _____

Gas exchange occurs in both photosynthesis and cellular respiration. Contrast these two processes by
completing the following table:

	Organelle	Gas Given Off	Gas Taken Up
7. Photosynthesis			
8. Cellular respiration			

9. What experimental conditions were used in this laboratory to test for cellular respiration in
   photosynthesizing plant cells? _____

10. Suppose you replaced *Elodea* with animal cells in an experimental test tube testing for photosynthesis.
    Would the results differ according to the use of a white light or no light? _____ Explain. _____
    _____

7

Cellular Respiration

Introduction

In this laboratory, you will study **cellular respiration** in germinating soybeans. Cellular respiration is an ATP-generating process that involves the complete breakdown most often of glucose to carbon dioxide and water. In eukaryotes, glucose breakdown begins in the cytoplasm, but is completed in mitochondria. Cellular respiration is aerobic and requires oxygen, and it results in a buildup of ATP (Fig. 7.1). This equation represents cellular respiration:

> 🕚 **Planning Ahead** You may wish to start the fermentation experiment on page 82 first, to allow time for incubation.

$$C_6H_{12}O_6 \ + \ 6\,O_2 \xrightarrow{\hspace{3cm}} 6\,CO_2 \ + \ 6\,H_2O \ + \ ATP$$

glucose   oxygen   carbon   water
                                     dioxide

Figure 7.1   Cellular respiration.
Both plant cells and animal cells have mitochondria and carry on cellular respiration, a process that utilizes the equation above.

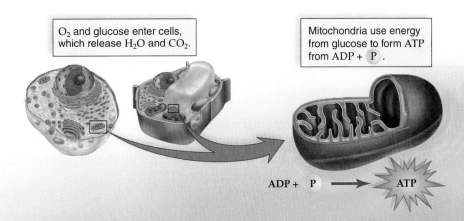

O$_2$ and glucose enter cells, which release H$_2$O and CO$_2$.

Mitochondria use energy from glucose to form ATP from ADP + P .

ADP + P ⟶ ATP

In this laboratory, you will also study **ethanol fermentation,** which is an ATP-generating process. When an organism, such as yeast, breaks down glucose to ethanol and carbon dioxide, only 2 ATP result but the process is anaerobic and does not require oxygen. Fermentation occurs in the cytoplasm and mitochondria are not involved. This reaction represents yeast fermentation:

$$C_6H_{12}O_6 \longrightarrow 2\,CO_2 + 2\,C_2H_5OH + 2\,ATP$$

$$\text{glucose} \qquad\qquad \underset{\text{dioxide}}{\text{carbon}} \quad \text{ethanol}$$

When animals, such as humans, ferment, they produce lactate instead of ethanol and carbon dioxide.

7.1 Cellular Respiration

When germination occurs and plants begin to grow, cellular respiration can provide them with the ATP they need to produce all the molecules that allow them to grow. Consider, for example, Figure 7.2, which depicts soybean germination.

In the Experimental Procedure that follows, we are going to measure the amount of oxygen uptake as evidence that germinating soybeans are carrying on cellular respiration. The need for oxygen by a germinating soybean will be compared to the need by nongerminating soybeans. State a hypothesis for this experiment here:

Hypothesis: _____

Figure 7.2 Germination of a soybean seed.

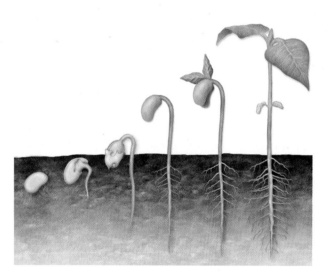

Virtual Lab Energy in a Cell   A virtual lab, called Energy in a Cell is available on the *Biology* website **www.mhhe.com/maderbiology11.** It will help you test your knowledge of photosynthesis and cellular respiration. Follow the directions given but be aware that in this virtual lab the Celvin cycle reactions are called dark reactions.

Experimental Procedure: Cellular Respiration

1. Obtain a volumeter, an apparatus that measures changes in gas volumes. Remove the three vials from the volumeter. Remove the stoppers from the vials and the vials from the volumeter. Label the vials 1, 2, and 3.
2. Using the same amounts, place a small wad of absorbent cotton in the bottom of each vial. Without getting the sides of the vials wet, use a dropper to saturate the cotton with 15% potassium hydroxide (KOH). The KOH absorbs CO_2 as it is given off by the soybeans. Place a small wad of dry cotton on top of the KOH-soaked absorbent cotton (Fig. 7.3).

Figure 7.3 Vials.

In this experiment, three vials are filled as noted.

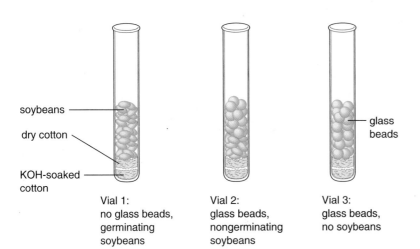

soybeans

dry cotton

KOH-soaked cotton

glass beads

Vial 1:
no glass beads,
germinating
soybeans

Vial 2:
glass beads,
nongerminating
soybeans

Vial 3:
glass beads,
no soybeans

> ⚠ **Potassium hydroxide (KOH)** is a strong, caustic base. Exercise care in using this chemical. If any KOH spills on your skin, rinse immediately with clear water. Follow your instructor's directions for disposal of tubes that contain KOH.

3. Count 25 germinating soybean seeds and add to vial 1. Count 25 dry (nongerminating) soybean seeds and add to vial 2. Add glass beads to vial 2 so that the volume occupied in vial 2 is approximately the same as in vial 1. Add only glass beads to vial 3 to bring to approximately the same volume.

4. Replace the vials in the volumeter and firmly place the stoppers in the vials (Fig. 7.4). Each stopper has a vent (outlet tube) ending in rubber tubing held shut with a clamp. Remove the clamps. Adjust the graduated side arm until only about 5 mm to 1 cm protrudes through the stopper.

5. Use a Pasteur pipe to inject *Brodie manometer fluid* (or water colored with vegetable dye and a small amount of detergent) into each graduated side arm so that approximately 1 cm of dye is drawn into each side arm.

Figure 7.4 Volumeter containing three respirometers.

In this experiment, the respirometers are vials filled as per Figure 7.3 with graduated side arms attached. Oxygen uptake is measured by movement of a marker drop in each side arm.

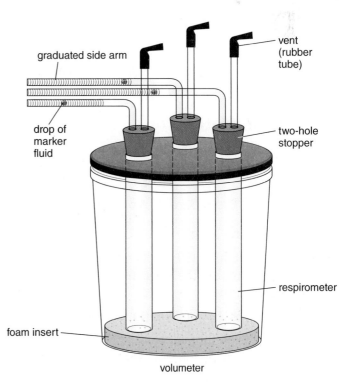

graduated side arm

vent (rubber tube)

drop of marker fluid

two-hole stopper

respirometer

foam insert

volumeter

6. After waiting one minute for equilibrium, reattach the pinch clamp to the outlet rubber tube and mark the position of the dye with a wax pencil. The dye should be near the outlet and as O_2 is taken up, the marker drop will move toward the vials.

7. Record in Table 7.1 for each vial:
   - the initial position of the marker drop to the nearest 0.01 mL.
   - the position of the marker drop after ten minutes.
   - the position of the marker drop after ten more minutes.
   - the net change of position. (The distance the marker drop moved from the initial reading.)

8. Did the marker drop change in vial 3 (glass beads)? _____ By how much? _____ Enter this number in the "Correction" column of Table 7.1, and use this number to correct the net change you observed in vials 1 and 2. (This is a correction for any change in volume due to atmospheric pressure changes or temperature changes.) This will complete Table 7.1.

Table 7.1	Cellular Respiration						
Vial	Contents	Initial Reading	Reading After 10 Minutes	Reading After 20 Minutes	Net Change	Correction	(Corrected) Net Change
1	Germinating soybeans						
2	Dry (nongerminating) soybeans Glass beads						
3	Glass beads						

Conclusions: Cellular Respiration

- Do your results support or fail to support the hypothesis? _____
  Explain. _____
- Why was it necessary to absorb the carbon dioxide? _____

  _____

- Explain which respirometer in the soybean experiment was a negative control? _____

7.2 Fermentation

When you study the ability of yeast to ferment sugar, you will need to use a respirometer to measure the amount of CO_2 given off. This experimental procedure shows you how to prepare your **respirometer.**

Experimental Procedure: Respirometer Practice

1. Completely fill a small tube (15 × 125 mm) with *water only* (Fig. 7.5).
2. Invert a large tube (20 × 150 mm) over the small tube, and with your finger or a pencil, push the small tube up into the large tube until the upper lip of the small tube is in contact with the bottom of the large tube.
3. Quickly invert both tubes. Do not permit the small tube to slip away from the bottom of the large tube. A little water will leak out of the small tube and be replaced by an air bubble.
4. Practice this inversion until the bubble in the small tube is as small as you can make it.

Figure 7.5 Respirometer for fermentation.
Place a small tube inside a large tube. Hold the small tube in place as you rotate the entire apparatus, and an air bubble will form in the small tube.

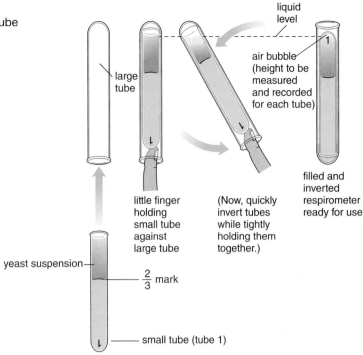

Ethanol Fermentation

Yeast fermentation to produce ethanol (C_2H_5OH) and carbon dioxide (CO_2) has long been utilized by humans to produce wine and bread. During the production of wine, it is the ethanol that is desired while bread and other baked goods rise when yeast gives off carbon dioxide (Fig. 7.6). Recently, there has been a great deal of interest in using ethanol produced by yeast fermentation of corn as a substitute for gasoline in cars.

Testing Sugars

In the Experimental Procedure that follows, we are going to test which of several sugars is a better food source for yeast when they ferment. It will be assumed that the ease of sugar fermentation correlates with the amount of carbon dioxide given off within a certain time limit. As background data, observe the structure of the three sugars involved:

Figure 7.6 Products of fermentation.

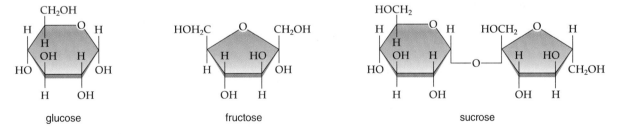

glucose   fructose   sucrose

State a hypothesis for your experiment in which you sequence the sugars according to how easily you expect yeast to ferment them:

Hypothesis: _____

Laboratory 7   Cellular Respiration   **81**

Have four large test tubes ready. With a wax pencil, label and mark off a small test tube at the 2/3-full level. Use this tube to mark off three other small tubes at the same level.

1. Label and fill the small tubes as directed, and record the contents in Table 7.2.

   Tube 1   Fill to the mark with *glucose solution.*

   Tube 2   Fill to the mark with *fructose solution.*

   Tube 3   Fill to the mark with *sucrose solution.*

   Tube 4   Fill to the mark with *distilled water.*

2. Resuspend a yeast solution each time, and fill all four tubes to the top with yeast suspension (Fig. 7.5).

3. Slide the large tubes over the small tubes, and invert them in the way you practiced. This will mix the yeast and sugar solutions.

4. Place the respirometers in a tube rack, and measure the initial height of the air space in the rounded bottom of the small tube. Record the height in Table 7.2.

5. Place the respirometers in an incubator or in a warm water bath maintained at 37°C. Note the time, and allow the respirometers to incubate about 20 minutes (incubator) or one hour (water bath). However, watch your respirometers and if they appear to be filling with gas quite rapidly, stop the incubation when appropriate.

6. At the end of the incubation period, measure the final height of the gas bubbles, and record it in Table 7.2. Calculate the net change, and record it in Table 7.2.

Table 7.2   Fermentation by Yeast

Tube	Sugar	Initial Gas Height	Final Gas Height	Net Change	Ease of Fermentation
1					
2					
3					
4				_____	

Conclusions: Yeast Fermentation

- From your results, evaluate how the sugars tested compare as an energy source for yeast fermentation. Enter your evaluation in Table 7.2.

- Do your data support or fail to support your hypothesis? _____

  Explain. _____

- Can your results be correlated with the comparative structure of the sugars? (See page 81.) _____

  Explain. _____

- Explain which respirometer was a negative control. _____

1. In the Experimental Procedure: Cellular Respiration, what gas was taken up by the soybeans? _____

2. Explain the role of each of the following components in the cellular respiration experiment:

   a. KOH _____

   b. Germinating soybeans _____

   c. Dry (nongerminating) soybeans _____

3. If you performed the cellular respiration experiment without soaking the cotton with KOH, what results would you predict? _____

   Why? _____

4. Soybeans don't move around as animals do. For what purpose do they need energy? _____

   _____

5. Both cellular respiration and fermentation ordinarily begin with what molecule? _____

6. How do the overall equations for these processes indicate that cellular respiration is aerobic and that fermentation is anaerobic? _____

   _____

7. Glucose breakdown results in the breaking of C—H bonds and stored energy is released. Contrast the end products of fermentation and cellular respiration in terms of their energy content. _____

   _____

8. Fermentation results in the net production of only 2 ATP, while cellular respiration results in the production of many more ATP. Explain these results with reference to the end products of both of these processes. _____

   _____

9. In the Experimental Procedure: Yeast Fermentation, the gas bubble got larger. What gas was causing this increase in bubble size? _____

10. Why might you hypothesize that, of the three sugars (glucose, fructose, and sucrose), glucose would result in the most activity during the fermentation experiment? _____

    _____

8

Mitosis and Meiosis

Learning Outcomes

8.1 The Cell Cycle
- Name and describe the stages of the cell cycle.   85
- Describe how the cell prepares for mitosis, and identify the phases of mitosis in models and microscopic slides: explain how the chromosome number stays constant.   86–90
- Describe animal and plant cell cytokinesis.   91–92

8.2 Meiosis
- Name and describe the phases of meiosis I and meiosis II with attention to the movement of chromosomes: explain how the chromosome number is reduced.   93–97

8.3 Mitosis Versus Meiosis
- Compare the outcomes of mitosis and meiosis.   98
- Contrast the steps of mitosis with those of meiosis I and meiosis II. State two critical events during meiosis I.   98–99

8.4 Karyotype Anomalies
- Recognize that abnormalities in chromosome number and structure can occur when cells divide.   100

8.5 Gametogenesis in Animals
- Contrast spermatogenesis and oogenesis using models and slides.   101–3

Introduction

Dividing cells experience nuclear division, cytoplasmic division, and a period of time between divisions called interphase. During **interphase,** the nucleus appears normal, and the cell is performing its usual cellular functions. Also, the cell is increasing all of its components, including such organelles as the mitochondria, ribosomes, and centrioles, if present. DNA replication (making an exact copy of the DNA) occurs toward the end of interphase. Thereafter, the chromosomes, which contain DNA, are duplicated and contain two chromatids held together at a **centromere.** These chromatids are called **sister chromatids.**

During nuclear division, called **mitosis,** the new nuclei receive the same number of chromosomes as the parental nucleus. When the cytoplasm divides, a process called **cytokinesis,** two daughter cells are produced. In multicellular organisms, mitosis permits growth and repair of tissues. In eukaryotic, unicellular organisms, mitosis is a form of asexual reproduction. Sexually reproducing organisms utilize another form of nuclear division, called **meiosis.** In animals, meiosis is a part of gametogenesis, the production of gametes (sex cells). The **gametes** are sperm in male animals and eggs in female animals. As a result of meiosis, the daughter cells have half the number of chromosomes as the parental cell. As we shall see, meiosis contributes to recombination of genetic material and to diversity among sexually reproducing organisms.

This laboratory examines both mitotic and meiotic cell division to show their similarities and differences and it also discusses chromosome anomalies that can occur during mitosis and meiosis.

8.1 The Cell Cycle

The **cell cycle** is an orderly series of stages that includes preparation to divide and the division process (Fig. 8.1). During interphase (between cell divisions), the cell carries on its normal activities and gets ready to divide. After cell division, during a stage called G_1, the cell increases in size as growth occurs. During the S stage, DNA replication occurs in addition to growth, and during the G_2 stage cytoplasmic organelle duplication occurs as does synthesis of the proteins involved in regulating cell division. Then, mitosis occurs and the cell divides. State the event of each stage on the line provided.

G_1 _____

S _____

G_2 _____

M _____

Explain why the term "cell cycle" is appropriate.

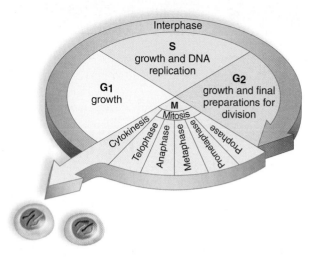

Figure 8.1   The cell cycle.
Immature cells go through a cycle that consists of four stages: G_1, S (for synthesis), G_2, and M (for mitosis). Eventually, some daughter cells "break out" of the cell cycle and become specialized cells.

The time required for the entire cell cycle varies according to the organism, but 18 to 24 hours is typical for animal cells. Mitosis (including cytokinesis, if it occurs) lasts less than an hour to slightly more than 2 hours; for the rest of the time, the cell is in interphase.

Mitosis

Mitosis is nuclear division that results in two new nuclei, each having the same number of chromosomes as the original nucleus. The **parent cell** is the cell that divides, and the resulting cells are called **daughter cells.** Mitosis allows the body to grow and repair itself. When you suffer a cut and it heals, the cell cycle, including mitosis, is occurring.

When cell division is about to begin, chromatin starts to condense and compact to form visible, rodlike sister chromatids held together at the centromere. *Consult Table 8.1 and label the sister chromatids, centromere, and kinetochore in the drawing of a duplicated chromosome in Figure 8.2.* This illustration represents a chromosome as it would appear just before nuclear division occurs.

Figure 8.2   Duplicated chromosomes.
DNA replication results in a duplicated chromosome that consists of two sister chromatids held together at a centromere.

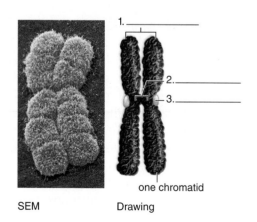

1. _____

2. _____

3. _____

one chromatid

SEM   Drawing

Table 8.1 Structures Associated with Mitosis

Structure	Description
Nucleus	A large organelle containing the chromosomes and acting as a control center for the cells
Chromosome	Rod-shaped body in the nucleus that is seen during mitosis and meiosis and that contains DNA and, therefore, the hereditary units, or genes
Nucleolus	An organelle found inside the nucleus; composed largely of RNA for ribosome formation
Spindle	Microtubule structure that brings about chromosome movement during cell division
Metaphase plate	The center of the fully formed spindle
Chromatids	The two identical parts of a chromosome following DNA replication
Centromere	A constriction where duplicates (sister chromatids) of a chromosome are held together
Kinetochore	A body that appears during cell division on either side of the centromere; it attaches a chromatid to a spindle fiber
Centrosome	The central microtubule-organizing center of cells; consists of granular material; in animal cells, contains two centrioles
Centrioles*	Short, cylindrical organelles in animal cells that contain microtubules and are associated with the formation of the spindle during cell division
Aster*	Short, radiating fibers produced by the centrioles

*Animal cells only

Spindle

The **spindle** is a structure that appears and brings about an orderly distribution of chromosomes to the daughter cell nuclei. A spindle has fibers that stretch between two poles (ends). Spindle fibers are bundles of microtubules, protein cylinders found in the cytoplasm that can assemble and disassemble. The **centrosome,** the main microtubule-organizing center of the cell, divides before mitosis so that during mitosis, each pole of the spindle has a centrosome. Animal cells contain two barrel-shaped organelles called **centrioles** in each centrosome and asters, arrays of short microtubules radiating from the poles (see Fig. 8.3). The fact that plant cells have a centrosome but lack centrioles suggests that centrioles are not required for spindle formation.

Observation: Animal Mitosis

Animal Mitosis Models

1. Using Figure 8.3 as a guide, identify the phases of mitosis in animal mitosis models.
2. Each species has its own chromosome number. Counting the number of centromeres tells you the number of chromosomes in the models. What is the number of chromosomes in each of the cells

   in this model series? _____

Whitefish Blastula Slide

The blastula is an early embryonic stage in the development of animals. The **blastomeres** (blastula cells) shown are in different phases of mitosis (see Fig. 8.3).

1. Examine a prepared slide of whitefish blastula cells undergoing mitotic cell division.
2. Try to find a cell in each phase of mitosis. Have a partner or your instructor check your identification.

Figure 8.3 Phases of mitosis in animal and plant cells.
The colors signify that the chromosomes were inherited from different parents.

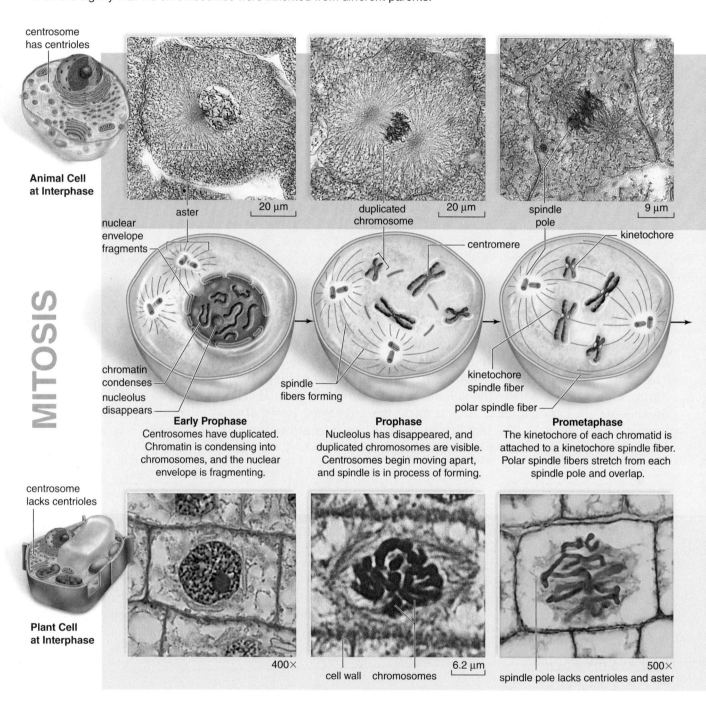

MITOSIS

centrosome has centrioles

Animal Cell at Interphase

aster

20 μm

duplicated chromosome

20 μm

spindle pole

9 μm

nuclear envelope fragments

centromere

kinetochore

chromatin condenses

nucleolus disappears

spindle fibers forming

kinetochore spindle fiber

polar spindle fiber

Early Prophase
Centrosomes have duplicated. Chromatin is condensing into chromosomes, and the nuclear envelope is fragmenting.

Prophase
Nucleolus has disappeared, and duplicated chromosomes are visible. Centrosomes begin moving apart, and spindle is in process of forming.

Prometaphase
The kinetochore of each chromatid is attached to a kinetochore spindle fiber. Polar spindle fibers stretch from each spindle pole and overlap.

centrosome lacks centrioles

Plant Cell at Interphase

400×

cell wall   chromosomes

6.2 μm

500×

spindle pole lacks centrioles and aster

Beginning on page 88 the phases of mitosis are shown in Figure 8.3. Mitosis is the type of nuclear division that (1) occurs in the body (somatic) cells; (2) results in two daughter cells because there is only one round of division; and (3) keeps the chromosome number constant (same as the parent cell).

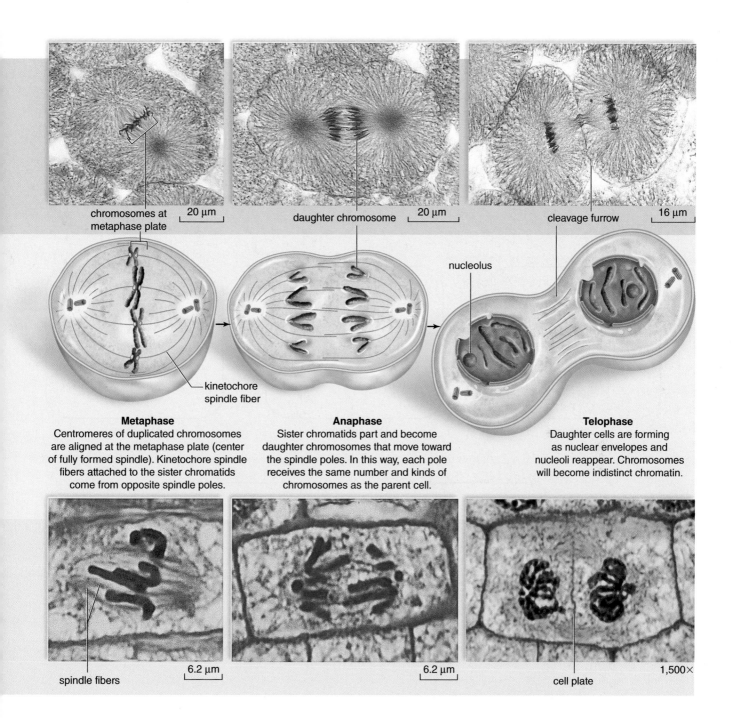

chromosomes at metaphase plate | 20 μm

daughter chromosome | 20 μm

cleavage furrow | 16 μm

nucleolus

kinetochore spindle fiber

Metaphase
Centromeres of duplicated chromosomes are aligned at the metaphase plate (center of fully formed spindle). Kinetochore spindle fibers attached to the sister chromatids come from opposite spindle poles.

Anaphase
Sister chromatids part and become daughter chromosomes that move toward the spindle poles. In this way, each pole receives the same number and kinds of chromosomes as the parent cell.

Telophase
Daughter cells are forming as nuclear envelopes and nucleoli reappear. Chromosomes will become indistinct chromatin.

spindle fibers | 6.2 μm

6.2 μm

cell plate | 1,500×

Plant Mitosis Models

1. Identify the phases of plant cell mitosis using models of plant cell mitosis and Figure 8.3 as a guide.

2. Notice that plant cells do not have centrioles and asters. Plant cells do have centrosomes and this accounts for the formation of a spindle.

3. What is the number of chromosomes in each of the cells in this model series? _____

Onion Root Tip Slide

1. In plants, the root tip contains tissue that is continually dividing and producing new cells. Examine a prepared slide of onion root tip cells undergoing mitotic cell division. Try to find the phases that correspond to those shown in Figure 8.3.

2. Using high power, focus up and down on a cell in telophase. You may be able to just make out the cell plate, the region where a plasma membrane is forming between the two prospective daughter cells. Later, cell walls appear in this area.

3. In the boxes provided, draw and label the stages of mitosis as observed in the onion root tip slide.

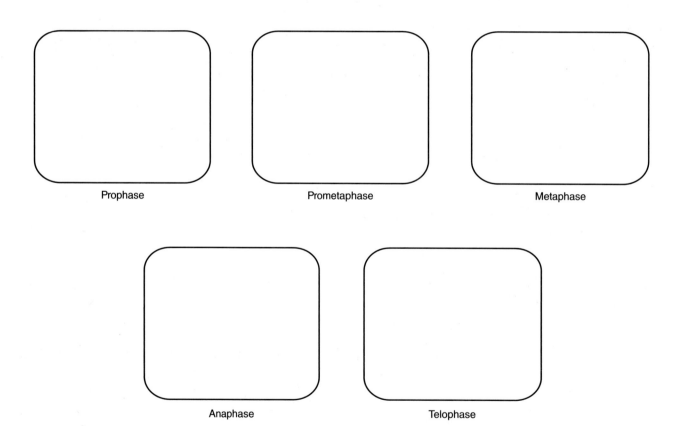

Prophase   Prometaphase   Metaphase

Anaphase   Telophase

Cytokinesis

Cytokinesis, division of the cytoplasm, usually accompanies mitosis. During cytokinesis, each daughter cell receives a share of the organelles that duplicated during interphase. Cytokinesis begins in anaphase, continues in telophase, and reaches completion by the start of the next interphase.

Cytokinesis in Animal Cells

In animal cells, a **cleavage furrow,** an indentation of the membrane between the daughter nuclei, begins as anaphase draws to a close (Fig. 8.4). The cleavage furrow deepens as a band of actin filaments called the contractile ring slowly constricts the cell, forming two daughter cells.

 Were any of the cells of the whitefish

blastula slide undergoing cytokinesis? ____

How do you know? _____

_____

_____

Virtual Lab Cell Reproduction
A virtual lab called Cell Reproduction is available on the *Biology* website **www.mhhe .com/maderbiology11**. Follow the directions given to review mitosis and compare normal tissues to cancerous tissues.

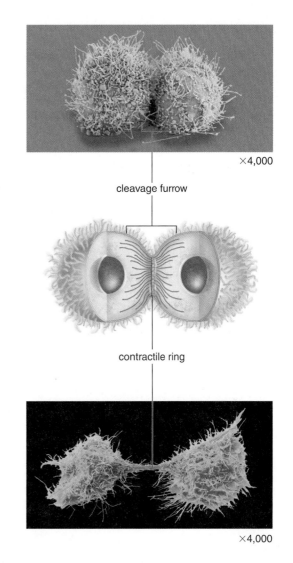

×4,000

cleavage furrow

contractile ring

×4,000

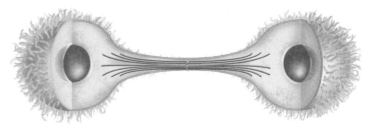

Figure 8.4 Cytokinesis in animal cells.
A single cell becomes two cells by a furrowing process. A contractile ring composed of actin filaments gradually gets smaller, and the cleavage furrow pinches the cell into two cells.

Cytokinesis in Plant Cells

After mitosis, the cytoplasm divides by cytokinesis. In plant cells, membrane vesicles derived from the Golgi apparatus migrate to the center of the cell and form a **cell plate** (Fig. 8.5), the location of a new plasma membrane for each daughter cell. Later, individual cell walls appear in this area. Were any of the cells of the onion root tip slide undergoing cytokinesis? _____

How do you know? _____

Figure 8.5   Cytokinesis in plant cells.
During cytokinesis in a plant cell, a cell plate forms midway between two daughter nuclei and extends to the plasma membrane. Vesicles containing cell wall components fuse to form the cell plate.

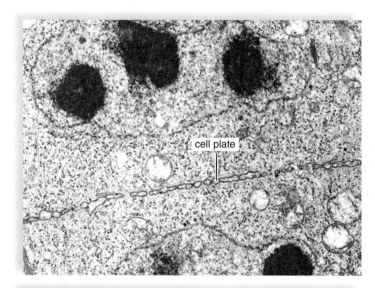

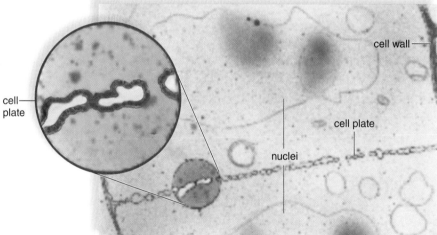

Summary of Mitotic Cell Division

1. The nuclei in the daughter cells have the _____ number of chromosomes as the parent cell had.

2. Mitosis is cell division in which the chromosome number _____

3. If a parent cell has 16 chromosomes, how many chromosomes do the daughter cells have following mitosis? _____

8.2 Meiosis

Meiosis is a form of nuclear division in which the chromosome number is reduced by half during the production of gametes (sperm or eggs). See Section 8.5 for more details. The nucleus of the parent cell about to undergo meiosis has the **diploid (2n) number** of chromosomes—that is, two sets of the same type chromosomes. After meiosis is complete, the daughter nuclei have the **haploid (n) number**, or one set of these chromosomes. For example, diploid fruit fly cells have 8 chromosomes (2n = 8) but undergo meiosis to produce gametes with 4 chromosomes (n = 4). In sexually reproducing species, meiosis must occur or the chromosome number would double with each generation.

A diploid nucleus contains **homologues,** also called homologous chromosomes. Homologues look alike and carry the genes for the same traits. Before meiosis begins, the chromosomes are already double stranded—that is, they contain sister chromatids.

Experimental Procedure: Meiosis

In this exercise, you will use pop beads to construct homologues and move the chromosomes to simulate meiosis.

Building Chromosomes to Simulate Meiosis

1. Obtain the following materials: 48 pop beads of one color (e.g., red) and 48 pop beads of another color (e.g., blue) for a total of 96 beads; eight magnetic centromeres; and four centriole groups.
2. Build a homologous pair of duplicated chromosomes using Figure 8.6a as a guide. Each chromatid will have 16 beads. Be sure to bring the centromeres of two units of the same color together so that they attract and link to form one duplicated chromosome. (One member of the pair will be red, and the other will be blue.)
3. Build another homologous pair of duplicated chromosomes using Figure 8.6b as a guide. Each chromatid will have eight beads. Be sure to bring the centromeres of two units of the same color together so that they attract. (One member of the pair will be red, and the other will be blue.)
4. Note that your chromosomes are the same as those in Figure 8.7. The red chromosomes were inherited from one parent, and the blue chromosomes were inherited from the other parent.

Figure 8.6 Two pairs of homologues.
The chromosomes of these homologous pairs are duplicated.

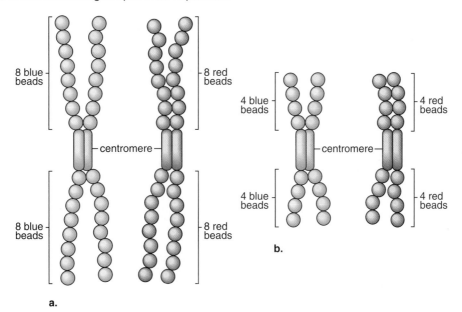

Simulating Meiosis I

Meiosis requires two nuclear divisions, and the first round during meiosis is called **meiosis I** (see Fig. 8.7).

Meiosis I has the same phases as mitosis. Gametes differ genetically for two reasons. (1) During prophase I, homologues line up next to one another during a process called **synapsis**. Then during **crossing over**, the nonsister chromatids of a homologue pair exchange genetic material. (2) Homologues pair, and each pair independently goes to the metaphase plate without regard to the orientation of other pairs. This means that either homologue, red or blue, can face a particular pole. During anaphase I, homologues separate from each other. With the completion of telophase I, all possible combinations of the haploid number of chromosomes (still composed of two chromatids) can be in the daughter nuclei.

Prophase I

1. Using Figure 8.7 as a guide, put all four of the chromosomes you built in the center of your work area, which represents the nucleus. Place two pairs of centrioles outside the nucleus.
2. Separate the pairs of centrioles, and move one pair to opposite poles of a simulated spindle.
3. Simulate synapsis (described above) by bringing the homologues together, side by side.
4. Simulate crossing over (described above) by exchanging the exact segments of two nonsister chromatids of a single homologous pair. Why use nonsister chromatids and not sister chromatids?

_____

Metaphase I

5. Position the homologues at the metaphase plate in such a way that the homologues are prepared to move apart toward the poles. For each pair, blue or red can face either pole.

Anaphase I

6. Separate the homologues, and move the homologues toward the opposite poles.

Telophase I

7. During telophase I, the chromosomes are at the poles and daughter nuclei form. Fill in the following blanks with the words *red-long, red-short, blue-long,* and *blue-short* to show what chromosomes are in the daughter nuclei.

   Daughter nucleus 1: _____ and _____

   Daughter nucleus 2: _____ and _____

8. What other combinations would have been possible? (*Hint:* Alternate the colors of one pair at metaphase I.)

   Daughter nucleus 1: _____ and _____

   Daughter nucleus 2: _____ and _____

Conclusions: Meiosis I

- Do the chromosomes inherited from the mother or father always remain together following

  meiosis I? _____ Explain. _____

- Name two ways that meiosis contributes to genetic variation among offspring:

  1. _____

  2. _____

Interkinesis

Interkinesis is the period between meiosis I and meiosis II. In some species, daughter nuclei form but daughter cells do not form, and daughter nuclei undergo meiosis II right after meiosis I. Does DNA

replication occur during interkinesis? _____ Explain. _____

Simulating Meiosis II

The second round of nuclear division during meiosis is called **meiosis II** (see Fig. 8.7). Each daughter cell from meiosis I undergoes meiosis II and therefore meiosis produces four daughter cells altogether.

The chromosomes proceed individually during prophase II and metaphase II. During anaphase II, the centromeres divide and the chromatids separate, becoming daughter chromosomes that move toward the poles. Following telophase II the daughter nuclei and cells form. Notice that meiosis II is exactly like mitosis except that the nucleus of the parent cell and the daughter nuclei are haploid.

Prophase II

1. Using Figure 8.7 as a guide, choose one daughter nucleus from step 7, page 94, to be the parent nucleus undergoing meiosis II.
2. Place a pair of centrioles at the simulated poles of a new spindle.

Metaphase II

3. Move each duplicated chromosomes to the metaphase plate. How many duplicated chromosomes are at the metaphase plate? _____

Anaphase II

4. Pull the magnets of each duplicated chromosome apart. What does this action represent? _____

_____

Telophase II

5. Put the chromosomes—each having one chromatid—at the poles near the centrioles. At the end of telophase, the daughter nuclei reform.

Conclusions: Meiosis II

• You chose only one daughter nucleus from meiosis I to be the parent nucleus that divides during meiosis II. In reality both daughter nuclei go on to divide again. Therefore, how many nuclei are usually present when meiosis II is complete? _____

• In this exercise, how many chromosomes were in the parent cell nucleus undergoing meiosis II?

_____

• How many chromosomes are in the daughter nuclei following meiosis II? _____
Explain. _____

Summary of Meiotic Cell Division

1. The parent cell has the diploid (2n) number of chromosomes, and the daughter cells have the _____ (n) number of chromosomes.
2. Meiosis is cell division in which the chromosome number _____
3. If a parent cell has 16 chromosomes, the daughter cells will have how many chromosomes following meiosis? _____
4. Whereas meiosis reduces the chromosome number, **fertilization** restores the chromosome number. A zygote (fertilized egg) contains the same number of chromosomes as the parent, but are these exactly the same chromosomes as either parent had? _____ Why not? _____
Fertilization is another way that sexual reproduction results in genetic variation. Explain. _____

_____

Meiosis Phases

Figure 8.7 Meiosis I and II in plant cell micrographs and animal cell drawings.

2n is the diploid number of chromosomes and **n** is the haploid number of chromosomes.

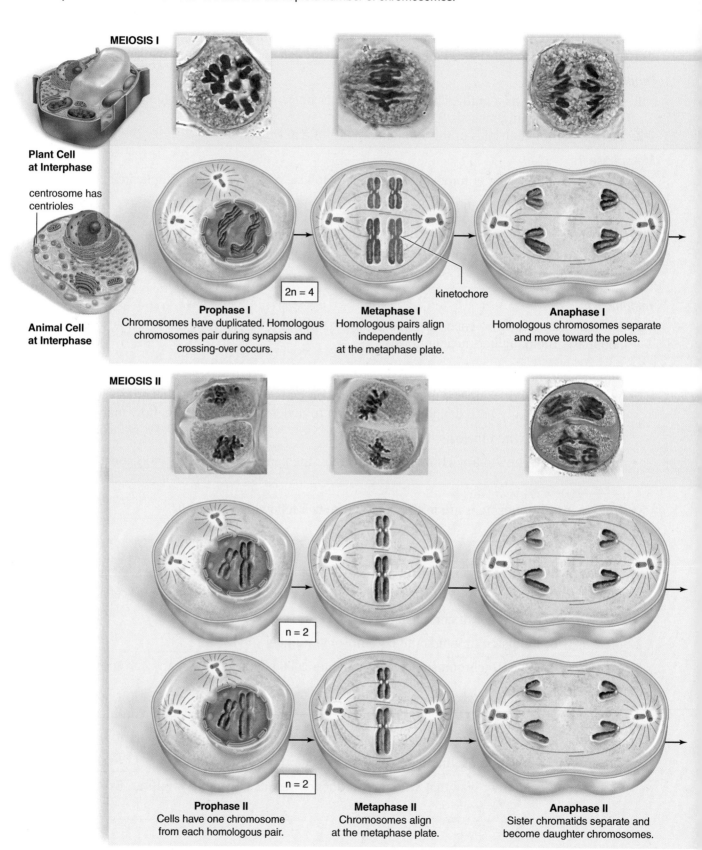

MEIOSIS I

Plant Cell at Interphase

centrosome has centrioles

Animal Cell at Interphase

2n = 4

kinetochore

Prophase I
Chromosomes have duplicated. Homologous chromosomes pair during synapsis and crossing-over occurs.

Metaphase I
Homologous pairs align independently at the metaphase plate.

Anaphase I
Homologous chromosomes separate and move toward the poles.

MEIOSIS II

n = 2

n = 2

Prophase II
Cells have one chromosome from each homologous pair.

Metaphase II
Chromosomes align at the metaphase plate.

Anaphase II
Sister chromatids separate and become daughter chromosomes.

Beginning on page 96, the phases of meiosis are shown in Figure 8.7. Meiosis is the type of nuclear division that (1) occurs in the sex organs to produce sperm or eggs; (2) results in four cells because there are two rounds of cell division; and (3) reduces the chromosome number to half that of the parent cell.

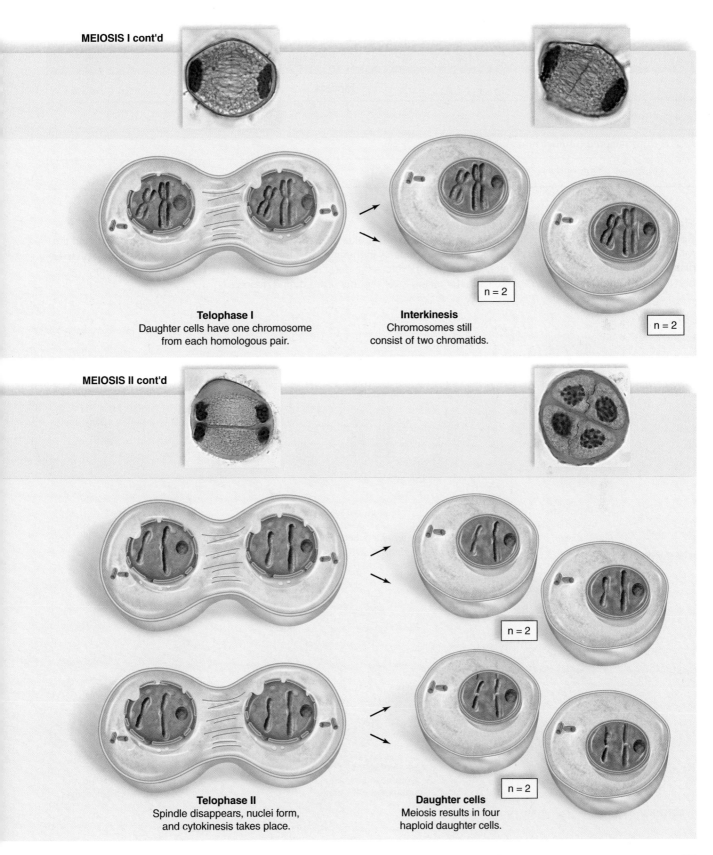

MEIOSIS I cont'd

Telophase I
Daughter cells have one chromosome from each homologous pair.

Interkinesis
Chromosomes still consist of two chromatids.

n = 2

n = 2

MEIOSIS II cont'd

Telophase II
Spindle disappears, nuclei form, and cytokinesis takes place.

Daughter cells
Meiosis results in four haploid daughter cells.

n = 2

n = 2

Laboratory 8   Mitosis and Meiosis   **97**

8.3 Mitosis Versus Meiosis

In comparing mitosis to meiosis it is important to note that meiosis requires two nuclear divisions but mitosis requires only one nuclear division. Therefore mitosis produces two daughter cells and meiosis produces four daughter cells. Following mitosis, the daughter cells are still diploid but following meiosis, the daughter cells are haploid. Figure 8.8 explains why. Fill in Table 8.2 to indicate general

Table 8.2 Differences Between Mitosis and Meiosis	Mitosis	Meiosis
1. Number of divisions		
2. Chromosome number in daughter cells		
3. Number of daughter cells		

Figure 8.8 Meiosis I compared to mitosis.

Compare metaphase of meiosis I to metaphase of mitosis. The blue chromosomes were inherited from one parent, and the red chromosomes were inherited from the other parent. Only in metaphase I are the homologues paired at the metaphase plate. Homologues separate during anaphase I, and therefore the daughter cells are haploid. The exchange of color between nonsister chromatids represents the crossing-over that occurred during meiosis I.

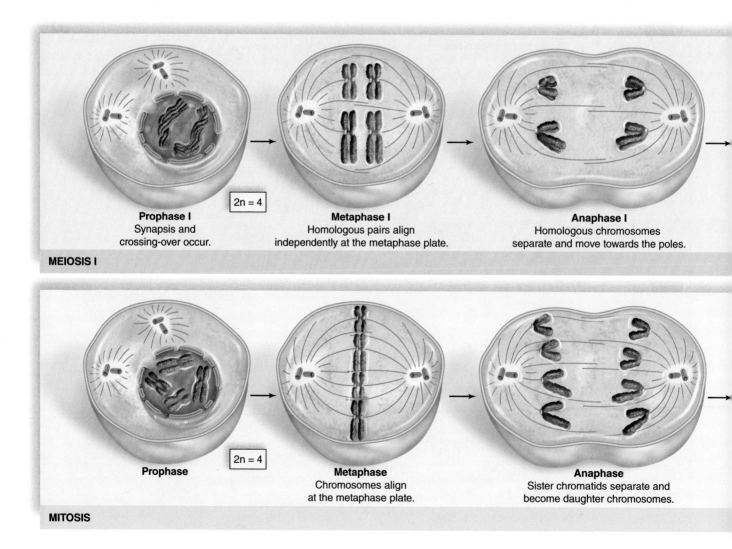

MEIOSIS I

| Prophase I | Metaphase I | Anaphase I |
| Synapsis and crossing-over occur. | Homologous pairs align independently at the metaphase plate. | Homologous chromosomes separate and move towards the poles. |

2n = 4

MITOSIS

| Prophase | Metaphase | Anaphase |
| | Chromosomes align at the metaphase plate. | Sister chromatids separate and become daughter chromosomes. |

2n = 4

differences between mitosis and meiosis, and complete Table 8.3 to indicate specific differences between mitosis and meiosis I. Mitosis need only be compared with meiosis I because the exact events occur during both mitosis and meiosis II except that the cells are diploid during mitosis and haploid during meiosis II.

Table 8.3 Mitosis Compared with Meiosis I	
Mitosis	**Meiosis I**
Prophase: No pairing of chromosomes	Prophase I: _____
Metaphase: Duplicated chromosomes at metaphase plate	Metaphase I: _____
Anaphase: Sister chromatids separate	Anaphase I: _____
Telophase: Chromosomes have one chromatid	Telophase I: _____

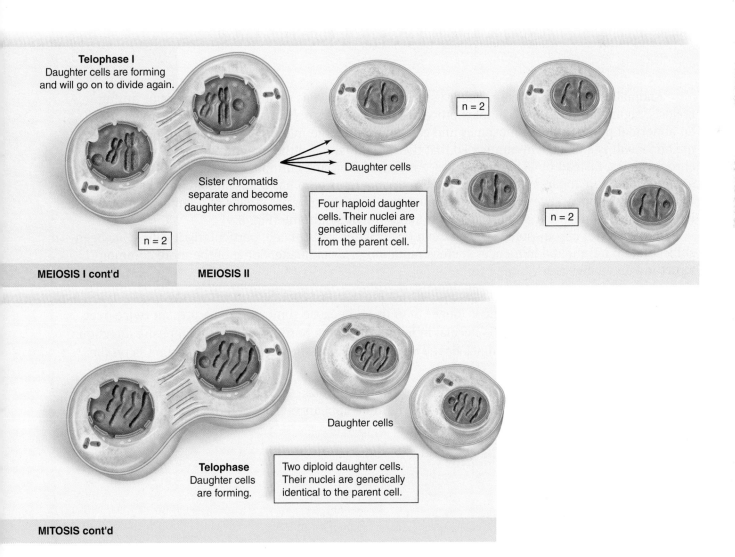

Telophase I
Daughter cells are forming and will go on to divide again.

Sister chromatids separate and become daughter chromosomes.

Daughter cells

n = 2

n = 2

n = 2

Four haploid daughter cells. Their nuclei are genetically different from the parent cell.

MEIOSIS I cont'd   **MEIOSIS II**

Daughter cells

Telophase
Daughter cells are forming.

Two diploid daughter cells. Their nuclei are genetically identical to the parent cell.

MITOSIS cont'd

8.4 Karyotype Anomalies

In a karyotype, the chromosomes of an organism are arranged so that the pairs of chromosomes can be seen (Fig. 8.9). At that time, it is possible to observe any possible anomalies in chromosome number and structure.

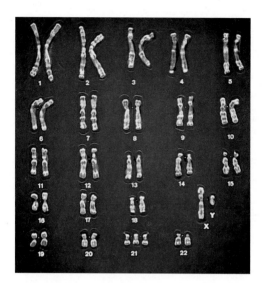

Figure 8.9 Karyotype of a person with Down syndrome.
Note the three number 21 chromosomes.

Anomalies of chromosome number usually arise when cells divide during mitosis or during meiosis, signifying that these complex processes don't always occur as expected. Cancer cells have an abnormal number of chromosomes and constantly undergo mitosis. The most common abnormal meiotic result is a change in number so that the individual has either 45 chromosomes or 47 chromosomes, instead of 46 chromosomes. An extra or missing chromosome can cause a fetus and/or a child to develop apparent anomalies. For example, a fetus who has a missing X chromosome may fail to develop the appearance of a female and may not have the internal organs of a female. Either sex that has an extra chromosome 21 has the symptoms of Down syndrome (Fig. 8.9).

Anomalies of chromosome structure can occur when cells divide, particularly if cells have been subject to environmental influences such as radiation or drug intake. Some of the more common structural anomalies are:

Deletion: The chromosome is shorter than usual because some portion is missing.
Duplication: The chromosome is longer than usual because some portion is present twice over.
Inversion: The chromosome is normal in length but some portion runs in the opposite direction.
Translocation: Two chromosomes have switched portions and each switched portion is on the wrong chromosome.

Anomalies of chromosome structure or number can result in recognized syndromes, a collection of symptoms that always occur together. Which syndrome appears depends on the particular anomaly. This topic is studied in more detail in laboratory 10.

8.5 Gametogenesis in Animals

Gametogenesis is the formation of **gametes** in animals. In humans and other mammals, the gametes are sperm and eggs. **Fertilization** occurs when the nucleus of a sperm fuses with the nucleus of an egg.

Gametogenesis in Mammals

Gametogenesis occurs in the testes of males, where **spermatogenesis** produces sperm. Gametogenesis occurs in the ovaries of females, where **oogenesis** produces oocytes (eggs).

A diploid (2n) nucleus contains the full number of chromosomes, which is 46 in humans, and a haploid (n) nucleus contains half as many, which is 23 in humans. Gametogenesis involves meiosis, the process that reduces the chromosome number from 2n to n. In sexually reproducing species, if meiosis did not occur, the chromosome number would double with each generation. Meiosis consists of two divisions: the first meiotic division (meiosis I) and the second meiotic division (meiosis II). Therefore, you would expect four haploid cells at the end of the process. Indeed, there are four sperm as a result of spermatogenesis (Fig. 8.10). However, in females, meiosis I results in a secondary oocyte and one polar body. A **polar body** is a nonfunctioning cell that dies. A secondary oocyte does not undergo meiosis II unless fertilization (fusion of egg and sperm) occurs. At the completion of oogenesis, there is a single egg (Fig. 8.10). Fertilization restores the full chromosome number, which is 46 in humans. Now the individual has two of every type of chromosome.

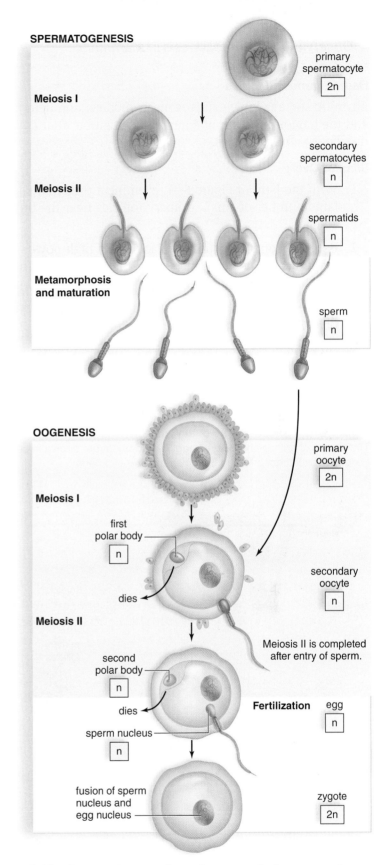

Figure 8.10 Spermatogenesis and oogenesis in mammals.
Spermatogenesis produces four viable sperm, whereas oogenesis produces one egg and two polar bodies. In humans, both sperm and egg have 23 chromosomes each; therefore, following fertilization, the zygote has 46 chromosomes.

Gametogenesis Models

Examine any available gametogenesis models, and determine the diploid number of the parent cell and the haploid number of a gamete. Remember that counting the number of centromeres tells you the number of chromosomes.

Slide of Ovary

1. With the help of Figure 8.11, examine a prepared slide of an ovary. Under low power, you will see a large number of small, primary follicles near the outer edge. A primary follicle contains a primary oocyte.

2. Find a secondary follicle, and switch to high power. Note the secondary oocyte (egg), surrounded by numerous cells, to one side of the liquid-filled follicle.

3. Also look for a large, fluid-filled vesicular (Graafian) follicle, which contains a mature secondary oocyte to one side. The vesicular follicle will be next to the outer surface of the ovary because this type of follicle releases the egg during ovulation.

4. How many secondary follicles can you find on your slide? _____ How many vesicular follicles can you find? _____ How does this number compare with the number of sperm cells seen in the testis cross section (see Fig. 8.12)? _____

Figure 8.11 Microscopic ovary anatomy.

The stages of follicle and oocyte (egg) development are shown in sequence. Each follicle goes through all the stages. Following ovulation, a follicle becomes the corpus luteum.

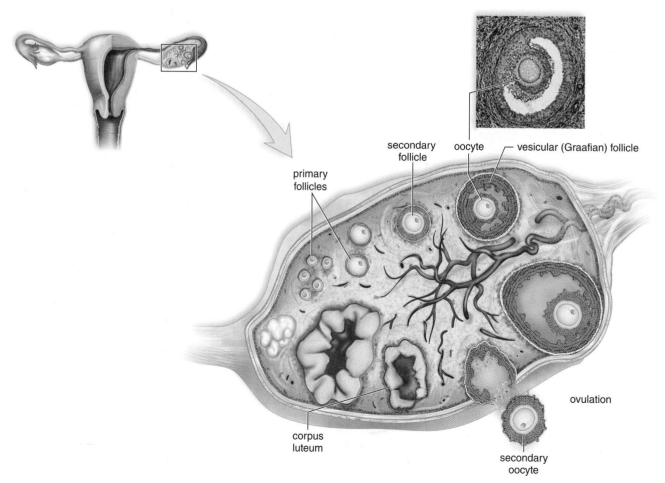

Figure 8.12   Microscopic testis anatomy.
a. A testis contains many seminiferous tubules. **b.** Scanning electron micrograph of a cross section of the seminiferous tubules, where spermatogenesis occurs. Note the location of interstitial cells in clumps among the seminiferous tubules in this light micrograph.

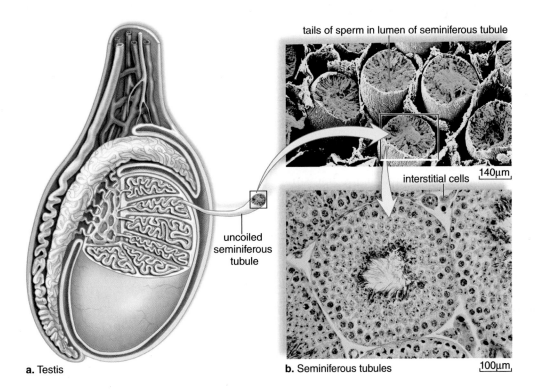

tails of sperm in lumen of seminiferous tubule

interstitial cells   140µm

uncoiled
seminiferous
tubule

a. Testis

b. Seminiferous tubules   100µm

Slide of Testis

1. With the help of Figure 8.12, examine a prepared slide of a testis. Under low power, note the many circular structures. These are the **seminiferous tubules,** where sperm formation takes place.

2. Switch to high power, and observe one tubule in particular. Find mature sperm (which look like thin, fine, dark lines) in the middle of the tubule. **Interstitial cells,** which produce the male sex hormone testosterone, are between the tubules.

Summary of Gametogenesis

1. What is gametogenesis? _____

   In general, how many chromosomes are in a gamete?* _____

2. What is spermatogenesis? _____

   How many chromosomes does a human sperm have?* _____

3. What is oogenesis? _____

   How many chromosomes does a human egg have?* _____

4. Following fertilization, how many chromosomes does the zygote, the first cell of the new

   individual, have?* _____

   * Use either diploid or haploid to answer these questions.

1. During anaphase of mitosis in humans or other 2n organisms, do the chromosomes have one or two chromatids as they move toward the poles? _____

2. During anaphase of meiosis I, do the chromosomes have one or two chromatids as they move toward the poles? _____

3. During anaphase of meiosis II, do the chromosomes have one or two chromatids as they move toward the poles? _____

4. Asexual reproduction of a haploid protozoan can be described as n → n. Explain. _____

   _____

5. Explain why furrowing is a suitable mechanism for cytokinesis of animal cells but not plant cells.

   _____

6. A student is simulating meiosis I with chromosomes that are red-long and yellow-long; red-short and yellow-short. Why would you *not* expect to find both red-long and yellow-long in one resulting daughter cell? _____

   _____

7. If there are 13 pairs of homologues in a primary spermatocyte, how many chromosomes are there in a sperm? _____

8. Assume that you have built a pair of homologues, each having two chromatids. One homologue contains red beads, while the other contains yellow beads. Describe the appearance of two nonsister chromatids following crossing-over. _____

   _____

9. Give one major difference between mitosis and meiosis. _____

   _____

10. A person with Down syndrome has what type of chromosome anomaly? _____

Biology Website

Enhance your study of the text and laboratory manual with study tools, practice tests, and virtual labs. Also ask your instructor about the resources available through ConnectPlus, including the media-rich eBook, interactive learning tools, and animations.

www.mhhe.com/maderbiology11

McGraw-Hill Access Science Website

An Encyclopedia of Science and Technology Online which provides more information including videos that can enhance the laboratory experience.

www.accessscience.com

9

Mendelian Genetics

Learning Outcomes

9.1 One-Trait Crosses
- State Mendel's law of segregation, and relate this law to laboratory exercises. 106–11
- Describe and predict the results of a one-trait cross in both tobacco seedlings and *Drosophila*. 106–11
- Describe the life cycle of *Drosophila* (fruit fly), and recognize these stages in a culture bottle. 108–10

9.2 Two-Trait Crosses
- State Mendel's law of independent assortment, and relate this law to laboratory exercises. 111–15
- Explain and predict the results of a two-trait cross in corn plants and *Drosophila*. 112–15

9.3 X-Linked Crosses
- Explain and predict the results of an X-linked cross in *Drosophila*. 116–17

9.4 Chi-Square Analysis
- Utilize the chi-square statistical test to help determine whether data do or do not support a hypothesis. 118–19

Introduction

Gregor Mendel, sometimes called the "father of genetics," formulated the basic laws of genetics examined in this laboratory. He determined that individuals have two alternate forms of a gene (now called **alleles**) for each trait in their body cells. Today, we know that alleles are on the chromosomes. An individual can be **homozygous dominant** (two dominant alleles, *GG*), **homozygous recessive** (two recessive alleles, *gg*), or **heterozygous** (one dominant and one recessive allele, *Gg*). **Genotype** refers to an individual's genes, while **phenotype** refers to an individual's appearance (Fig. 9.1). Homozygous dominant and heterozygous individuals show the dominant phenotype; homozygous recessive individuals show the recessive phenotype.

Allele Key
T = tall plant
t = short plant
Phenotype
Genotype
tall
TT
tall
Tt
short
tt
Allele Key
L = long wings
l = short wings
Phenotype
Genotype
long wings
LL
long wings
Ll
short wings
ll

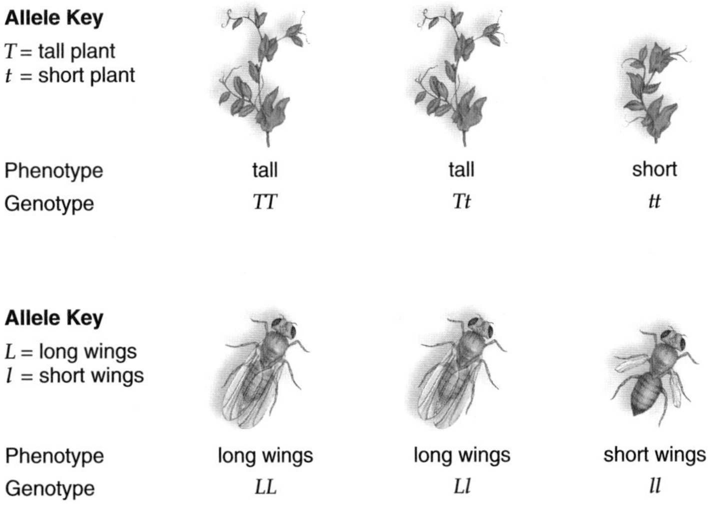

Figure 9.1 Genotype versus phenotype. Only with homozygous recessive do you immediately know the genotype.

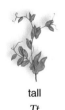

Punnett Squares

Punnett squares named after the man who first used them, allow you to easily determine the results of a cross between individuals whose genotypes are known. Consider that when fertilization occurs two gametes such as a sperm and an egg join together. Whereas individuals have two alleles for every trait, gametes have only one allele because alleles are on the chromosomes and homologues separate during meiosis. Heterozygous parents with the genotype *Aa* produce two types of gametes: 50% of the gametes contain an *A* and 50% contain an *a*. A Punnett square allows you to vertically line up all possible types of sperm and to horizontally line up all possible types of eggs. Every possible combination of gametes occurs within the squares and these combinations indicate the genotypes of the offspring. In Figure 9.2, one offspring is *AA* = homozygous dominant, two are *Aa* = heterozygous, and one is *aa* = homozygous recessive. Therefore, three of the offspring will have the dominant phenotype and one individual will have the recessive phenotype. This is said to be a **phenotypic ratio** of 3:1.

A Punnett square can be used for any cross regardless of the trait(s) and the genotypes of the parents. All you need to do is use the correct letters for the particular trait(s), and make sure you have given the parents the correct genotypes and correct proportion of each type gamete. Then you can determine the genotypes of the offspring and the resulting phenotypic ratio among the offspring.

Figure 9.2   What are the expected results of a cross?
A Punnett square allows you to determine the expected phenotypic ratio for a cross.

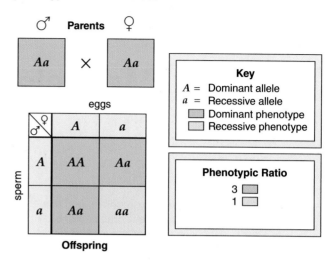

Offspring

Virtual Lab Punnett Squares   A virtual lab called Punnett Squares is available on the *Biology* website **www.mhhe.com/maderbiology11**. It will allow you to practice filling in Punnett squares and determining your results.

9.1   One-Trait Crosses

A single pair of alleles is involved in one-trait crosses. Mendel found that reproduction between two heterozygous individuals *(Aa)*, called a **monohybrid cross**, results in both dominant and recessive phenotypes among the offspring. In Figure 9.2, the expected phenotypic ratio among the offspring is 3:1. Three offspring have the dominant phenotype for every one that has the recessive phenotype.

Mendel realized that these results are obtainable only if the alleles of each parent segregate (separate from each other) during meiosis. Therefore, Mendel formulated his first law of inheritance:

Law of Segregation

Each organism contains two alleles for each trait, and the alleles segregate during the formation of gametes. Each gamete then contains only one allele for each trait. When fertilization occurs, the new organism has two alleles for each trait, one from each parent.

Inheritance is a game of chance. Just as there is a 50% probability of heads or tails when tossing a coin, there is a 50% probability that a sperm or egg will have an *A* or an *a* when the parent is *Aa*. The chance of an equal number of heads or tails improves as the number of tosses increases. In the same way, the chance of an equal number of gametes with *A* and *a* improves as the number of gametes increases. Therefore, the 3:1 ratio among offspring is more likely the more offspring you count for the same type cross.

Color of Tobacco Seedlings

In tobacco plants, a dominant allele *(C)* for chlorophyll gives the plants a green color, and a recessive allele *(c)* for chlorophyll causes a plant to appear white. If a tobacco plant is homozygous for the recessive allele *(c)*, it cannot manufacture chlorophyll and thus appears white (Fig. 9.3).

Figure 9.3 Monohybrid cross.
These tobacco seedlings are growing on an agar plate.
The white plants cannot manufacture chlorophyll.

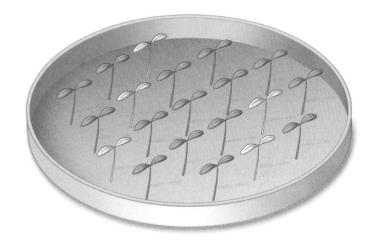

Experimental Procedure: Color of Tobacco Seedlings

1. Obtain a numbered agar plate on which tobacco seedlings are growing. They are the offspring of a cross between heterozygous parents: the cross *Cc* × *Cc*. Complete the Punnett square to determine the expected phenotypic ratio.

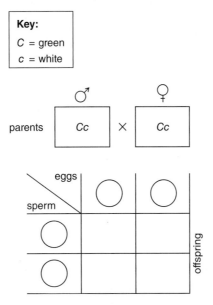

What is the expected phenotypic ratio? _____

2. Using a stereomicroscope, view the seedlings, and count the number that are green and the number that are white. Record the plate number and your results in Table 9.1.

3. Repeat steps 1 and 2 for two additional plates. Total the number that are green and the number that are white.

4. Complete Table 9.1 by recording the class data. Total the number that are green and the number that are white per class.

Table 9.1 Color of Tobacco Seedlings		
	Number of Offspring	
	Green Color	**White Color**
Plate # _____		
Plate # _____		
Plate # _____		
Totals		
Class data		

Conclusions: Color of Tobacco Seedlings

- Calculate the actual phenotypic ratio you observed. _____ Does your ratio differ from the expected ratio? _____ Explain. _____

  _____

- If directed by your instructor, use the chi-square test (see Section 9.4, page 118 of this laboratory) to determine if the deviation from the expected results can be accounted for by chance alone. Chi-square value: _____

- Repeat these steps using the class data. What is the phenotypic ratio per the class? _____ Do your class data give a ratio that is closer to the expected ratio, and is the chi-square deviation insignificant? _____ Explain. _____

Drosophila Melanogaster Culture

In a culture vial, female flies lay **eggs** on the medium provided. After a day or two, the eggs hatch into small **larvae** that feed and grow for about eight days, depending on the temperature. During this period, they **molt** (shed their external skeleton) twice. Therefore, three larval periods of growth, called **instars**, occur between molts. When fully grown, the third-stage instar larvae cease feeding and **pupate.** During pupation, which lasts about four days, the larval tissues are reorganized to form those of the adult. The life cycle is summarized in Figure 9.4.

Observation: Drosophila melanogaster Culture

Obtain a culture vial of *Drosophila melanogaster* to examine and answer these questions:

1. Where in the vial are the adult flies? _____ the eggs? _____

2. Where are the larva and what do they look like? _____

3. Where are the pupae and what do they look like? _____

Figure 9.4   Life cycle of the fruit fly, showing differences between adult sexes.

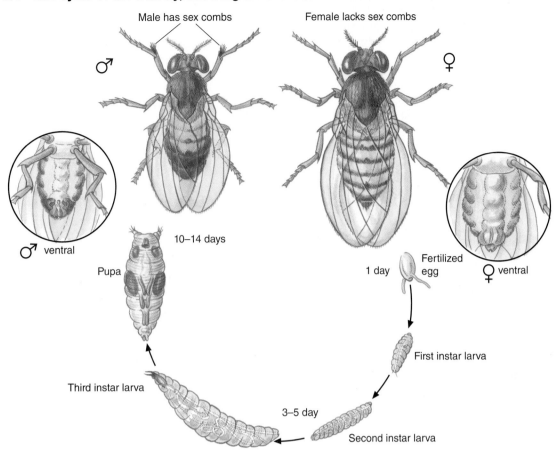

Observation: *Drosophila* Flies

You will be provided with either slides, frozen flies, or live flies to examine. If the flies are alive, follow the directions on page 110 for using FlyNap™ to anesthetize them. Put frozen or anesthetized flies on a white card, and use a camel-hair brush to move them around. Use a stereomicroscope or a hand lens to see the flies clearly. If you are looking at slides, you may use the scanning lens of your compound light microscope to view the flies.

1. Examine wild-type flies. **Wild-type** flies have the common characteristics of flies found in nature or in your kitchen on ripe fruit. You will find that they have the characteristics illustrated in Figure 9.4 for wing length, eye color, and body color. Complete Table 9.2 for wild-type flies.

2. Examine mutant flies. **Mutant flies** exhibit a **mutation**—a genetic change has occured that causes them to have a characteristic that differs from the wild-type flies. Two of the characteristics are the same as the wild type and one will be different. A mutant fly is homozygous recessive for that characteristic; therefore, as long as mutant flies reproduce only with mutant flies of the same type, they can be maintained in culture. Complete Table 9.2 for the mutant flies you examine.

3. Distinguish males from females. Examine any flies you have available and also Figure 9.4 in order to see that male flies can be distinguished from female flies.
   - The male is generally smaller.
   - The male has a more rounded abdomen than the female. The female has a pointed abdomen.
   - Dorsally, the male is seen to have a black-tipped abdomen, whereas the female appears to have dark lines only at the tip.
   - Ventrally, the abdomen of the male has a dark region at the tip due to the presence of claspers; this dark region is lacking in the female.

Table 9.2 Characteristics of Wild-Type and Mutant Flies

	Wild-Type	Ebony Body	Short-Wing	Sepia-Eye	White-Eye
Wing length					
Color of eyes					
Color of body					

Anesthetizing Flies

The use of FlyNap™ allows flies to be anesthetized for at least 50 minutes without killing them.

1. Dip the absorbent end of a swab into the FlyNap™ bottle as shown.
2. Tap the bottom of the culture vial on the tabletop to knock the flies to the bottom of the vial. (If the medium is not firm, you may wish to transfer the flies to an empty bottle before anesthetizing them. Ask your instructor for assistance, if this is the case.)
3. With one finger, push the plug slightly to one side. Remove the swab from the FlyNap™, and quickly stick the anesthetic end into the culture vial beside the plug so that the anesthetic tip is below the plug. Keep the culture vial upright with the swab in one place.

4. Remove the plug and the swab immediately after the flies are anesthetized (approximately 2 minutes in an empty vial, 4 minutes in a vial with medium), and spill the flies out onto a white file card. The length of time the flies remain anesthetized depends on the amount of FlyNap™ on the swab and on the number and age of the flies in the culture vial.
5. Transfer the anesthetized flies from the white file card onto the glass plate of a stereomicroscope for examination, or use a hand lens.

Wing Length in *Drosophila*

In fruit flies, the allele for long wings (*L*) is dominant over the allele for short (vestigial) wings (*l*). In this laboratory, we will consider a cross between heterozygous flies: *Ll* × *Ll*.

Complete this Punnett square to determine the expected phenotypic ratio. What is the expected phenotypic ratio among the offspring? _____

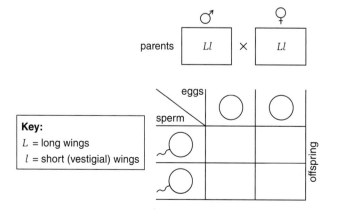

Experimental Procedure: Wing Length in Drosophila

Either you will be examining the results of a monohybrid cross or your professor will provide you with the data for such a cross. If the latter occurs, then enter the data in Table 9.3 and proceed directly to the Conclusions on this page. If you are observing live flies follow the directions on page 110 for anesthetizing and removing flies. When counting, use the stereomicroscope or a hand lens.

Divide your flies into those with long wings and those with short (vestigial) wings. Record your data and the class data in Table 9.3. For the class data, total the number of long wing and short wing flies per class.

Table 9.3 Wing Length in *Drosophila*		
	Number of Offspring	
	Long Wings	**Short Wings**
Your data		
Class data		

Conclusions : Wing Length in Drosophila

- Calculate the phenotypic ratio based on your data. _____

  Do your results differ from the expected ratio? _____ Explain. _____

- If so directed, do a chi-square test (see Section 9.4 of this laboratory) to determine if the deviation from the expected results can be accounted for by chance alone. Chi-square value: _____

- If your class did the experimental procedure what is the phenotypic ratio per the class? _____

  Regardless, do you expect the class data to give a ratio that is closer to the expected ratio, and an

  insignificant chi-square deviation? _____ Explain. _____

  _____

- Using the Punnett square provided, determine the phenotypic ratio expected for the cross *Ll* × *ll*.

  _____

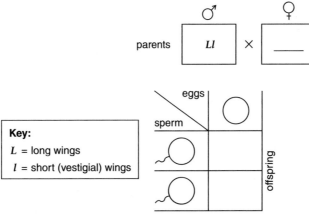

Key:
L = long wings
l = short (vestigial) wings

9.2 Two-Trait Crosses

Two-trait crosses involve two pairs of alleles. Mendel found that during a **dihybrid cross,** when two dihybrid individuals *(AaBb)* reproduce, the phenotypic ratio among the offspring is 9:3:3:1, representing four possible phenotypes. He realized that these results could only be obtained if the alleles of the parents segregated independently of one another when the gametes were formed. From this, Mendel formulated his second law of inheritance:

Law of Independent Assortment

Members of an allelic pair segregate (assort) independently of members of another allelic pair. Therefore, all possible combinations of alleles can occur in the gametes.

Color and Texture of Corn

In corn plants, the allele for purple kernel *(P)* is dominant over the allele for yellow kernel *(p)*, and the allele for smooth kernel *(S)* is dominant over the allele for rough kernel *(s)* (Fig. 9.5).

Figure 9.5   Dihybrid cross.
Four types of kernels are seen on an ear of corn following a dihybrid cross: purple smooth, purple rough, yellow smooth, and yellow rough.

20 mm

Experimental Procedure: Color and Texture of Corn

1. Obtain an ear of corn from the supply table. You will be examining the results of the cross *PpSs* × *PpSs*. Complete the Punnett square on page 113.

2. What is the expected phenotypic ratio among the offspring? _____

3. Count the number of kernels of each possible phenotype listed in Table 9.4. Record the sample number and your results in Table 9.4. Use three samples, and total your results for all samples. Also record the class data. Total the number of kernels that are the four phenotypes per class.

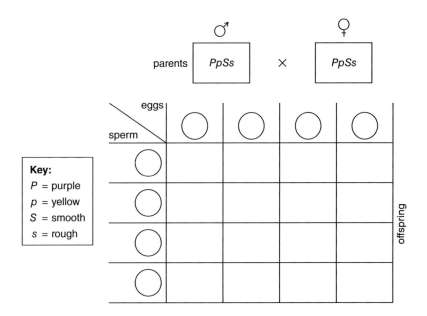

parents   PpSs   ×   PpSs

Key:
P = purple
p = yellow
S = smooth
s = rough

Table 9.4 Color and Texture of Corn

	Number of Kernels			
	Purple Smooth	**Purple Rough**	**Yellow Smooth**	**Yellow Rough**
Sample # _____				
Sample # _____				
Sample # _____				
Totals				
Class data				

Conclusions: Color and Texture of Corn

- From your data, which two traits seem dominant? _____ and _____
  Which two traits seem recessive? _____ and _____
- Calculate the actual phenotypic ratio you observed. _____ Does your ratio differ from the
  expected ratio? _____
- Use the chi-square test (see Section 9.4 of this laboratory) to determine if the deviation from the
  expected results can be accounted for by chance alone. Chi-square value:_____
- Repeat these steps using the class data. What is the phenotypic ratio per the class? _____
  Do your class data give a ratio that is closer to the expected ratio, and is the chi-square deviation
  insignificant? _____ Explain._____
  _____

Wing Length and Body Color in *Drosophila*

In *Drosophila*, long wings *(L)* are dominant over short (vestigial) wings *(l)*, and gray body *(G)* is dominant over ebony (black) body *(g)*. Consider the dihybrid cross *LlGg* × *LlGg* and complete this Punnett square:

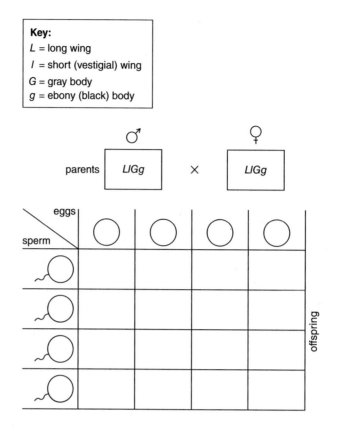

Key:
L = long wing
l = short (vestigial) wing
G = gray body
g = ebony (black) body

parents ♂ *LlGg* × ♀ *LlGg*

eggs / sperm

offspring

What is the expected phenotypic ratio for this cross? _____

Experimental Procedure: Wing Length and Body Color in Drosophila

Either you will be examining the results of a dihybrid cross or your professor will provide you with the data for such a cross. If the latter occurs, then enter the data in Table 9.5 and proceed directly to the Conclusions on page 115. If you are observing live flies follow the standard directions (see page 110) for anesthetizing and removing flies. Find one fly of each phenotype, and check with the instructor that you have identified them correctly before proceeding. When counting, use the stereomicroscope or a hand lens. Divide your flies into the groups indicated in Table 9.5, and record your results. Also record the class data. Total the number of flies that are the four phenotypes per class.

Table 9.5 Wing Length and Body Color in *Drosophila*			
Phenotypes			
Long Wings Gray Body	**Long Wings Ebony Body**	**Short Wings Gray Body**	**Short Wings Ebony Body**
Number of offspring			
Class data			

- Calculate the phenotypic ratio based on your data. _____ Do your results differ from the expected ratio? _____ Explain. _____

  _____

- If directed by your instructor, use the chi-square test (see Section 9.4) to determine if the deviation from the expected results can be accounted for by chance alone. Chi-square value: _____

- If the class did the experimental procedure what is the phenotypic ratio per the class? _____ Regardless, do you expect the class data to give a ratio that is closer to the expected ratio, and an insignificant chi-square deviation? _____ Explain. _____

  _____

- Use the Punnett square provided to calculate the expected phenotypic results for the cross *L1Gg × l1gg*.

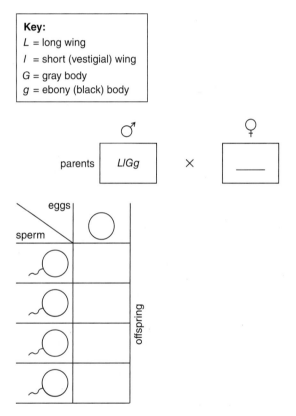

Key:
L = long wing
l = short (vestigial) wing
G = gray body
g = ebony (black) body

parents ♂ *LlGg* × ♀

eggs / sperm — offspring

What is the expected phenotypic ratio for this cross? _____

Summary

Using the alleles *Aa*, and if necessary *Bb*, make a list of the four types of crosses you have studied so far and the expected phenotypic results. These results will be obtained for these crosses no matter what the trait.

9.3 X-Linked Crosses

In animals such as fruit flies, chromosomes differ between the sexes. All but one pair of chromosomes in males and females are the same; these are called **autosomes** because they do not actively determine sex. The pair that is different is called the **sex chromosomes**. In fruit flies and humans, the sex chromosomes in females are XX and those in males are XY.

Some alleles on the X chromosome have nothing to do with gender and these genes are said to be X-linked. The Y chromosome does not carry these genes and indeed carries very few genes. Males with a normal chromosome inheritance are never heterozygous for X-linked alleles and if they inherit a recessive X-linked allele, it will be expressed.

Red/White Eye Color in *Drosophila*

In fruit flies, red eyes (X^R) are dominant over white eyes (X^r). You will be examining the results of the cross $X^RY \times X^RX^r$. Complete this Punnett square and state the expected phenotypic ratio for this cross?

Females: _____ Males: _____

Experimental Procedure: Red/White Eye Color in Drosophila

Either you will be examining the results of a sex-linked cross or your professor will provide you with the data for such a cross. If so, enter the data in Table 9.6 and proceed directly to the Conclusions on page 117. If you are observing live flies follow the standard directions (see page 110) for anesthetizing and removing flies. When counting, use the stereomicroscope or a hand lens. Record your results for each sex separately in Table 9.6. Also record the class data. Total the number of flies that have red eyes or white eyes per class.

Table 9.6 Red/White Eye Color in *Drosophila*		
	Number of Offspring	
Your Data:	**Red Eyes**	**White Eyes**
Males		
Females		
Class Data:		
Males		
Females		

Conclusions: Red/White Eye Color in Drosophila

Virtual Lab Sex-linked Traits   A virtual lab called Sex-linked Traits is available on the *Biology* website **www.mhhe.com/maderbiology11**. It will allow you to practice filling in Punnett squares and determining your results for X-linked traits.

- Calculate the phenotypic ratio based on your data for males and females separately.

  Males: _____

  Females: _____

- Does this ratio differ from the expected ratio? _____

- If directed by your instructor, use the chi-square test (see Section 9.4) to determine if the deviation from the expected results can be accounted for by chance alone. Chi-square value: _____

- Repeat these steps using the class data, if students in your lab did the experimental procedure. Would you expect class data to give a ratio that is closer to the expected ratio, and an insignificant chi-square deviation? _____ Explain. _____

  _____

- Using the Punnett square provided, calculate the expected phenotypic results for the cross $X^RY \times X^rX^r$.

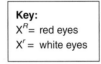

Key:
X^R = red eyes
X^r = white eyes

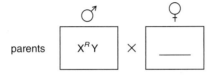

parents

♂ X^RY × ♀ _____

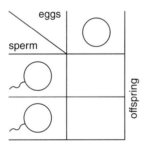

eggs

sperm

offspring

What is the expected phenotypic ratio among the offspring? _____

9.4 Chi-Square Analysis

Your experimental results can be evaluated using the **chi-square** (χ^2) test. This is the statistical test most frequently used to determine whether data obtained experimentally provide a "good fit," or approximation, to the expected data. Basically, the chi-square test can determine whether any deviations from the expected values are due to chance. Chance alone can cause the actual observed ratio to vary somewhat from the calculated ratio for a genetic cross. For example, a ratio of exactly 3:1 for a monohybrid cross is only rarely observed. Actual results will differ, but at some point, the difference is so great as to be unexpected. The chi-square test indicates this point. After this point, the original hypothesis (for example, that a monohybrid cross gives a 3:1 ratio) is not supported by the data.

The formula for the test is $\chi^2 = \Sigma(d^2/e)$

where χ^2 = chi-square

Σ = sum of

d = difference between expected and observed results (most often termed the *deviation*)

e = expected results

For example, in a monohybrid cross involving fruit flies with long wings (dominant) and fruit flies with short wings (recessive), a 3:1 ratio is expected. Therefore, if you count 160 flies, 40 are expected to have short wings, and 120 are expected to have long wings. But if you count 44 short-winged flies and 116 long-winged flies, then the value for the chi-square test would be as calculated in Table 9.7.

Table 9.7	Calculation of Chi-Square (Example)				
Phenotype	Observed Number	Expected Results (e)	Difference (d)	d^2	Partial Chi-Square (d^2/e)
Short wings	44	40	4	16	16/40 = 0.400
Long wings	116	120	4	16	16/120 = 0.133
					Chi-square = $\chi^2 = \Sigma(d^2/e)$ = 0.400 + 0.133 = 0.533

Now look up this chi-square value (χ^2) in a table that indicates whether the probability (p) is that the differences noted are due only to chance in the form of random sampling error or whether the results should be explained on the basis of a different prediction (hypothesis). In Table 9.8, the notation C refers to the number of "classes," which in this laboratory would be determined by the number of phenotypic traits studied. The example involves two classes: short wings and long wings. However, as indicated in the table by $C-1$, it is necessary to subtract 1 from the total number of classes. In the example, $C-1 = 1$. Therefore, the χ^2 value (0.533) would fall in the first line of Table 9.8 between 0.455 and 1.074. These correspond with p values of 0.50 and 0.30, respectively. This means that, by random chance, this difference between the actual count and the expected count would occur between 30% and 50% of the time. In biology, it is generally accepted that a p value greater than 0.10 is acceptable and that a p value lower than 0.10 indicates that the results cannot be due to random sampling and therefore do not support (do not "fit") the original prediction (hypothesis). A chi-square analysis is used to *refute* (falsify) a hypothesis, not to *prove* it.

Table 9.8 Values of Chi-Square

	Hypothesis Is Supported						Hypothesis Is Not Supported			
	Differences Are Insignificant						Differences Are Significant			
p	0.99	0.95	0.80	0.50	0.30	0.20	0.10	0.05	0.02	0.01
$C-1$										
1	0.00016	0.0039	0.064	0.455	1.074	1.642	2.706	3.841	5.412	6.635
2	0.0201	0.103	0.446	1.386	2.408	3.219	4.605	5.991	7.824	9.210
3	0.115	0.352	1.005	2.366	3.665	4.642	6.251	7.815	9.837	11.345
4	0.297	0.711	1.649	3.357	4.878	5.989	7.779	9.488	11.668	13.277
5	0.554	1.145	2.343	4.351	6.064	7.289	9.236	11.070	13.388	15.086

Source: Data from W. T. Keeton, et al., *Laboratory Guide for Biological Science*, 1968, p. 189.

Use Table 9.9 for performing a chi-square analysis of your results from a previous Experimental Procedure in this laboratory. If you performed a one-trait (monohybrid) cross, you will use only the first two lines. If you performed a two-trait (dihybrid) cross, you will use four lines.

$\chi^2 =$ _____

$C-1 =$ _____

p (from Table 9.8) = _____

Table 9.9 Calculation of Chi-Square

Phenotype	Observed Number	Expected Results (e)	Difference (d)	d^2	Partial Chi-Square (d^2/e)
					=
					=
					=
					=
				Chi-square $= \chi^2 = \Sigma(d^2/e) =$	

Conclusions: Chi-Square Analysis

- Do your results support your original prediction? _____
- If not, how can you account for this? _____

_____

1. If offspring exhibit a 3:1 phenotypic ratio, what are the genotypes of the parents? _____

2. In fruit flies, which of the characteristics that you studied was X-linked? _____

3. If offspring exhibit a 9:3:3:1 phenotypic ratio, what are the genotypes of the parents?

   _____

   _____

4. What was the phenotype of the parents if you counted 90 long-winged flies to 30 short-winged flies as the offspring? _____

   _____

5. Briefly describe the life cycle of *Drosophila*. _____

   _____

   _____

6. When doing a genetic cross, why is it necessary to remove parent flies before the pupae have hatched?

   _____

7. What is the genotype of a white-eyed male fruit fly? _____

8. What may have been the genotypes of parents if you counted 20 green tobacco seedlings and 20 white tobacco seedlings in one agar plate? _____

9. Suppose you counted tobacco seedlings for a monohybrid cross in six agar plates, and your data were as follows: 125 green plants and 39 white plants. According to the chi-square test, are your deviations from the expected values due to chance? _____

10. Suppose that students in the laboratory periods before yours removed some of the purple and yellow corn kernels on the ears of corn as they were performing the Experimental Procedure. What effect would this have on your results?

    _____

    _____

10
Human Genetics

Learning Outcomes

Introduction

It's possible to view the chromosomes of an individual because cells can be microscopically examined and photographed just before cell division occurs (Fig. 10.1). A computer is then used to arrange the chromosomes by pairs. The resulting pattern of chromosomes is called a **karyotype**. Most individuals have a normal karyotype because they inherited 22 pairs of autosomes and one pair of sex chromosomes. Occasionally individuals inherit an abnormal number of chromosomes and therefore they have a karyotype anomaly. We will have the opportunity to observe a few karyotype anomalies in this laboratory.

Next, we consider autosomal dominant and recessive traits in human beings before moving on to recessive X-linked traits. We will have the opportunity to observe that Mendel's laws apply not only to peas and fruit flies, but also to humans. This laboratory provides an opportunity to practice genetics problems. Finally, we will learn how to read and construct a **pedigree**, which is a diagram showing the inheritance of a genetic disorder within a family. Genetic counselors use pedigrees to counsel couples about the chances of passing a genetic disorder to their children.

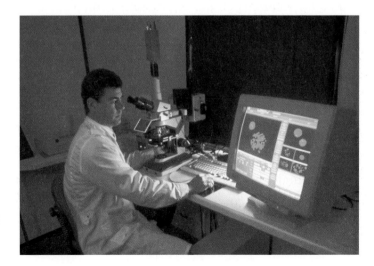

Figure 10.1   Human karyotype preparation.
Cells are microscopically examined and photographed and then a computer arranges the chromosomes by pairs.

10.1 Chromosomal Inheritance

Usually, gametogenesis and fertilization occur normally, and an individual inherits 22 pairs of autosomes and one pair of sex chromosomes. Gametogenesis involves meiosis, the type of cell division that reduces the chromosome number by one-half because the chromosome pairs separate during meiosis I. When homologues fail to separate during meiosis I, called **nondisjunction**, gametes with too few (n − 1) or too many (n + 1) chromosomes result (Fig. 10.2).

Nondisjunction can also occur during meiosis II if the chromatids fail to separate and the daughter chromosomes go into the same daughter cell. Again, the gametes will be either n − 1 or n + 1. Nondisjunction leads to various syndromes. A **syndrome** is a group of symptoms that occur together.

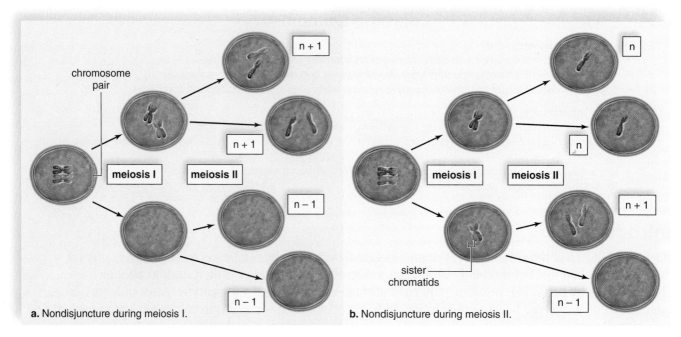

a. Nondisjuncture during meiosis I.　　**b.** Nondisjuncture during meiosis II.

Figure 10.2　Nondisjunction during oogenesis.
Recall that size and not color indicates the particular human chromosome. **a.** Nondisjunction involving one homologue pair has occurred during meiosis I, and all eggs after meiosis II have an abnormal number of chromosomes; in humans either 24 or 22 chromosomes. **b.** Nondisjunction involving one daughter chromosome has occurred during meiosis II, and only two eggs have an abnormal number of chromosomes. n = haploid number

Observation: Autosome Chromosome Anomalies

A trisomy occurs when the individual has three chromosomes instead of two chromosomes at one karyotype location.

A trisomy arises when an egg has two chromosomes of the same size because the sperm contributes another chromosome of that type to the offspring.

Three trisomies are seen among human newborns:

Down syndrome

Trisomy 21 (Down syndrome) is the most common autosomal trisomy in humans. Survival to adulthood is common. Characteristic facial features include an eyelid fold, a flat face, and a large fissured tongue. Some degree of mental retardation is common as is early-onset Alzheimers disease. Sterility due to sexual underdevelopment may be present. The karyotype change that signifies Down syndrome is indicated in Figure 10.3. With the assistance of Figure 10.2, explain how trisomy 21 comes about?

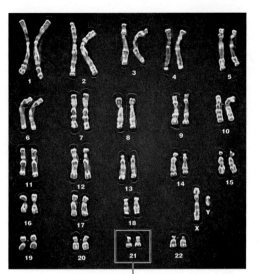

Figure 10.3 Normal human male karyotype.
Banding patterns appear when the chromosomes are stained. In a karyotype, the pairs of chromosomes are arranged by size and banding patterns. The sex chromosomes of a male are X and Y. A female has two X chromosomes. The karyotype change that indicates Down syndrome has been indicated. Students do the same for Patau syndrome and Edwards syndrome.

Down syndrome has three #21 chromosomes

Trisomy 13 (Patau syndrome) is characterized by polydactyly (extra fingers and toes) plus overlapping fingers. Cleft palate and other skeletal defects are common as well as eye, brain, and circulatory defects. Survival is rare beyond a few months. Indicate in Figure 10.3 the karyotype change that signifies Patau syndrome, i.e., which chromosome should have three copies?

Patau syndrome

_____

Trisomy 18 (Edwards syndrome) is characterized by clinched fists and overlapping fingers. Other skeletal defects include a small, odd-shaped head and a small mouth and jaw. The newborn has a low birth weight and most infants die within the first month due to internal organ defects. Indicate in Figure 10.3 the karyotype change that signifies Edwards syndrome, i.e., which chromsome should have three copies? _____

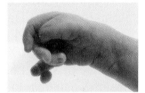

Edwards syndrome

Observation: Sex Chromosomes Anomalies

A female with **Turner syndrome** (XO) has only one sex chromosome, an X chromosome; the O signifies the absence of the second sex chromosome. Because the ovaries never become functional, these females do not undergo puberty or menstruation, and their breasts do not develop. Generally, females with Turner syndrome have a short build, folds of skin on the back of the neck, difficulty recognizing various spatial patterns, and normal intelligence. With hormone supplements, they can lead fairly normal lives.

When an egg having two X chromosomes is fertilized by an X-bearing sperm, an individual with **poly-X syndrome** results. The body cells have three X chromosomes and therefore 47 chromosomes. Although they tend to have learning disabilities, poly-X females have no apparent physical anomalies, and many are fertile and have children with a normal chromosome count.

Turner syndrome

Poly-X syndrome

When an egg having two X chromosomes is fertilized by a Y-bearing sperm, a male with **Klinefelter syndrome** results. This individual is male in general appearance, but the testes are underdeveloped, and the breasts may be enlarged. The limbs of XXY males tend to be longer than average, muscular development is poor, body hair is sparse, and many XXY males have learning disabilities. Fill in the following diagram to show that nondisjunction during meiosis can lead to Turner, poly-X, or Klinefelter syndrome:

Klinefelter syndrome

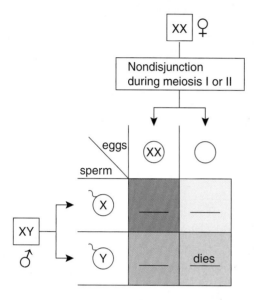

Jacob syndrome can be due to nondisjunction during meiosis II of spermatogenesis. These males are usually taller than average, suffer from persistent acne, and tend to have speech and reading problems. At one time, it was suggested that XYY males were likely to be criminally aggressive, but the incidence of such behavior has been shown to be no greater than that among normal XY males. Fill in the following diagram to show that nondisjunction during meiosis II in a male can lead to this syndrome:

Jacob syndrome

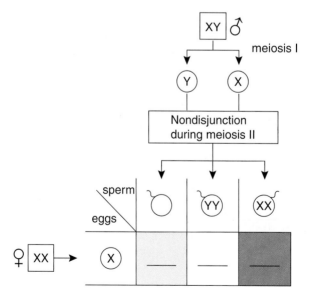

Conclusions: Sex Chromosome Anomalies

Complete Table 10.1 to show how a physician would recognize each of these syndromes from a karyotype. In Figure 10.4 use this information to correctly label each karyotype.

Table 10.1 Numerical Sex Chromosome Anomalies	
Syndrome Number	**Comparison with Normal Number**
Turner:	Normal Female:
Poly X:	Normal Female:
Klinefelter:	Normal Male:
Jacob:	Normal Male:

Figure 10.4   Sex chromosome anomalies.
The karyotypes of various sex chromosome anomalies are shown. Label each one with the name of the resulting syndrome.

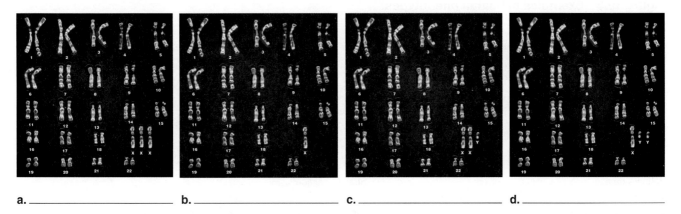

a. _____   b. _____   c. _____   d. _____

10.2   Genetic Inheritance

Just as we inherit pairs of chromosomes, we inherit pairs of alleles, alternate forms of a gene. As usual, a **dominant allele** is assigned a capital letter, while a **recessive allele** is given the same letter lowercased.

The **genotype** tells the alleles of the individual, while the **phenotype** describes the appearance of the individual.

Autosomal Dominant and Recessive Traits

The alleles for autosomal traits are carried on the autosomes. If individuals are homozygous dominant (*AA*) or heterozygous (*Aa*), their phenotype is the dominant trait. If individuals are homozygous recessive (*aa*), their phenotype is the recessive trait.

For example, if W = widow's peak and w = straight hairline, what would be homozygous dominant? _____ What would be heterozygous? _____ and what would be homozygous recessive? _____ Which one of these would have a straight hairline? _____

1. For this Experimental Procedure, you will need a lab partner to help you determine your phenotype for the traits listed in the first column of Table 10.2. Figure 10.5 illustrates some of these traits. Record your phenotypes by circling them in the first column of the table.

2. Determine your probable genotype. If you have the recessive phenotype, you know your genotype. If you have the dominant phenotype, you may be able to decide whether you are homozygous dominant or heterozygous by recalling the phenotype of your parents, siblings, or children. Circle your probable genotype in the second column of Table 10.2.

3. Your instructor will tally the class's phenotypes for each trait so that you can complete the third column of Table 10.2.

4. Complete Table 10.2 by calculating the percentage of the class with each trait. Are dominant phenotypes always the most common in a population? _____ Explain. _____

Figure 10.5 Examples of human phenotypes.

Widow's peak

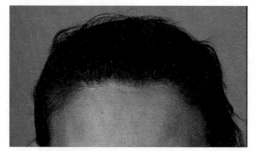

Straight hairline

Bent little finger

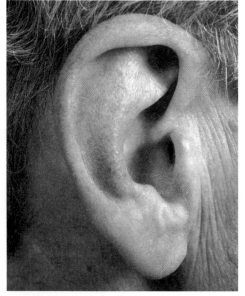

Unattached earlobes

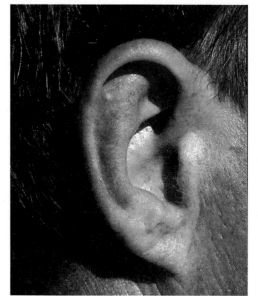

Attached earlobes

Straight little finger

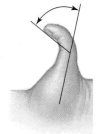

Hitchhiker's thumb

Table 10.2 Autosomal Human Traits

Trait: d = Dominant r = Recessive	Possible Genotypes	Number in Class	Percentage of Class with Trait
Hairline: Widow's peak (d) Straight hairline (r)	WW or Ww ww		
Earlobes: Unattached (d) Attached (r)	EE or Ee ee		
Skin pigmentation: Freckles (d) No freckles (r)	FF or Ff ff		
Hair on back of hand: Present (d) Absent (r)	HH or Hh hh		
Thumb hyperextension—"hitchhiker's thumb": Last segment cannot be bent backward (d) Last segment can be bent back to 60° (r)	TT or Tt tt		
Bent little finger: Little finger bends toward ring finger (d) Straight little finger (r)	LL or Ll ll		
Interlacing of fingers: Left thumb over right (d) Right thumb over left (r)	II or Ii ii		

Genetics Problems Involving the Traits in Table 10.2

1. Nancy and the members of her immediate family have attached earlobes. Her maternal grandfather has unattached earlobes. What is the genotype of her maternal grandfather?_____ Nancy's maternal grandmother is no longer living. What could have been the genotype of her maternal grandmother? _____ or _____

    Do two Punnett squares—one for each possible genotype for Nancy's maternal grandmother. Compare the chances her maternal grandparents could have had a child with attached earlobes as Nancy's mother has. Compare the chances.*_____

    *one out of four = 25% chance
     one out of two = 50% chance

Figure 10.6 Two common patterns of autosomal inheritance in humans.

Left: Both parents are heterozygous. *Right:* One parent is heterozygous and the other is homozygous recessive.

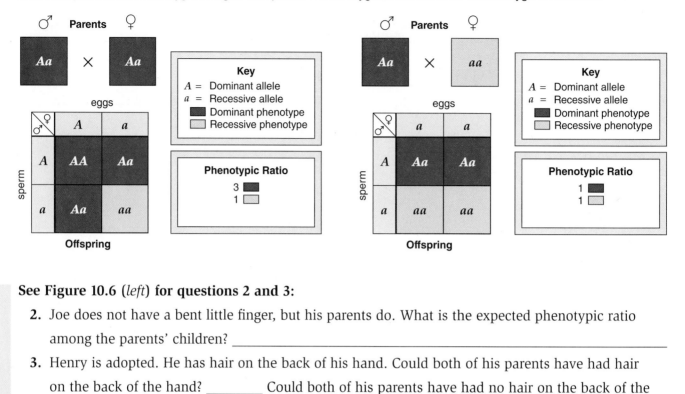

See Figure 10.6 (*left*) for questions 2 and 3:

2. Joe does not have a bent little finger, but his parents do. What is the expected phenotypic ratio among the parents' children? _____

3. Henry is adopted. He has hair on the back of his hand. Could both of his parents have had hair on the back of the hand? _____ Could both of his parents have had no hair on the back of the hand? _____ Explain. _____

_____

_____

Genetics Problems Involving Genetic Disorders

See Figure 10.6 (*left*) for questions 1 and 2:

1. Cystic fibrosis is an autosomal recessive disorder. If both of Sally's parents are heterozygous for cystic fibrosis, what are her chances of inheriting cystic fibrosis? _____

2. Nancy has cystic fibrosis, but neither parent has cystic fibrosis. What is the genotype of all people involved? _____

See Figure 10.6 (*right*) for questions 3 and 4:

3. Huntington disease is an autosomal dominant disorder. If only one of Sam's parents is heterozygous for Huntington disease and the other is homozygous recessive, what are his chances of inheriting Huntington disease? _____

4. In Henry's family, only his father has Huntington disease. What are the genotypes of Henry, his mother, and his father? _____

Sex Linkage

The sex chromosomes carry genes that affect traits other than the individual's sex. Genes on the sex chromosomes are called **sex-linked genes.** The vast majority of sex-linked genes have alleles on the X chromosome and are called **X-linked genes.** Most often, the abnormal condition is recessive.

Color blindness is an X-linked, recessive trait. The possible genotypes and phenotypes are as follows:

Females
$X^B X^B$ = normal vision
$X^B X^b$ = normal vision (carrier)
$X^b X^b$ = color blindness

Males
$X^B Y$ = normal vision
$X^b Y$ = color blindness

Experimental Procedure: X-Linked Traits

1. Your instructor will provide you with a color blindness chart. Have your lab partner present the chart to you. Write down the words or symbols you see, but do not allow your partner to see what you write, and do not discuss what you see. This is important because color-blind people see something different than do people who are not color blind.

2. Now test your lab partner as he or she has tested you.

3. Are you color blind? _____ If so, what is your genotype? _____

4. If you are a female and are not color blind, you can judge whether you are homozygous or heterozygous by knowing if any member of your family is color blind. If your father is color blind,

   what is your genotype? _____ If your mother is color blind, what is your genotype? _____

   If you know of no one in your family who is color blind, what is your probable genotype? _____

Genetics Problems Involving X-Linked Genetic Disorders

See Figure 10.7 (*left*) for questions 1 and 2:

1. If a father is color blind, but the mother does not have a color-blind allele, what are the chances his

   daughters will be color blind? _____ Be carriers? _____

2. Mary Jo is a carrier for hemophilia, an X-linked recessive disorder. Her mother is perfectly normal.

   What is her father's genotype? _____

Figure 10.7 Two common patterns of X-linked inheritance in humans.
Left: A color-blind father has carrier daughters. *Right:* The sons of a carrier mother have a 50% chance of being color blind.

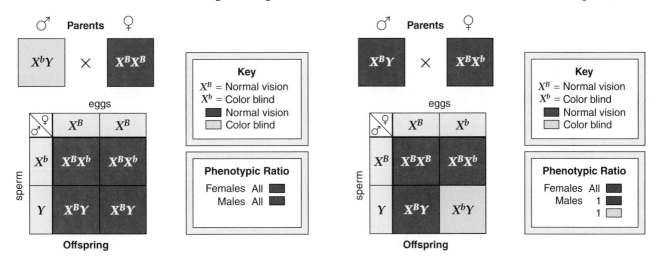

See Figure 10.7 (*right*) for questions 3 and 4 :

3. If a boy is color blind, from which parent did he inherit the defective allele? _____

4. Mary has a color-blind son but Mary and both of Mary's parents have normal vision. Give the genotype of all people involved. _____

5. A person with Klinefelter syndrome (XXY) is color blind. Both his father and mother have normal vision, but his maternal uncle is color blind. In which parent and at what meiotic division did sex chromosome nondisjunction occur? _____

6. A person with Turner syndrome has hemophilia. Her mother does not have hemophilia, but her father does. In which parent did nondisjunction occur, considering that the single X came from the father? _____ Is it possible to tell if nondisjunction occurred during meiosis I or meiosis II? _____ Explain. _____

_____

Pedigrees

A **pedigree** shows the inheritance of a genetic disorder within a family and can help determine whether any particular individual has an allele for that disorder. Then a Punnett square can be done to determine the chances of a couple producing an affected child.

In a pedigree, Roman numerals indicate the generation, and Arabic numerals indicate particular individuals in that generation. The symbols used to indicate normal and affected males and females, reproductive partners, and siblings are shown in Figure 10.8.

Figure 10.8   Pedigree symbols.

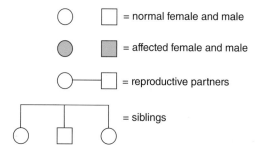

Pedigree Analyses

For each of the following pedigrees, determine how a genetic disorder is inherited. Use Table 10.3 to help with this determination. Is the inheritance pattern autosomal dominant, autosomal recessive, or X-linked recessive? Also, decide the genotype of particular individuals in the pedigree. Remember that the *genotype* indicates the dominant and recessive alleles present and the *phenotype* is the actual physical appearance of the trait in the individual. A pedigree indicates the phenotype, and you can reason out the genotype.

Table 10.3 Pedigree Solution Chart

Inheritance Pattern	Notes	Clues	Possible Genotypes
Autosomal dominant (any chromosome except X or Y)	If at least one chromosome has the allele, the individual will be affected.	One or both parents are affected. Many of the children are affected.	AA or Aa = affected aa = normal
Autosomal recessive (any chromosome except X or Y)	Both chromosomes must have the recessive allele for the individual to be affected.	If neither parent is affected, few of the children are affected.	AA or Aa = normal Aa = carrier* aa = affected
X-linked recessive (only the X chromosome)	The trait is only carried on the X chromosome. There must be a recessive allele on the X chromosome for the trait to be expressed.	Trait is primarily found in males. It is often passed from grandfather to grandson.	X^AX^A and X^AX^a = normal female; X^AX^a = carrier female* X^AY = normal male X^aX^a = affected female X^aY = affected male

*A carrier is one who does not show the trait but has the ability to pass it on to his or her offspring.

1. Study the following pedigree:

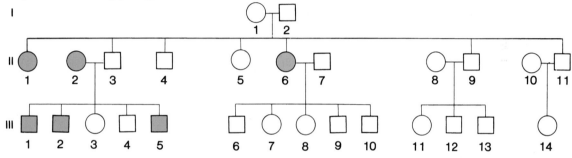

   a. What is the inheritance pattern for this genetic disorder? _____

   b. What is the genotype of the following individuals? Use *A* for the dominant allele and *a* for the recessive allele.

   Generation I, individual 1: _____

   Generation II, individual 1: _____

   Generation III, individual 8: _____

2. Study the following pedigree:

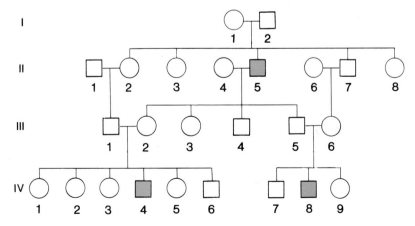

a. What is the inheritance pattern for this genetic disorder? _____

b. What is the genotype of the following individuals?

Generation I, individual 1: _____

Generation II, individual 8: _____

Generation III, individual 1: _____

Construction of a Pedigree

You are a genetic counselor who has been given the following information from which you will construct a pedigree.

1. Your data: Henry has a double row of eyelashes, which is a dominant trait. Both his maternal grandfather and his mother have double eyelashes. Their spouses are normal. Henry is married to Isabella and their first child Polly has normal eyelashes. The couple wants to know the chances of any child having a double row of eyelashes.

2. *Construct two blank pedigrees.* Begin with the maternal grandfather and grandmother and end with Polly.

Pedigree 1   **Pedigree 2**

3. Pedigree 1: Try out a pattern of autosomal dominant inheritance by assigning appropriate genotypes for an autosomal dominant pattern of inheritance to each person in this pedigree. Pedigree 2: Try out a pattern of X-linked dominant inheritance by assigning appropriate genotypes for this pattern of inheritance to each person in your pedigree. Which pattern is correct? _____

4. What is your key for this trait?

Key: _____ normal eyelashes   _____ double row of eyelashes

5. Use correct genotypes to show a cross between Henry and Isabella and from experience with crosses state the expected phenotypic ratio among the offspring:

Cross: Henry   Isabella   **Phenotypic ratio:**

_____ X _____   _____

6. What are the percentage chances of Henry and Isabella having a child with double eyelashes? _____

Laboratory Review 10

1. Name one pair of chromosomes not homologous in a normal karyotype. _____

2. If nondisjunction occurs in a heterozygous woman, would nondisjunction during meiosis I or meiosis II produce an egg in which the two X chromosomes carry the same alleles? _____
   Explain. _____

3. Which one could produce a sperm with two X chromosomes: nondisjunction during meiosis I or nondisjunction during meiosis II? _____
   Explain. _____

4. If an individual exhibits the dominant trait, do you know the genotype? _____
   Why or why not? _____
   _____

5. A son is color blind, but both parents are normal. Give the genotype of the mother _____ and the father _____. Explain the pattern of inheritance. _____
   _____

6. What pattern of inheritance in a pedigree would allow you to decide that a trait is X-linked?
   _____

7. What pattern of inheritance in a pedigree would allow you to decide that a trait is autosomal recessive?
   _____

8. What is the difference between a Punnett square and a pedigree?_____
   _____

9. What is the probability that two individuals with an autosomal recessive trait will have a child with the same trait? _____

10. What is the probability that a woman whose father was color blind will have a son who is color blind?
    _____

11

DNA Biology and Technology

Learning Outcomes

Introduction

This laboratory pertains to molecular genetics and biotechnology. Molecular genetics is the study of the structure and function of **DNA (deoxyribonucleic acid),** the genetic material. Biotechnology is the manipulation of DNA for the benefit of human beings and other organisms.

First we will study the structure of DNA and see how that structure facilitates DNA replication in the nucleus of cells. DNA replicates prior to cell mitosis; following mitosis, each daughter nucleus has a complete copy of the genetic material. DNA replication also precedes gametogenesis.

Then we will study the structure of **RNA (ribonucleic acid)** and how it differs from that of DNA, before examining how DNA, with the help of RNA, specifies protein synthesis. The linear construction of DNA, in which nucleotide follows nucleotide, is paralleled by the linear construction of the primary structure of protein, in which amino acid follows amino acid. Essentially, we will see that the sequence of nucleotides in DNA codes for the sequence of amino acids in a protein. We will also review the role of three types of RNA in protein synthesis. DNA's code is passed to messenger RNA (mRNA), which moves to the ribosomes. Transfer RNA (tRNA) brings the amino acids to the ribosomes, and they become linked by peptide bonds in the order directed by mRNA. In this way a particular polypeptide forms.

We now understand that a mutated gene has an altered DNA base sequence, and altered sequences can cause genetic disorders. You will have an opportunity to carry out a laboratory procedure that detects whether an individual is normal, has sickle-cell disease, or is a carrier.

11.1 DNA Structure and Replication

The structure of DNA lends itself to **replication,** the process that makes a copy of a DNA molecule. DNA replication is a necessary part of chromosome duplication, which precedes mitosis and meiosis. Therefore, DNA replication is needed for growth and repair and for sexual reproduction.

DNA Structure

DNA is a polymer of nucleotide monomers (Fig. 11.1). Each nucleotide is composed of three molecules: deoxyribose (a 5-carbon sugar), a phosphate, and a nitrogen-containing base.

Figure 11.1   Overview of DNA structure.
Diagram of DNA double helix shows that the molecule resembles a twisted ladder. Sugar-phosphate backbones make up the sides of the ladder, and hydrogen-bonded bases make up the rungs of the ladder. Complementary base pairing dictates that A is bonded to T and G is bonded to C and vice versa. *Label the boxed nucleotide pair as directed in the next Observation.*

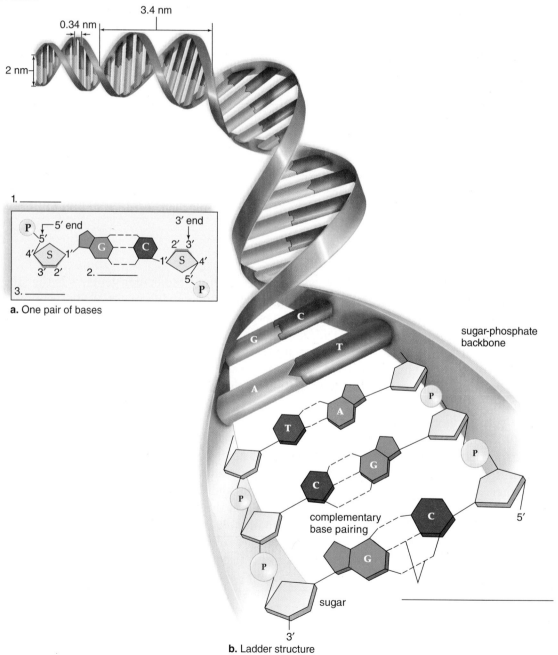

a. One pair of bases

b. Ladder structure

1. A boxed nucleotide pair is shown in Figure 11.1*a*. If you are working with a kit, draw a representation of one of your nucleotides here. *Label phosphate, base pair, and deoxyribose in your drawing and 1–3 in Figure 11.1a.*

2. Notice the four types of bases: cytosine (C), thymine (T), adenine (A), and guanine (G). What is the color of each of the four types of bases in Figure 11.1? In your kit? Complete Table 11.1 by writing in the colors of the bases.

Table 11.1 Base Colors		
	In Figure 11.1*b*	**In Your Kit**
Cytosine		
Thymine		
Adenine		
Guanine		

3. Using Figure 11.1 as a guide, join several nucleotides together. Observe the entire DNA molecule. What type of molecules make up the backbone (uprights of ladder) of DNA (Fig. 11.1*b*)? _____ and _____ In the backbone, the phosphate of one nucleotide is bonded to a sugar of the next nucleotide.

4. Using Figure 11.1 as a guide, join the bases together with hydrogen bonds. Label a hydrogen bond in Figure 11.1. Dashes are used to represent hydrogen bonds in Figure 11.1*b* because hydrogen bonds are (strong or weak). _____

5. Notice in Figure 11.1*b* and in your model, that the base A is always paired with the base _____, and the base C is always paired with the base _____. This is called **complementary base pairing**.

6. In Figure 11.1*b*, what molecules make up the rungs of the ladder? _____

7. Each half of the DNA molecule is a DNA strand. Why is DNA called a double helix (Fig. 11.1*b*)?

_____

DNA Replication

During replication, the DNA molecule is duplicated so that there are two DNA molecules. We will see that complementary base pairing makes replication possible.

1. Before replication begins, DNA is unzipped. Using Figure 11.2a as a guide, break apart your two DNA strands. What bonds are broken in order to unzip the DNA strands? _____

2. Using Figure 11.2b as a guide, attach new complementary nucleotides to each strand using complementary base pairing.

3. Show that you understand complementary base pairing by completing Table 11.2. You now have two DNA molecules (Fig. 11.2c). Are your molecules identical? _____

4. Because of complementary base pairing, each new double helix is composed of an _____ strand and a _____ strand. *Write old or new in 1–10, Figure 11.2a, b, and c. Conservative means to save something from the past.* Why is DNA replication called **semiconservative**?

   _____

   _____

   _____

   _____

Figure 11.2   DNA replication.
Use of the ladder configuration better illustrates how replication takes place. **a.** The parental DNA molecule. **b.** The "old" strands of the parental DNA molecule have separated. New complementary nucleotides available in the cell are pairing with those of each old strand. **c.** Replication is complete.

1. _____   2. _____

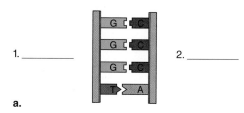

a.

3. _____   4. _____   6. _____

5. _____

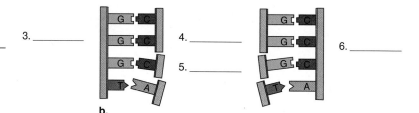

b.

7. _____   8. _____   10. _____

9. _____

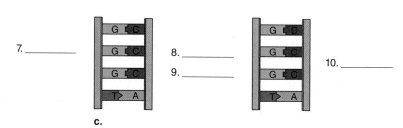

c.

5. Genetic material has to be inherited from cell to cell and organism to organism. Consider that because of DNA replication, a chromosome is composed of two chromatids and each chromatid is a DNA double helix. The chromatids separate during cell division so that each daughter cell receives a copy of each chromosome. In your own words, how does replication provide a means for passing DNA from cell to cell and organism to organism?

_____

_____

Table 11.2 DNA Replication																												
Old strand	G	G	G	T	T	C	C	A	T	T	A	A	A	T	T	C	C	A	G	A	A	A	T	C	A	T	A
New strand																											

11.2 RNA Structure

Like DNA, RNA is a polymer of nucleotides (Fig. 11.3). In an RNA nucleotide, the sugar ribose is attached to a phosphate molecule and to a nitrogen-containing base, C, U, A, or G. In RNA, the base uracil replaces thymine as one of the pyrimidine bases. RNA is single stranded, whereas DNA is double stranded.

Figure 11.3 Overview of RNA structure.
RNA is a single strand of nucleotides. *Label the boxed nucleotide as directed in the next Observation.*

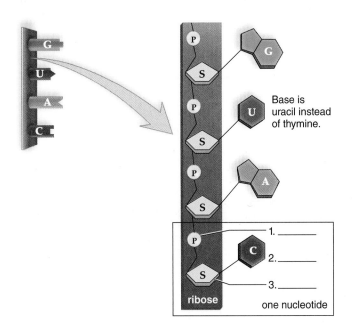

1. Describe the backbone of an RNA molecule. _____

2. Where are the bases located in an RNA molecule? _____

3. Complete Table 11.3 to show the complementary DNA bases for these RNA bases.

Table 11.3 DNA and RNA Bases				
RNA Bases	C	U	A	G
DNA Bases				

Observation: RNA Structure

1. If you are using a kit, draw a nucleotide for the construction of mRNA. *Label the ribose (the sugar in RNA), the phosphate, and the base in your drawing and in 1–3, Figure 11.3.*

2. Complete Table 11.4 by writing in the colors of the bases for Figure 11.3 and for your kit.

Table 11.4	Base Colors	
	In Figure 11.3	**In Your Kit**
Cytosine		
Uracil		
Adenine		
Guanine		

3. The base uracil substitutes for the base thymine in RNA. Complete Table 11.5 to show the several other ways RNA differs from DNA.

Table 11.5	DNA Structure Compared with RNA Structure	
	DNA	**RNA**
Sugar	Deoxyribose	
Bases	Adenine, guanine, thymine, cytosine	
Strands	Double stranded with base pairing	
Helix	Yes	

11.3 DNA and Protein Synthesis

Protein synthesis requires the processes of transcription and translation. During **transcription,** which takes place in the nucleus, an RNA molecule called **messenger RNA (mRNA)** is made complementary to one of the DNA strands. This mRNA leaves the nucleus and goes to the ribosomes in the cytoplasm. Ribosomes are composed of **ribosomal RNA (rRNA)** and proteins in two subunits. They provide a location for protein synthesis to occur.

During **translation,** RNA molecules called **transfer RNA (tRNA)** bring amino acids to the ribosome, and the amino acids join in the order prescribed by mRNA. In this way the sequence of amino acids in a new polypeptide was originally specified by DNA. This is the information that DNA, the genetic material, stores. Explain the role DNA, mRNA, and tRNA have in protein synthesis here:

DNA _____

mRNA _____

tRNA _____

Transcription

During **transcription**, complementary RNA is made from a DNA template (Fig. 11.4). A portion of DNA unwinds and unzips at the point of attachment of the enzyme RNA polymerase. A strand of mRNA is produced when complementary nucleotides join in the order dictated by the sequence of bases in DNA. Transcription occurs in the nucleus, and the mRNA passes out of the nucleus to enter the cytoplasm.

Label Figure 11.4. For number 1, note the name of the enzyme that carries out mRNA synthesis. For number 2, note the name of this polynucleotide molecule.

Observation: Transcription

1. If you are using a kit, unzip your DNA model so that only one strand remains. This strand is called the **template strand** because it will be used as a template to construct an mRNA molecule. The **gene strand** is preferred terminology for the complementary DNA strand you discarded, because as you can verify in Figure 11.4 it has the same sequence of bases as the mRNA molecule, except uracil substitutes for thymine.*

2. Using Figure 11.4 as a guide, construct a messenger RNA (mRNA) molecule by first lining up RNA nucleotides complementary to the template strand of your DNA molecule. Join the nucleotides together to form mRNA.

3. A portion of DNA has the sequence of bases shown in Table 11.6. *Complete Table 11.6 to show the sequence of bases in mRNA.*

4. If you are using a kit, unzip mRNA transcript from the DNA. Locate the end of the strand that will move to the _____ in the cytoplasm.

Figure 11.4   Messenger RNA (mRNA).
Messenger RNA, which is complementary to a section of DNA forms during transcription.

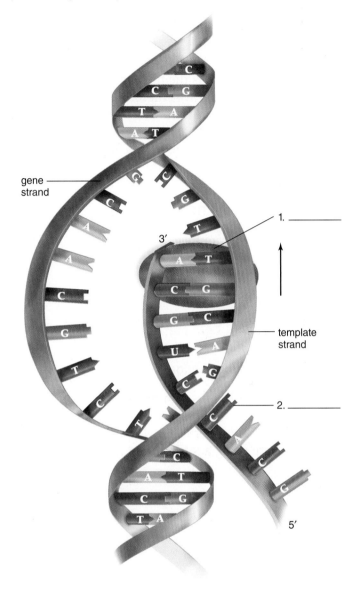

gene strand

1. _____

3′

template strand

2. _____

5′

Table 11.6 Transcription	
DNA	T A C A C G A G C A A C T A A C A T
mRNA	

*Sometimes the template strand is called the *sense strand* and, if so, the gene strand is called the *antisense strand*.

Translation

During **translation**, a polypeptide is made. DNA specifies the sequence of amino acids in a polypeptide because every three bases stand for an amino acid. Therefore, DNA is said to have a **triplet code.** The bases in mRNA are complementary to the bases in DNA. Every three bases in mRNA are called a **codon.** One codon of mRNA represents one amino acid. Thus, the sequence of DNA bases serves as the blueprint for the sequence of amino acids assembled to make a protein. The correct sequence of amino acids in a polypeptide is the message that mRNA carries.

Messenger RNA leaves the nucleus and proceeds to the ribosomes, where protein synthesis occurs. As previously mentioned, transfer RNA (tRNA) molecules transfer amino acids to the ribosomes. Each tRNA has one particular amino acid at one end and a specific **anticodon** at the other end (Fig. 11.5). *Label Figure 11.5,* where the amino acid is represented as a colored ball, the tRNA is green, and the anticodon is the sequence of three bases. (The anticodon is complementary to an mRNA codon.)

Figure 11.5 Transfer RNA (tRNA).
Each type of transfer RNA with a specific anticodon carries a particular amino acid to the ribosomes.

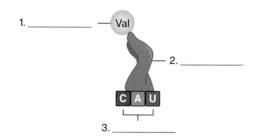

Observation: Translation

1. Figure 11.6 shows seven tRNA–amino acid complexes. Every amino acid has a name; in the figure, only the first three letters of the name are inside the ball. Using the mRNA sequence given in Table 11.7, number the tRNA–amino acid complexes in the order they will come to the ribosome.
2. If you are using a kit, arrange your tRNA–amino acid complexes in the order consistent with Table 11.7. Complete Table 11.7. Why are the codons and anticodons in groups of three? _____

Figure 11.6 Transfer RNA diversity.
Each type of tRNA carries only one particular amino acid, designated here by the first three letters of its name.

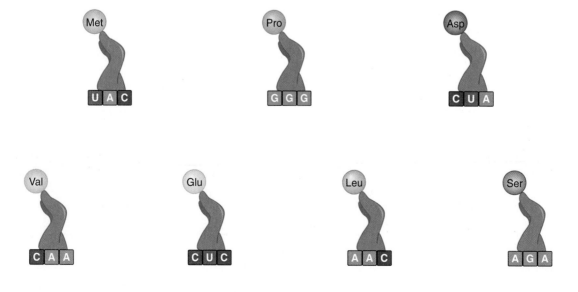

Table 11.7 Translation

mRNA codons	AUG	CCC	GAG	GUU	GAU	UUG	UCU
tRNA anticodons							
Amino acid*							

*Use three letters only. See Table 11.8 for the full names of these amino acids.

Table 11.8 Names of Amino Acids

Abbreviation	Name
Met	methionine
Pro	proline
Asp	aspartate
Val	valine
Glu	glutamate
Leu	leucine
Ser	serine

3. Figure 11.7 shows the manner in which the polypeptide grows. A ribosome has three binding sites. From left to right, they are the A (amino acid) site, the P (peptide) site, and the E (exit) site. A tRNA leaves from the E site after it has passed its amino acid or peptide to the newly arrived tRNA–amino acid complex. Then the ribosome moves forward, making room for the next tRNA–amino acid. This sequence of events occurs over and over until the entire polypeptide is borne by the last tRNA to come to the ribosome. Then a release factor releases the polypeptide chain from the ribosome. *In Figure 11.7, label the ribosome, the mRNA, and the peptide.*

Figure 11.7 Protein synthesis.

1. A ribosome has a site for two tRNA–amino acid complexes. 2. Before a tRNA leaves, an RNA passes its attached peptide to newly arrived tRNA–amino acid complex. 3. The ribosome moves forward, and the next tRNA–amino acid complex arrives.

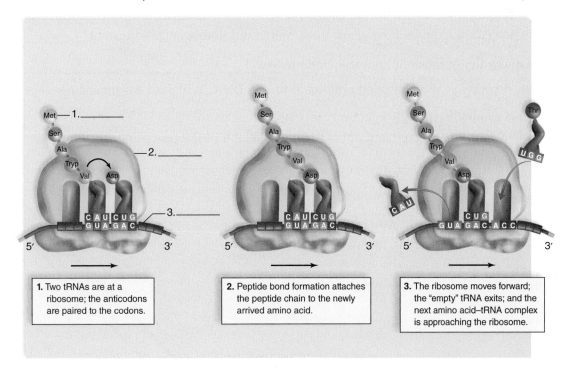

1. Two tRNAs are at a ribosome; the anticodons are paired to the codons.

2. Peptide bond formation attaches the peptide chain to the newly arrived amino acid.

3. The ribosome moves forward; the "empty" tRNA exits; and the next amino acid–tRNA complex is approaching the ribosome.

11.4 Isolation of DNA

In the following Experimental Procedure, you will isolate DNA from the cells of an organism using a modified procedure like that used worldwide in biotechnology laboratories. You will extract DNA from a vegetable or fruit filtrate that contains DNA in solution. To prepare the filtrate, your instructor homogenized the vegetable or fruit with a detergent. The detergent emulsifies and forms complexes with the lipids and proteins of the plasma membrane, causing them to precipitate out of solution. Cell contents, including DNA, become suspended in solution. The cellular mixture is then filtered to produce the *filtrate* that contains DNA and its adhering proteins.

The DNA molecule is easily degraded (broken down), so it is important to closely follow all instructions. Handle glassware carefully to prevent nucleases in your skin from contaminating the glassware.

Experimental Procedure: Isolating DNA

1. Obtain a pair of gloves and wear them when doing this procedure.
2. Obtain a large, clean test tube, and place it in an ice bath. Let stand for a few minutes to make sure the test tube is cold. Everything must be kept very cold.
3. Obtain approximately 4 mL of the *filtrate,* and add it to your test tube while keeping the tube in the ice bath.
4. Obtain and add 2 mL of cold *meat tenderizer solution* to the solution in the test tube, and mix the contents slightly with a stirring rod or Pasteur pipette. Let stand for 10 minutes so the enzyme has time to strip the DNA of protein.
5. Use a graduated cylinder or pipette to slowly add an equal volume (approximately 6 mL) of ice-cold *95% ethanol* along the inside of the test tube. Keep the tube in the ice bath, and tilt it to a 45° angle. You should see a distinct layer of ethanol over the white precipitate. The white precipitate is the DNA. Let the tube sit for 2 to 3 minutes.
6. Insert a glass rod or a Pasteur pipette into the tube until it reaches the bottom of the tube. To collect the DNA, gently swirl the glass rod or pipette, always in the same direction. (You are not trying to mix the two layers; you are trying to wind the DNA onto the glass rod like cotton candy.) This process is called "spooling" the DNA. The stringy, slightly gelatinous material that attaches to the pipette is DNA (Fig. 11.8). If the DNA has been damaged, it will still precipitate, but as white flakes that cannot be collected on the glass rod.
7. Answer the following questions:

   a. This procedure requires homogenization. When did homogenization occur? _____

   What was the purpose of homogenization? _____

   b. Next, deproteinization stripped proteins from the DNA. Which of the preceding steps represents deproteinization? _____

   c. Finally, DNA was precipitated out of solution. Which of the preceding steps represents the precipitation of DNA? _____

Figure 11.8   Isolation of DNA.
The addition of ethanol causes DNA to come out of solution so that it can be spooled onto a glass rod.

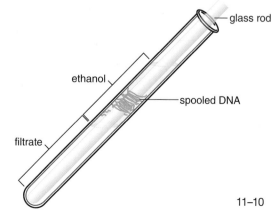

11.5 Genetic Disorders

The base sequence of DNA in all the chromosomes is an organism's genome. The Human Genome Project was a monumental effort to determine the normal order of all the 3.6 billion nucleotide bases in the human genome. Someday it will be possible to sequence anyone's genome within a relatively short time, and thereby determine what particular base sequence alterations signify that he or she has a disorder or will have one in the future. In this laboratory, you will study the alteration in base sequence that causes a person to have sickle-cell disease.

In persons with sickle-cell disease, the red blood cells aren't biconcave disks like normal red blood cells—they are sickle shaped. Sickle-shaped cells can't pass along narrow capillary passageways. They clog the vessels and break down, causing the person to suffer from poor circulation, anemia, and poor resistance to infection. Internal hemorrhaging leads to further complications, such as jaundice, episodic pain in the abdomen and joints, and damage to internal organs.

Sickle-shaped red blood cells are caused by an abnormal hemoglobin (Hb^S). Individuals with the $Hb^A Hb^A$ genotype are normal; those with the $Hb^S Hb^S$ genotype have sickle-cell disease, and those with the $Hb^A Hb^S$ have sickle-cell trait. Persons with sickle-cell trait do not usually have sickle-shaped cells unless they experience dehydration or mild oxygen deprivation.

Genomic Sequence for Sickle-Cell Disease

Examine Figures 11.9a and b, which show the DNA base sequence, the mRNA codons, and the amino acid sequence for a portion of Hb^A and the same portion for Hb^S.

Figure 11.9   Sickle-cell disease.
a. When red blood cells are normal, the base sequence (in one location) for Hb^A alleles is CTC. **b.** In sickle-cell disease at these locations, it is CAC.

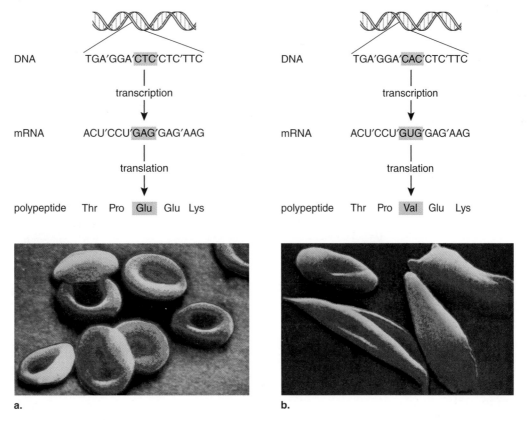

1. In what three-DNA-base sequence does Hb^A differ from Hb^S? Hb^A _____ Hb^S _____

2. What are the codons for these three bases? Hb^A _____ Hb^S _____

3. What is the amino acid difference? Hb^A _____ Hb^S _____

This one amino acid difference causes the polypeptide chain in sickle-cell hemoglobin to pile up as firm rods that push against the plasma membrane and deform the red blood cell into a sickle shape:

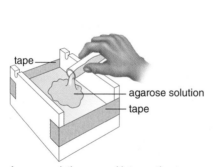

glutamate
(polar R group)

valine
(nonpolar R group)

Gel Electrophoresis

The two most widely used techniques for separating molecules in biotechnology are chromatography and gel electrophoresis. Chromatography separates molecules on the basis of their solubility and size. **Gel electrophoresis** separates molecules on the basis of their charge and size (Fig. 11.10).

During gel electrophoresis, charged molecules migrate across a span of gel (gelatinous slab) because they are placed in a powerful electrical field. In the present experiment, the fragment mixture for each DNA sample is placed in a small depression in the gel called a well. The gel is placed in a powerful electrical field. The electricity causes equal-length DNA fragments, which are negatively charged, to move through the gel to the positive pole at a faster rate than those that have no charge.

Almost all gel electrophoresis is carried out using horizontal gel slabs. First, the gel is poured onto a glass plate, and the wells are formed. After the samples are added to the wells, the gel and the glass plate are put into an electrophoresis chamber, and buffer is added. The fragments begin to migrate after the electrical current is turned on. With staining, the fragments appear as a series of bands spread from one end of the gel to the other.

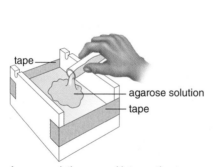

a. Agarose solution poured into casting tray

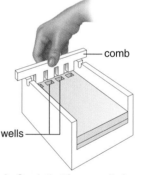

b. Comb that forms wells for samples

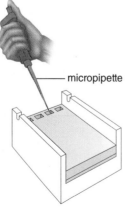

c. Wells that can be loaded with samples

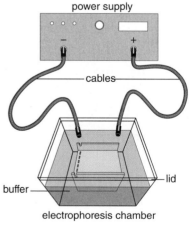

d. Electrophoresis chamber and power supply

Figure 11.10 Equipment and procedure for gel electrophoresis.

In this procedure, you will perform gel electrophoresis, if instructed to do so, and analyze your data to come to a conclusion.

Performing Gel Electrophoresis

1. Obtain three samples of hemoglobin provided by your kit. They are labeled sample A, B, and C.
2. If so directed by your instructor, carry out gel electrophoresis of these samples (see Fig. 11.10).

> ⚠ **Gel electrophoresis** Students should wear personal protective equipment: safety goggles and smocks or aprons while loading gels and during electrophoresis and protective gloves while staining.

Analyzing the Electrophoresed Gel

1. Sickle-cell hemoglobin (Hb^S) migrates slower toward the positive pole than normal hemoglobin (Hb^A) because the amino acid valine has no polar R groups, whereas the amino acid glutamate does have a polar R group.
2. In Figure 11.11, label the lane that contains only Hb^S, signifying that the individual is Hb^SHb^S.
3. Label the lane that contains only Hb^A, signifying that the individual is Hb^AHb^A.
4. Label the lane that contains both Hb^S and Hb^A, signifying that the individual is Hb^AHb^S.

Figure 11.11   Gel electrophoresis of hemoglobins.

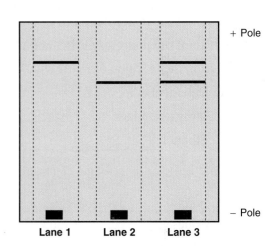

Conclusion: Genomic Sequence for Sickle-Cell Disease

- You are a genetic counselor. A young couple seeks your advice because sickle-cell disease occurs among the family members of each. You order DNA base sequencing to be done. The results come back that at one of the two loci for normal Hb^A, each has the abnormal sequence CAC instead of CTC. The other locus is normal. What genotype do they each have? _____
What are the chances that this couple will have a child with sickle-cell disease (Hb^A is dominant and Hb^S is recessive)? _____

1. Explain why DNA is said to have a structure that resembles a ladder.

   _____

2. How is complementary base pairing different when pairing DNA to DNA than when pairing DNA to mRNA?

   _____

   _____

3. Explain why the genetic code is called a triplet code. _____

4. What role does each of the following molecules play in protein synthesis?

   a. DNA _____

   b. mRNA _____

   c. tRNA _____

   d. Amino acids _____

5. Which of the molecules listed in question 4 are involved in transcription? _____

6. Which of the molecules listed in question 4 are involved in translation? _____

7. During the isolation of DNA, what role was played by these substances?

   a. Detergent _____

   b. Meat tenderizer _____

   c. Ethanol _____

8. What is the purpose of gel electrophoresis? _____

   _____

9. Why does sickle-cell hemoglobin (Hb^S) migrate slower than normal hemoglobin (Hb^A) during gel

   electrophoresis? _____

   _____

10. Why are red blood cells sickle shaped in a person with sickle-cell disease? _____

    _____

12

Evidences of Evolution

Introduction

Evolution is the process by which life has changed through time. A **species** is a group of similarly constructed organisms that share common genes, and a **population** is all the members of a species living in a particular area. When new variations arise that allow certain members of a population to capture more resources, these individuals tend to survive and to have more offspring than the other, unchanged members. Therefore, each successive generation will include more members with the new variation. Eventually, most members of a population and then the species will have the same **adaptations,** i.e., structures, physiology, and behavior that make an organism suited to its environment.

Evolution, which has been ongoing since the origin of life, is also an explanation for the unity of life. All organisms share the same characteristics of life because they can trace their ancestry to the first cell or cells. In this laboratory, you will study three types of data that support the concept of **common descent:** (1) the fossil record, (2) comparative anatomy (embryological and adult), and (3) molecular differences. **Fossils** are the remains or evidence of some organism that lived long ago. Fossils can be used to trace the history of life on Earth. A comparative study of the anatomy of different groups of organisms from insects to vertebrates has shown that each group has structures of similar construction called **homologous structures.** For example, all vertebrate animals have essentially the same type of skeleton. Homologous structures signify relatedness through evolution because they can be traced to a **common ancestor** of the group. Living organisms use the same basic molecules including ATP, the carrier of energy in cells; DNA, which makes up genes; and proteins, such as enzymes and antibodies. In this laboratory, you will use an antigen-antibody reaction to show the degree of evolutionary relatedness between different vertebrates.

Table 12.1 The Geological Timescale: Major Divisions of Geological Time and Some of the Major Evolutionary Events that Occurred

Era	Period	Epoch	Millions of Years Ago	Plant Life	Animal Life
Cenozoic*	Quaternary	Holocene	0.01–present	Human influence on plant life.	Age of Homo sapiens.
				Significant Mammalian Extinction	
		Pleistocene	1.8–0.01	Herbaceous plants spread and diversify.	Presence of ice age mammals. Modern humans appear.
	Tertiary	Pliocene	5.3–1.8	Herbaceous angiosperms flourish.	First hominids appear.
		Miocene	23–5.3	Grasslands spread as forests contract.	Apelike mammals and grazing mammals flourish; insects flourish.
		Oligocene	33.9–23	Many modern families of flowering plants evolve.	Browsing mammals and monkeylike primates appear.
		Eocene	55.8–33.9	Subtropical forests with heavy rainfall thrive.	All modern groups of mammals are represented.
		Paleocene	65.5–55.8	Flowering plants continue to diversify.	Primitive primates, herbivores, carnivores, and insectivores appear.
Mesozoic				**Mass Extinction: Dinosaurs and Most Reptiles**	
	Cretaceous		145.5–65.5	Flowering plants, conifers persist.	Placental mammals appear; modern insect groups appear.
	Jurassic		199.6–145.5	Flowering plants appear.	Dinosaurs flourish; birds appear.
				Mass Extinction	
	Triassic		251–199.6	Forests of conifers and cycads dominate.	First mammals appear; first dinosaurs appear; corals and molluscs dominate seas.
Paleozoic				**Mass Extinction**	
	Permian		299–251	Gymnosperms diversify.	Reptiles diversify; amphibians decline.
	Carboniferous		359.2–299	Age of great coal-forming forests: Ferns, club mosses, and horsetails flourish.	Amphibians diversify; first reptiles appear; first great radiation of insects.
				Mass Extinction	
	Devonian		416–359.2	First seed plants appear. Seedless vascular plants diversify.	Jawed fishes diversify and dominate the seas; first insects and first amphibians appear.
	Silurian		443.7–416	Seedless vascular plants appear.	First jawed fishes appear.
				Mass Extinction	
	Ordovician		488.3–443.7	Nonvascular land plants appear. Marine algae flourish.	Invertebrates spread and diversify; jawless fishes (first vertebrates) appear.
	Cambrian		542–488.3	First plants appear on land. Marine algae flourish.	All types of ancestral invertebrates present; first chordates appear.
Precambrian Time			600	Oldest soft-bodied invertebrate fossils.	
			1,400–700	Protists evolve and diversify.	
			2,200	Oldest eukaryotic fossils.	
			2,700	O_2 accumulates in atmosphere.	
			3,500	Oldest known fossils (prokaryotes).	
			4,600	Earth forms.	

*Many authorities divide the Cenozoic era into the Paleogene period (contains the Paleocene. Eocene, and Oligocene epochs) and the Neogene period (contains the Miocene, Pliocene, Pleistocene, and Holocene epochs).

12.1 Evidence from the Fossil Record

Because all life-forms evolved from the first cell or cells, life has a history, and this history is revealed by the fossil record. The geologic timescale, which was developed by both geologists and paleontologists, depicts the history of life based on the fossil record (Table 12.1). In this section, we will study the geologic timescale and then examine some fossils.

Geologic Timescale

Divisions of the Timescale

Notice that the geological timescale or simply the **timescale** divides the history of Earth into eras, then periods, and then epochs. The three eras span the greatest amounts of time, and the epochs are the shortest time frames. Notice that only the periods of the Cenozoic era are divided into epochs, meaning that more attention is given to the evolution of primates and flowering plants than to the earlier evolving organisms. Modern civilization is given its own epoch, despite the fact that humans have only been around about

0.04% of the history of life. List the four eras in the timescale starting with Precambrian time: _____

How to Read the Timescale

1. Using the timescale, you can trace the history of life by beginning with Precambrian time at the bottom. The timescale indicates that the first cells (the prokaryotes) arose some 3,500 million years ago (MYA). The prokaryotes evolved before any other group. Why do you read the timescale starting at the bottom? _____

2. The Precambrian time was very long, lasting from the time the Earth first formed until 542 MYA. The fossil record during the Precambrian time is meager, but the fossil record from the Cambrian period onward is rich (for reasons still being determined). This helps explain why the timescale usually does not show any periods until the Cambrian period of the Paleozoic era. You can also use the divisions of the timescale to check when certain groups evolved and/or flourished. Example: During the Ordovician period, the nonvascular plants appear on land, and the first jawless and jawed fishes appear in the seas.

   During the _____ era in the _____ period, the first flowering plants appear. How many millions of years ago was this period? _____

3. During the Paleozoic era, note the Carboniferous period. In this period great swamp forests covered the land. These are also called coal-forming forests because, with time, they became the coal we burn today. How do you know that the plants in this forest were not flowering trees, as most of our trees are today? _____

   What type animal was diversifying at this time? _____

4. You should associate the Cenozoic era with the evolution of humans. Among mammals, humans are primates. During what period and epoch did primates appear? _____
   Among primates, humans and certain of the apes are hominids. During what period and epoch did hominids appear?_____ The scientific name for humans is *Homo sapiens*.
   What period and epoch is the age of *Homo sapiens*? _____

Dating Within the Timescale

As you may realize, the timescale provides both relative dates and absolute dates. When you say, for example, "Flowering plants evolved during the Jurassic period," you are using relative time, because flowering plants evolved earlier or later than groups in other periods. If you use the dates that are given in MYA, you are using absolute time. Absolute dates are usually obtained by measuring the amount of a radioactive isotope in the rocks surrounding the fossils. Why wouldn't you expect to find human fossils and dinosaur fossils together in rocks dated similarly? _____

Limitations of the Timescale

Because the timescale tells when various groups evolved and flourished, it might seem that evolution has been a series of events leading only from the first cells to humans. This is not the case; for example, prokaryotes (bacteria and archaea) never declined and are still the most abundant and successful organisms on Earth. Even today, they constitute up to 90% of the total weight of living things.

Then, too, the timescale lists mass extinctions, but it doesn't tell when specific groups became extinct. **Extinction** is the total disappearance of a species or a higher group; **mass extinction** occurs when a large number of species disappear in a few million years or less. For lack of space, the geologic timescale can't depict in detail what happened to the members of every group mentioned. Figure 12.1 does show how mass extinction affected a few groups of animals. Which of the animals shown in

Figure 12.1 suffered the most during the P-T (Permian-Triassic) extinction? _____
The K-T extinction occurred between the Cretaceous and the Tertiary periods. Which animals shown in

Figure 12.1 became extinct during the K-T extinction? _____
Figure 12.1 shows only periods and no eras. *Fill in the eras* on the lines provided in the figure.

Figure 12.1 Mass extinctions.

Five significant mass extinctions and their effects on the abundance of certain forms of marine and terrestrial life. The width of the horizontal bars indicates the varying abundance of each life-form considered. Lines are provided for students to fill in the eras.

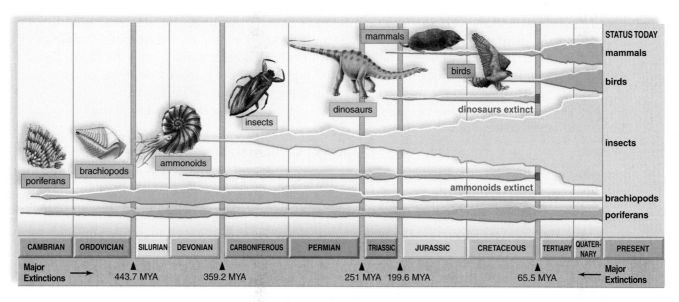

Era _____   _____   _____

Observation: Fossils

1. Obtain a box of selected fossils. If the fossils are embedded in rocks, examine the rock until you have found the fossil. Fossils are embedded in rocks because the sediment that originally surrounded them hardened over time. Most fossils consist of hard parts such as shells, bones, or teeth because these parts are not consumed or decomposed over time. One possible reason the Cambrian might be rich in fossils is that organisms before this time did not have

_____ .

2. The kit you are using, or your instructor, will identify which of the fossils are invertebrate animals. These fossils date back to which era and period? _____ Fill in the title of Table 12.2. List the names of these fossils in Table 12.2 and give a description of the fossil part that survived.

Table 12.2 Invertebrate Fossils from the _____ Era _____ Period	
Type of Fossil	Description of Hard Part

3. Which of the fossils available to you are vertebrates? _____
Use Table 12.1 (The Geological Timescale) to associate each fossil with the particular era and period when this type animal was most abundant. Fill in Table 12.3 according to sequence of the time frames from the latest (*top*) to earliest (*bottom*).

Table 12.3 Vertebrate Fossils		
Type of Fossil	Era, Period	Description of Hard Part

4. Which of the fossils available to you are plants? _____ Plants that have no hard parts become fossils when their impressions are filled in by minerals. Use Table 12.1 to associate each plant with a particular era and period. Assume trees are flowering plants and associate them with the era and period when flowering plants were most abundant. Fill in Table 12.4 according to the sequence of the time frames from the latest (*top*) to earliest (*bottom*).

Table 12.4 Plant Fossils		
Type of Fossil	Era, Period	Description of Fossil

12.2 Evidence from Comparative Anatomy

In the study of evolutionary relationships, parts of organisms are said to be **homologous** if they exhibit similar basic structures and embryonic origins. If these parts of organisms are similar in function only, they are said to be **analogous.** Only homologous structures indicate an evolutionary relationship and are used to determine evolutionary relationships.

Comparison of Adult Vertebrate Forelimbs

The limbs of vertebrates are homologous structures. Homologous structures share a basic pattern, although there may be specific differences. The similarity of homologous structures is explainable by descent from a **common ancestor**.

Observation: Vertebrate Forelimbs

1. The central diagram in Figure 12.2 represents the forelimb bones of the common ancestor. The basic components are the humerus (h), ulna (u), radius (r), carpals (c), metacarpals (m), and phalanges (p) in the five digits.
2. Carefully compare and label in Figure 12.2 the corresponding forelimb bones of the frog, the lizard, the bird, the bat, the cat, and the human. In particular, note the specific modifications that have occurred in some of the bones to meet the demands of a particular way of life.
3. Fill in Table 12.5 to indicate which bones in each specimen appear to most resemble the common ancestor and which most differ from the ancestral condition.
4. Relate the change in bone structure to mode of locomotion in two examples.

   Example 1: _____

   Example 2: _____

Table 12.5 Comparison of Vertebrate Forelimbs

Animal	Bones That Resemble Common Ancestor	Bones That Differ from Common Ancestor
Frog		
Lizard		
Bird		
Bat		
Cat		
Human		

Conclusion: Vertebrate Forelimbs

- Vertebrates are descended from a _____, but they are adapted

  to _____.

Figure 12.2 Vertebrate forelimbs.

Because all vertebrates evolved from a common ancestor, their forelimbs share homologous structures.

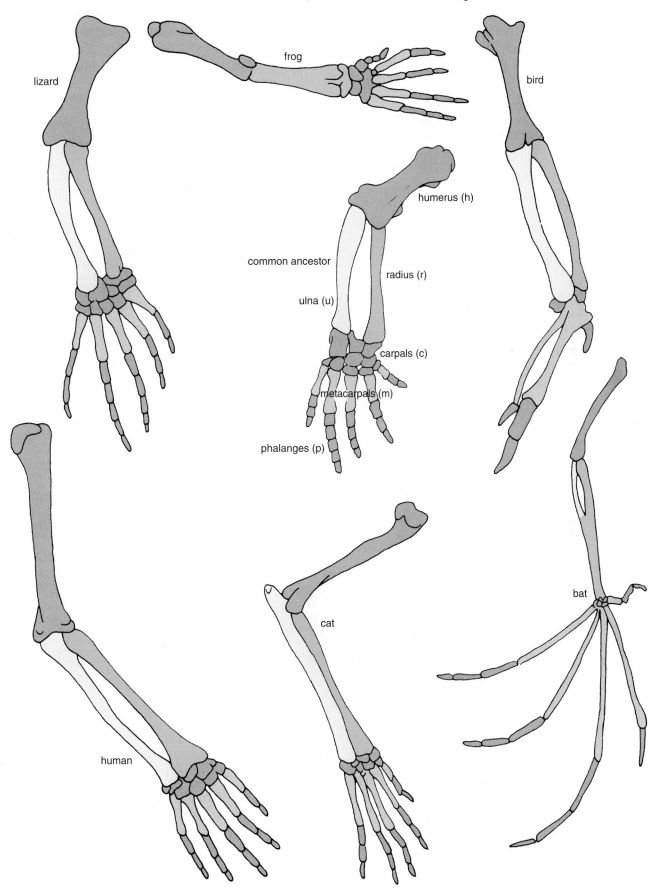

Comparison of Chimpanzee and Human Skeletons

Chimpanzees and humans are closely related, as is apparent from an examination of their skeletons (Fig. 12.3). However, they are adapted to different ways of life. Chimpanzees are adapted to living in trees and are herbivores—they eat mainly plants. Humans are adapted to walking on the ground and are omnivores—they eat both plants and meat.

Figure 12.3   Human and chimpanzee skeletons.
Differences in posture can be related to their adaptations to a different habitat. Humans now walk on land and chimpanzees primarily live in trees. The boxes identify particular parts of the skeleton.

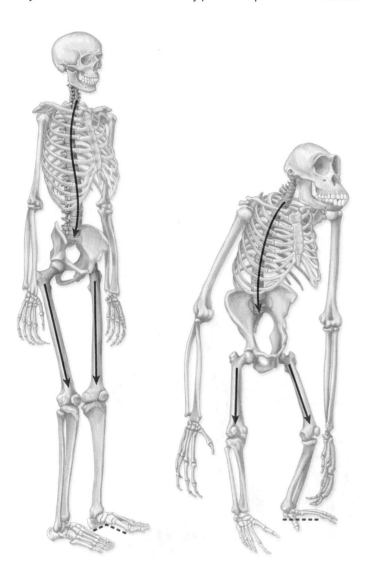

Human spine exits from the skull's center; ape spine exits from rear of skull.

Human spine is S-shaped; ape spine has a slight curve.

Human pelvis is bowl-shaped; ape pelvis is longer and more narrow.

Human femurs angle inward to the knees; ape femurs angle out a bit.

Human knee can support more weight than ape knee.

Human foot has an arch; ape foot has no arch.

Observation: Chimpanzee and Human Skeletons

Posture

Chimpanzees are arboreal and climb in trees. While on the ground, they tend to knuckle-walk, with their hands bent. Humans are terrestrial and walk erect. After noting that the boxes in Figure 12.3 correlate to parts of the skeleton, answer the following questions:

1. **Head and torso:** Where are the head and trunk with relation to the hips and legs—thrust forward over the hips and legs or balanced over the hips and legs? *Record your observations in Table 12.6.*
2. **Spine:** Which animal has a long and curved lumbar region, and which has a short and stiff lumbar region? *Record your observations in Table 12.6.*

   How does this contribute to an erect posture in humans? _____

   _____
3. **Pelvis:** Chimpanzees sway when they walk because lifting one leg throws them off balance. Which animal has a narrow and long pelvis, and which has a broad and short pelvis? *Record your observations in Table 12.6.*
4. **Femur:** In humans, the femur better supports the trunk. In which animal is the femur angled between articulations with the pelvic girdle and the knee? In which animal is the femur straight with no angle? *Record your observations in Table 12.6.*
5. **Knee joint:** In humans, the knee joint is modified to support the body's weight. In which animal is the knee joint larger? *Record your observations in Table 12.6.*
6. **Foot:** In humans, the foot is adapted for walking long distances and running with less chance of injury.

   In which animal is the big toe opposable? _____ How does an opposable toe assist chimpanzees? _____

   Which foot has an arch? _____ How does an arch assist humans? _____

   _____ *Record your observations in Table 12.6.*
7. How does the difference in the position of the foramen magnum, a large opening in the base of the skull for the spinal cord, correlate with the posture and stance of the two organisms?

Table 12.6 Comparison of Chimpanzee and Human Postures

Skeletal Part	Chimpanzee	Human
1. Head and torso		
2. Spine		
3. Pelvis		
4. Femur		
5. Knee joint		
6. Foot: Opposable toe		
Arch		

Conclusion: Chimpanzee and Human Skeletons

- Do your observations show that the skeletal differences between chimpanzees and humans can be related to posture? _____ Explain. _____

   _____

Figure 12.4 Chimpanzee and human skulls.

Differences in facial features can be related to a difference in diet. Chimpanzees are herbivorous, and humans are omnivores.

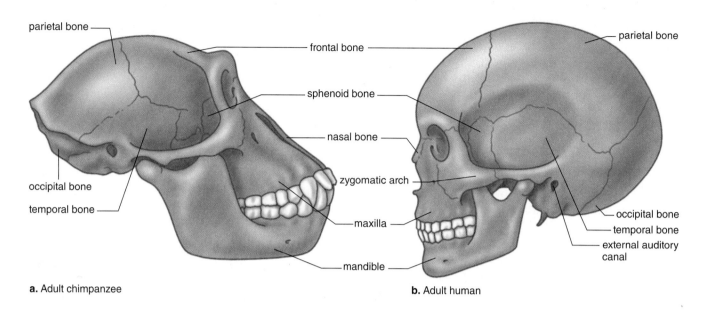

a. Adult chimpanzee   **b.** Adult human

Facial Features

Humans are omnivorous. A diet rich in meat does not require strong grinding teeth or well-developed facial muscles. Chimpanzees are herbivores, and a vegetarian diet requires strong teeth and strong facial muscles that attach to bony projections. Compare the skulls of the chimpanzee and the human in Figure 12.4 and answer the following questions:

1. **Supraorbital ridge:** For which skull is the supraorbital ridge (the region of frontal bone just above the eye socket) thicker? *Record your observations in Table 12.7.*
2. **Frontal bone:** Compare the slope of the frontal bones of the chimpanzee and human skulls. How are they different? *Record your observations in Table 12.7.*
3. **Teeth:** Examine the teeth in the adult chimpanzee and adult human skulls. Are the shapes and size of teeth similar in both? *Record your observations in Table 12.7.*
4. **Chin:** What is the position of the mouth and chin in relation to the profile for each skull? *Record your observations in Table 12.7.*

Table 12.7   Facial Features of Chimpanzees and Humans

Feature	Chimpanzee	Human
1. Supraorbital ridge		
2. Slope of frontal bone		
3. Teeth		
4. Chin		

Conclusion: Facial Features

• Do your observations show that diet can be related to the facial features of chimpanzees and humans? _____ Explain._____

_____

Comparison of Vertebrate Embryos

The anatomy shared by vertebrates extends to their embryological development. For example, as embryos, they all have a postanal tail, somites (segmented blocks of mesoderm lying on either side of the notochord), and paired pharyngeal pouches. In aquatic animals, these pouches become functional gills (Fig. 12.5). In humans, the first pair of pouches becomes the cavity of the middle ear and auditory tube, the second pair becomes the tonsils, and the third and fourth pairs become the thymus and parathyroid glands.

Figure 12.5   Vertebrate embryos.
During early developmental stages, vertebrate embryos have certain characteristics in common.

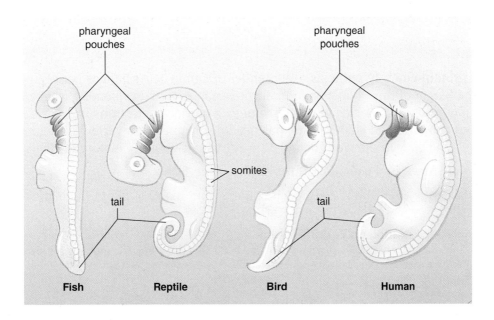

Observation: Chick and Pig Embryos

1. Obtain prepared slides of vertebrate embryos at comparable stages of development. Observe each of the embryos using a stereomicroscope.
2. List five similarities of the embryos:

   a. _____

   b. _____

   c. _____

   d. _____

   e. _____

Conclusion: Vertebrate Embryos

• Vertebrate embryos resemble one another because _____

_____.

12.3 Molecular Evidence

Almost all living organisms use the same basic biochemical molecules, including DNA, ATP, and many identical and nearly identical enzymes. In addition, living organisms utilize the same DNA triplet code and the same 20 amino acids in their proteins. There is no obvious functional reason these elements need to be so similar. Therefore, their similarity is best explained by descent from a common ancestor.

Molecular evidence for evolution includes many new findings. For example, using new kinds of microscopes and modern biotechnology techniques, scientists have discovered developmental genes, called *Hox* genes, that control the body's shape and form. Various data indicate that *Hox* genes must date back to a common ancestor that lived more than 600 MYA and that despite millions of years of divergent evolution, all animals share the same control switches for development. In general, the study of *Hox* genes has shown how animal diversity is due to variations in the expression of ancient genes present in animals from sponges to humans rather than to wholly new and different genes.

Protein Differences

According to the **protein clock hypothesis,** the number of amino acid changes between organisms is proportional to the length of time since two organisms began evolving separately from a common ancestor. The sequence of amino acids in **cytochrome *c*,** a carrier of electrons in the electron transport chain found in mitochondria and chloroplasts, has been determined in a variety of organisms. Are the number of amino acid changes shown in Figure 12.6 consistent with the history of life as revealed by the fossil record and shown in the geological timescale (Table 12.1)? _____ The theory of evolution has consistently been supported by varied data from many different fields of study.

Figure 12.6   Significance of molecular differences.
The branch points in this diagram indicate the number of amino acids that differ between human cytochrome *c* and the organisms depicted. These molecular data are consistent with those provided by a study of the fossil record and comparative anatomy.

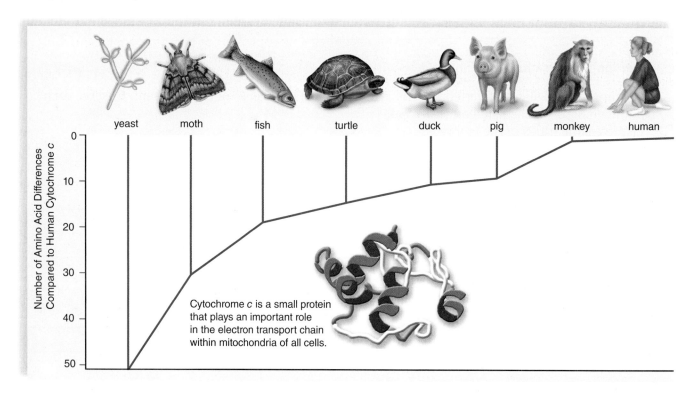

Cytochrome *c* is a small protein that plays an important role in the electron transport chain within mitochondria of all cells.

Protein Similarities

The immune system makes **antibodies** (proteins) that react with foreign proteins, termed **antigens.** Antigen-antibody reactions are specific. An antibody will react only with a particular antigen. In today's laboratory, this reaction can be observed when a precipitate, a substance separated from the solution, appears.

Experimental Procedure: Protein Similarities

In this procedure, it is assumed that human serum (containing human antigens) is injected into the bloodstream of a rabbit, and the rabbit makes antibodies against human antigens (Fig. 12.7). The experiment tests this sensitized rabbit serum against the antigens of other animals. The stronger the reaction (determined by the amount of precipitate), the more closely related the animal is to humans.

1. Obtain a chemplate (a clear glass tray with wells), one bottle of synthetic *human blood serum,* one bottle of synthetic *rabbit blood serum,* and five bottles (I–V) of *blood serum test solution.*
2. Put two drops of synthetic rabbit blood serum in each of the six wells in the chemplate. Label the wells 1–6. See yellow circles in Figure 12.8.
3. Add 2 drops of synthetic human blood serum to each well. See red circles in Figure 12.8. Stir with the plastic stirring rod that was attached to the chemplate. The rabbit serum has now been "sensitized" to human serum. (This simulates the production of antibodies in the rabbit's bloodstream in response to the human blood proteins.)
4. Rinse the stirrer. (The large cavity of the chemplate may be filled with water to facilitate rinsing.)
5. Add 4 drops of *blood serum test solution III* (tests for human blood proteins) to well 6.

   Describe what you see. _____

   _____

   This well will serve as the basis by which to compare all the other samples of test blood serum.
6. Now add 4 drops of *blood serum test solution I* to well 1. Stir and observe. Rinse the stirrer. Do the same for each of the remaining *blood serum test solutions (II–V)*—adding II to well 2, III to well 3, and so on. Be sure to rinse the stirrer after each use.
7. At the end of 10 and 20 minutes, record the amount of precipitate in each of the six wells in Figure 12.8. Well 6 is recorded as having + + + + amount of precipitate after both 10 and 20 minutes. Compare the other wells with this well (+ = trace amount; 0 = none). Holding the plate slightly above your head at arm's length and looking at the underside toward an overhead light source will allow you to more clearly determine the amount of precipitate.

Figure 12.7 Antigen-antibody reaction.
When antibodies react to antigens, a complex forms that appears as a precipitate.

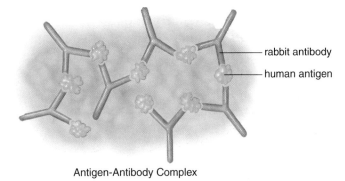

rabbit antibody

human antigen

Antigen-Antibody Complex

Figure 12.8 Molecular evidence of evolution.

The greater the amount of precipitate, the more closely related an animal is to humans.

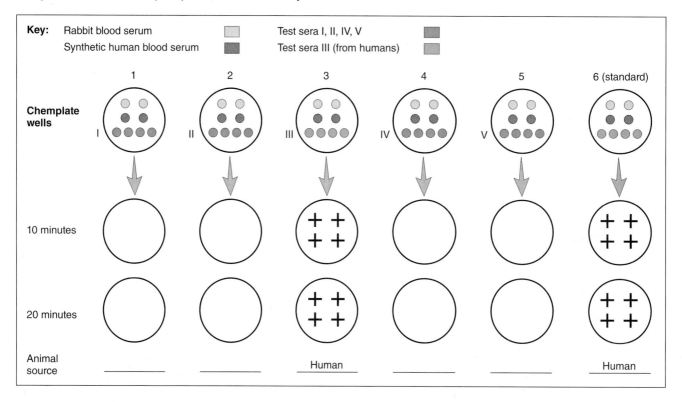

Conclusions: Protein Similarities

- The last row in Figure 12.8 tells you that the reaction in well 3 is that of a human. How do your
  test results confirm this? _____
- Aside from humans, the test sera (supposedly) came from a pig, a monkey, an orangutan, and a
  chimpanzee. Which is most closely related to humans—the pig or the chimpanzee?

  _____

- Based on how closely the amount of reaction (precipitate) matches that of a human, fill in the names of
  the animals in the last row of Figure 12.8. Use this chemical reaction to provide evidence of evolution.

  _____

  _____

- In Section 12.2, a comparison of bones in vertebrate forelimbs showed that vertebrates share a
  common ancestor. Molecular evidence shows us that of the vertebrates studied, _____ and
  _____ are most closely related. In evolution, closely related organisms share a recent common
  ancestor. Humans share a more recent common ancestor with chimpanzees than they do with pigs.

12.4 Summarizing the Evidences of Evolution

Work with a partner or in a group to conclude that your observations in this laboratory give evidences of evolution.

Evidence from the Fossil Record

In this section, you studied the geologic timescale (page 151). The geologic timescale gives powerful evidence of evolution because:

1. Fossils are _____.

2. Fossils can be arranged in a sequential manner because _____.

3. Younger fossils and not older fossils are more like _____.

4. In short, the fossil record shows _____.

Evidence from Comparative Anatomy

1. In this section, you compared vertebrate forelimbs (page 154). The similarity between the bones in all vertebrate forelimbs and a common ancestor shows that today's vertebrates

   _____.

   However, the various vertebrates are adapted to _____.

2. In this section, you also compared the chimpanzee and the human skeletons (page 156). A contrast in skeletons shows that the posture differences are due to _____

   _____.

   A contrast in skulls shows that the facial differences are due to _____.

3. In this section, you also compared vertebrate embryos (page 159). The similarity in the appearance of vertebrate embryos also shows that today's vertebrates _____

   _____.

Molecular Evidence

1. In this section (page 160) you learned that due to _____ descent, organisms share the same proteins (e.g., cytochrome c). However, distantly related organisms have

   _____ amino acid differences in these proteins than do _____

   related organisms.

2. In this section, you also used an antigen-antibody reaction to show how closely related certain vertebrates are to humans (page 161). Two closely related organisms share _____

   _____.

1. List three types of evidence that various types of organisms are related through common descent. _____

2. Why would you expect a fossil buried millions of years ago to not look exactly like a modern-day organism? _____

3. What accounts for many of the skeletal difference between chimpanzees and humans?_____

   _____

4. If a characteristic is found in bacteria, fungi, pine trees, snakes, and humans, when did it most likely evolve? _____ Why? _____

   _____

5. What are homologous structures, and what do they show about relatedness? _____

   _____

6. The development of reptiles, chicks, and humans shows similarities. What can we learn from this observation?_____

7. What do DNA mutations have to do with amino acid changes in a protein? _____

   _____

8. How did antigen-antibody reactions help determine the degree of relatedness between species in this laboratory? _____

   _____

9. Using plus (+) symbols, show the amount of reaction you would expect when a pig, monkey, and chimpanzee are tested for the same antibody-antigen reaction. _____

   _____

10. Define the following types of evidence for evolution:

    fossil _____

    common descent _____

    comparative anatomy _____

    adaptation _____

    molecular _____

13

Natural Selection

Introduction

Natural selection is the mechanism, first described by Charles Darwin, that brings about adaptation to the environment as evolution occurs. Organisms are remarkably adapted to their way of life. Through natural selection, birds have become variously adapted to acquiring food (Fig. 13.1). The beak of a honeycreeper can catch an insect while the talons of an osprey can catch a fish, for example. Today, Charles Darwin's theory of natural selection can be described in this way:

1. The members of a population—all the members of a species living in one locale at the same time— have heritable variations (phenotypic traits) that can be passed from one generation to the next.
2. There is a competition for resources (such as food, mates, shelter) and members of the population with traits that allow them to better capture resources will reproduce to a greater extent than those that lack these traits.
3. Across generations, a larger proportion of the population will have these adaptive traits. In this way, the environment has selected how the genotype and phenotype makeup of the population will change over time.

a.   b.

Figure 13.1 Adaptations.
a. The beak of a green honeycreeper is adapted to catching insects. **b.** The talons of an osprey are adapted to catching fish.

Figure 13.2 Agents of evolutionary change.

a. Natural Selection	b. Genetic Drift	c. Mutation	d. Gene Flow	e. Nonrandom Mating
		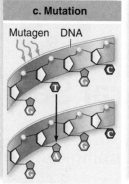		
The only agent to result in adaptation to the environment.	Chance results in gene pool frequency changes when population size decreases.	Mutation occurs infrequently but is the ultimate source of allele and phenotype variations.	Movement of new individuals or gametes into a population can cause gene pool frequency changes.	Inbreeding is a common reason for genotype frequency changes in a population.

In panel c, labels: Mutagen, DNA.
In panel e, label: Self-fertilization.

13.1 Hardy-Weinberg Law

The **Hardy-Weinberg law** gives us a way to know when evolution has occurred. It states that normally, allele frequencies in the gene pool of a population stay the same from generation to generation. If they change, microevolution has occurred. The agents that can cause a change in gene pool frequencies are natural selection, genetic drift, mutation, gene flow (the movement of alleles between populations), and nonrandom mating (Fig. 13.2). In today's laboratory we are studying natural selection, genetic drift, and mutation as causes of microevolution.

In order to understand and use the law, understand these terms:

Population: all the members of a species living in the same locale at the same time. You and your fellow lab members will be the population in today's laboratory.

Gene pool: the frequency of the alleles and genotypes of all the individuals in the population.

Frequency: the proportion of alleles and genotypes relative to the total number of individuals in the population.

Virtual Lab Natural Selection A Virtual Laboratory called Natural Selection is available on the *Biology* website **www.mhhe.com/maderbiology11**. After opening this virtual lab,

- Choose an environment (top left).
- Choose a frequency for alleles *A* and *a* (top right).
- Click the Generation 1 button and these frequencies plus the number of individuals out of 100 that are homozygous dominant, heterozygous, and homozygous recessive will appear in the environment and in the table below.
- Double-click Natural Selection and the new frequency data will be displayed. Note which allele apparently leads to a disadvantaged phenotype in this environment.
- Click on Generation 2 now lit and the Natural Selection button. Do the same for Generation 3–5 and watch the change in frequency data.

When we are dealing with a trait that has only two alleles, the Hardy-Weinberg equations allow us to calculate the gene pool frequencies in a population. The equations are:

$$p + q = 1$$
$$p^2 + 2pq + q^2 = 1$$

where
q = recessive allele frequency
q^2 = homozygous recessive frequency
p = dominant allele frequency
p^2 = homozygous dominant frequency
$2pq$ = heterzygous frequency

The Parent Population

In order to know if a change has occurred, we first need to know the allele and genotype frequencies for a particular trait in the gene pool of the present or **parent (P) population**. The trait under consideration will be the ability to taste PTC (phenylthiocarbamide). PTC is an antithyroid drug that prevents the thyroid gland from incorporating iodine into the thyroid hormone.

Observation: Gene Pool Frequencies of Parent Populations

Your Genotype

Taste a piece of paper impregnated with PTC. If you can taste the chemical, your genotype is either *TT* or *Tt*. If you cannot taste the chemical your genotype is *tt*. Are you homozygous recessive? _____

Homozygous Recessive Genotype (q^2) Frequency

Your instructor will determine what proportion of the class is *tt*. _____ = q^2 Frequencies are recorded as a decimal. For example, 25% becomes 0.25. Record the frequency of the P population that is homozygous recessive in Table 13.1.

Table 13.1 Parental (P) Generation	
Genotypes	**Genotype Frequencies**
Homozygous recessive (q^2)	
Homozygous dominant (p^2)	
Heterozygous ($2pq$)	

Homozygous Dominant Genotype (p^2) Frequency

If $q^2 =$ _____ , then $q =$ _____ . Note that $q + p = 1$. Therefore, what is p? _____
What is p^2? _____ Record this number as the frequency of the P population that is homozygous dominant in Table 13.1.

Heterozygous Genotype (2pq) Frequency

As indicated, calculate the frequency of the heterozygous genotype in the P population by calculating $2pq$: $2 \times p \times q =$ _____ . Record this number as the frequency of the P population that is heterozygous in Table 13.1.

13.2 Natural Selection and Genetic Drift

Natural Selection

When natural selection occurs, some aspect of the environment selects which traits are more likely or less likely to be passed on to the next generation. For example, light-colored peppered moths are more likely to survive predation by insects when the vegetation is light colored and less likely to survive predation by insects when the vegetation is dark colored. The situation is just the opposite for dark-colored peppered moths (Fig. 13.3a).

The population of field crickets on the island of Hawaii now contains many silent males because the cricket's chirping attracted a deadly parasitic fly that lays her eggs on the cricket's back. The eggs develop into maggots that feed on the cricket (Fig. 13.3b). This gave the silent males an advantage that allowed them to reproduce more than the other male members of the population.

Bacteria have an advantage when they can survive in the presence of antibiotics, and the greater reproduction of these bacteria under these conditions compared with other members of a population accounts for the reason why we are plagued by so many resistant forms of bacteria today.

a.   b.

Figure 13.3 Selective agents.
a. The selective agent in the environment for these peppered moths is the color of vegetation. Which colored moth is favored for greater reproductive success when the vegetation is dark? **b.** In Hawaii, parasitic crickets lay their eggs on chirping males and not silent ones; therefore, silent males are favored for greater reproductive success.

Experimental Procedure: Natural Selection

1. Using the frequencies calculated in Table 13.1 as a guide, your instructor will assign you a genotype for the trait. (Students who are homozygous recessive will be assigned this genotype.)
2. Write down your genotype, which is your initial parental (*P*) generation genotype. _____
3. To ensure random mating, be completely uninhibited and choose anyone in the class (male or female) as your mate, approaching him or her confidently. That person cannot refuse you.
4. Each couple will have two offspring. Since each partner contributes one allele to each offspring, follow this procedure for each offspring: Flip a coin. If it comes up heads, the allele on the left is passed on. If it comes up tails, the allele on the right is successful. Fill in Table 13.2.
5. In this experimental procedure we will assume that many modern commercial products now contain PTC and the inability to taste this chemical leads to the death (before reproductive age) of population members.

Table 13.2 Your F$_1$ Offspring

Offspring	Your Contribution	Your Partner's Contribution	Genotype
First offspring			
Second offspring			

6. Now discard each parental genotype (assume that the parents have passed away), and take one of the offsprings' genotypes and your partner should take the other. Record your new F$_1$ (first filial generation) genotype. _____ If this genotype is *tt*, take a seat because you are no longer able to reproduce because you do not survive.

7. Otherwise proceed to a new partner, and follow exactly the same procedure to generate new offspring. If possible, record your new genotype until you have completed five generations:
F$_2$ genotype _____ F$_3$ genotype _____ F$_4$ genotype _____ F$_5$ genotype _____

8. Now fill in Table 13.3 using the F$_5$ information for all *surviving adult* members of the population. This time, you complete the "Number of Students" column first and the "Genotype Frequencies" column second. Since you know the number of students with each genotype, it is only necessary to calculate the proportion of students with each genotype to know the genotype frequencies. For example,

$$\frac{\text{Number of students who are homozygous dominant}}{\text{Number of students remaining}} = \underline{\hspace{2cm}}$$

Table 13.3 F$_5$ Generation

Genotypes	Number of Students	Genotype Frequencies
Homozygous recessive (q^2)	0	0%
Homozygous dominant (p^2)		
Heterozygous ($2pq$)		

9. Compare Table 13.3 with Table 13.1. Do your results show that evolution occurred? _____ Was evolution due to natural selection? _____ Explain. _____

___

Genetic Drift

Genetic drift refers to changes in allelic frequencies of a gene that are due solely to chance. For example, a natural disaster such as a flood or fire, or hunting by humans, may eliminate many individuals from a population so that the few remaining reproduced more than they normally would have. The alleles of the surviving individuals might very well be passed on in greater measure than those of the individuals that died (Fig. 13.4). The bottleneck effect and founder effect are principal factors of genetic drift.

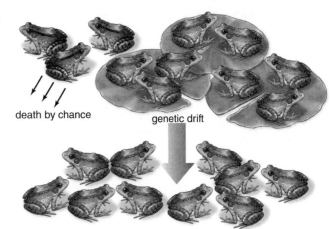

death by chance

genetic drift

Figure 13.4 Genetic drift.
Genetic drift occurs when by chance only certain members of a population (in this case, green frogs) reproduce and pass on their alleles to the next generation. The allele frequencies of the next generation's gene pool may be markedly different from those of the previous generation.

Bottleneck Effect

The **bottleneck effect** occurs when a majority of genotypes are prevented from participating in the production of the next generation due to extreme natural or human interference. For example, northern elephant seals (Fig. 13.5) have reduced genetic variation, probably because of a population bottleneck humans inflicted on them in the 1890s. Hunting reduced their population size to as few as 20 individuals at the end of the nineteenth century. Their population has since rebounded to over 30,000—but their genes still carry the marks of this bottleneck: The population has much less genetic variation than a population of southern elephant seals that was not so intensely hunted.

Figure 13.5 Bottleneck effect.
Members of the northern elephant seal population tend to be homozygous for a large proportion of traits, probably because they were hunted almost to extinction.

Founder Effect

When a few individuals by chance found a colony, only a fraction of the total genetic diversity of the original gene pool is represented. Therefore, the genotype frequencies in descendant populations will probably be quite different from those in the original large population. A small population is more apt to undergo genetic drift—high allele frequency changes by chance. This is called the **founder effect.**

Experimental Procedure: Founder Effect

1. To simulate the founder effect, your instructor will divide the original *P* population (class) described in Table 13.1 into two smaller populations. How many persons are in the new population? _____ Each member of the population will use their initial *P* genotype from page 168. Record your genotype here: _____ .

2. As before, simply calculate the proportion of students with each genotype to arrive at the current genotype frequencies.

Table 13.4 Parental (*P*) Generation		
Genotypes	**Number of Students**	**Genotype Frequencies**
Homozygous recessive (q^2)		
Homozygous dominant (p^2)		
Heterozygous ($2pq$)		

3. Compare Table 13.4 to Table 13.1. Do you expect the founder effect to occur? _____ Explain. _____

4. Now follow the same procedure as before until you have completed five generations. This time, (1) choose only members of your new population as mates and (2) all members of the population can reproduce.

Initial (P) genotype _____ F_1 genotype _____ F_2 genotype _____

F_3 genotype _____ F_4 genotype _____ F_5 genotype _____

5. After five generations, fill in Table 13.5. Simply calculate the proportion of students with each genotype to arrive at the genotype frequencies.

Table 13.5 F_5 Generation		
Genotypes	Number of Students	Genotype Frequencies
Homozygous recessive (q^2)		
Homozygous dominant (p^2)		
Heterozygous ($2pq$)		

6. Compare Table 13.5 with Table 13.1. Do the results show that the founder effect changes a Hardy-Weinberg equilibrium and that genetic drift has occurred? _____ Suppose p = a genetic disorder. Is this disorder now more or less prevalent compared to the original population? Explain.

_____

_____

Virtual Lab Role of Mutations in Natural Selection A Virtual Laboratory called Knocking Out Genes is available on the *Biology* website **www.mhhe.com/maderbiology11**.

After opening this virtual laboratory, follow these directions rather than the original ones. In this virtual laboratory you will be working with *Arabidopsis,* a small wildflower plant that serves as a model organism for plant research. This model organism can be easily grown in the laboratory because it is small and has a short generation time. Yet, anything learned about how genes work, in particular, can be applied to nearly all other organisms. If you wish to know more about *Arabidopsis*, see the *Arabidopsis* lab manual in the virtual lab.

This study will compare the growth pattern of wild-type plants (the control group) to the growth of mutant plants (the experimental group) under stressful conditions: UV exposure, high salinity of soil, and drought conditions. A mutant plant has at least one altered gene. Hypothesize whether you expect or do not expect the mutant plant to perform better than the wild-type plant under stressful environmental conditions.

Hypothesis _____

_____

Experiment 1 Optimum Conditions
This experiment compares the growth of wild-type plants (the control group) with mutant plants (the experimental group) under optimum growth conditions in a normal environment. This will establish a baseline: what the plants look like under optimum conditions in a normal environment. Optimum conditions means that all the needs of the plant in terms of macro and micro nutrients have been provided. Normal environment means that the normal amount of sunlight, water, and NaCl have also been provided.

1. While this lab is called Knocking Out Genes, we will assume that both gain of function mutations and loss of function mutations are possible.

2. Make sure you see "Optimum Growth Conditions" next to environment at the top of the screen. If not, click the reset button at the bottom of the screen until you have these conditions.

3. Click and drag wild-type seeds to the pots in the control group chamber. Click and drag mutant seeds to the pots in the experimental group chamber.

4. Click the Grow button and then the magnifying glass.

5. Examine any one of the wild-type plants (the other wild-type plants will look the same). Examine any one of the mutant plants (the other mutant plants will look the same). Confirm that both the wild-type plants and the mutant plants look the same—that is, normal: _____

Experiment 2 High Salinity

1. Click the reset button at the bottom and the clean pot button at the top of the screen. Change optimum growth conditions to High Salinity in Soil. We will assume that all nutrients and water have been provided for both sets of plants but the environment contains too much NaCl.

2. Click and drag wild-type seeds to the pots in the control group chamber. Click and drag mutant seeds to the pots in the experimental group chamber.

3. Click the Grow button and then the magnifying glass.

Results

4. Examine any one of the wild-type plants (the other wild-type plants will look the same). Examine any one of the mutant plants (the other mutant plants will look the same).

5. Describe the general appearance of both wild-type and mutant plants under High Salinity in Table 1.

Experiment 3 Drought Conditions and Experiment 4 UV Exposure

Repeat the directions given for High Salinity except change Environment to Drought Conditions for the third experiment and UV Exposure for the fourth experiment.

Results

Describe the appearance of both wild-type and mutant plants under Drought Conditions and UV Exposure, respectively, in Table 1.

Table 1   Appearance of Plants Under Various Environments

| | Environmental Conditions | | |
Plant	High Salinity	Drought	UV Exposure
Wild type			
Mutant			

Conclusion: Do your results (support or not support) your original hypothesis? My results _____

the hypothesis. Explain: _____

Questions

1. Why was it necessary to expose both the wild-type and the mutant plants to the various environments? Criticize the experimental design if you had kept the wild-type plants under optimum conditions of growth but had changed the environmental conditions for mutant plants. _____

2. Refer to Figure 13.2 and note that mutations are one cause of microevolution. Did this virtual laboratory show that mutations lead to adaptations when the environment changes and mutant plants are exposed to a new selective agent?

_____ If not, what advice do you have for the researchers based on the scientific method? _____

_____

Laboratory Summary

1. In this laboratory we learned that five conditions can cause microevolution to occur. How did we recognize microevolution had occurred? _____

2. Next we chose a selective agent that caused natural selection to occur. What was this selective agent? _____

3. Then we discovered that genetic drift, such as with the bottleneck and founder effect, occurs when (few or many) _____ of the original population reproduce together.

4. Finally, we carried out an experiment to show that mutations can lead to natural selection when organisms are placed in a _____ environment that contains a selective agent.

Laboratory Review 13

1. List the conditions necessary for a Hardy-Weinberg equilibium to occur. _____
   _____
   _____

2. What is the evidence that evolution is occurring in any given population? _____

3. Assume a Hardy-Weinberg equilibrium.

   a. If $p = 0.8$, what are the gene pool frequencies of a population?

   p = dominant allele = _____

   q = recessive allele = _____

   q^2 = _____ = _____ = _____ %

   p^2 = _____ = _____ = _____ %

   $2pq$ = _____ = _____ = _____ %

   b. What methodology was used in this laboratory to bring about natural selection? _____
   _____

   c. What methodology was used in this laboratory to bring about genetic drift? _____
   _____

   Explain. _____

4. Natural selection results in bacteria adapting (becoming resistant) to the presence of antibiotics.

   a. To what environment were the bacteria exposed? _____
   _____

   b. How do you know that at least some of the bacteria were preadapted to this environment?
   _____

5. How does the process of genetic drift differ from natural selection? _____

_____

6. How does the result of genetic drift differ from natural selection? _____

_____

7. Assume a Hardy-Weinberg equilibrium. If 49% of the population had a recessive phenotype for a trait, what does p equal? _____

8. How is a bottleneck effect like the founder effect? _____

_____

9. Even though it is the phenotype that is exposed to a particular environment, the genotype frequencies of a population can change over time. Explain. _____

_____

10. Natural selection does not cause mutations to occur, yet mutations do play a role in the natural selection process. Explain. _____

_____

Biology **Website**

Enhance your study of the text and laboratory manual with study tools, practice tests, and virtual labs. Also ask your instructor about the resources available through ConnectPlus, including the media-rich eBook, interactive learning tools, and animations.

www.mhhe.com/maderbiology11

McGraw-Hill Access Science Website

An Encyclopedia of Science and Technology Online which provides more information including videos that can enhance the laboratory experience.

www.accessscience.com

LABORATORY

14
Bacteria and Protists

Learning Outcomes

14.1 Bacteria
- Distinguish Gram-negative from Gram-positive bacteria.   176–77
- Recognize colonies of bacteria on agar plates.   178
- Identify the three commonly recognized shapes of bacteria.   179
- Recognize *Gloeocapsa, Oscillatoria,* and *Anabaena* as cyanobacteria.   182–83

14.2 Protists
- Identify and give examples of green, brown, and red algae.   184–87
- Describe the structure and importance of diatoms and dinoflagellates.   188
- Describe various types of locomotion used by protozoans.   189
- Describe sexual reproduction in plasmodial slime molds.   191

Introduction

The history of life began with the evolution of the prokaryotic cell. Although the prokaryotic cell contains genetic material, it is not located in a nucleus, and the cell also lacks any other type of membranous organelle. At one time prokaryotes were believed to be a unified group, but based on molecular data, they are now divided into two major groups—**domain Bacteria** and **domain Archaea.** Eukaryotes in **domain Eukarya** have a membrane-bounded nucleus and membranous organelles. Eukaryotes are more closely related to archaea than bacteria. Prokaryotes resemble each other structurally but are metabolically diverse. Eukaryotes, on the other hand, are structurally diverse and exist as protists, fungi, plants, and animals (Fig. 14.1).

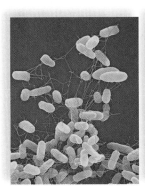    

Salmonella, a bacterium   *Paramecium,* a protist   Morel, a fungus   Sunflower, a plant   Snow goose, an animal

Figure 14.1   The world of living things.
Prokaryotes are represented in this illustration by the bacteria. The protists, fungi, plants, and animals are all eukaryotes.

14.1 Bacteria

Although two major groups of prokaryotes are now recognized, we will study the bacteria as a representative prokaryote. Most bacteria are **saprotrophic,** meaning that they send out digestive enzymes into the environment and thereafter take up the resulting nutrient molecules. Saprotrophic organisms are decomposers that play a major role in ecosystems by digesting the remains of dead organisms and returning inorganic nutrients to photosynthetic organisms. Some bacteria are **parasitic** and cause diseases, such as strep throat or gangrene. Other bacteria are **photosynthetic** or **chemosynthetic** and thus are able to make organic molecules utilizing inorganic molecules. Cyanobacteria are always photosynthetic. They contain chlorophyll, but the green color is often masked by other pigments. In fact, some cyanobacteria are red, brown, or even black.

Several techniques are used to identify bacteria; three methods are demonstrated in this laboratory: Gram stain, colony morphology, and shape of the bacterial cell.

Gram Stain

Most bacterial cells are protected by a **cell wall** that contains a unique molecule called **peptidoglycan.** Bacteria are commonly differentiated by using the Gram stain procedure, which distinguishes bacteria with a thick layer of peptidoglycan (Gram-positive) from those that have a thin layer of peptidoglycan (Gram-negative). Gram-positive bacteria retain a crystal violet-iodine complex and stain blue-purple, whereas Gram-negative bacteria decolorize and counterstain red-pink with safranin (Fig. 14.2).

Figure 14.2   Generalized structure of a bacterium.
a. Gram-positive cells have a thick layer of peptidoglycan. **b.** Gram-negative cells have a very thin layer of peptidoglycan. **c.** This difference causes Gram-positive cells to stain purple and Gram-negative cells to stain reddish-pink.

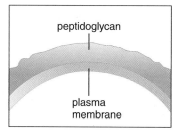

a. Gram-positive cell   b. Gram-negative cell

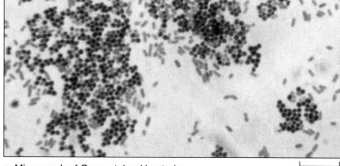

c. Micrograph of Gram stained bacteria

10 µm

Observation: Gram Staining

1. Examine prepared slides under high power. Note the color differences (Fig. 14.2c).

2. Study the directions for Gram staining on page 177.

3. Why do the Gram-positive cells in Figure 14.2c appear purple? _____

   Why do the Gram-negative cells appear reddish-pink? _____

Before you begin, fill in the first column of Table 14.1. Perform the Gram stain on at least two different bacteria, and complete the second column of Table 14.1.

Smear Preparation

1. Obtain an inoculating loop and place one loopful of water on a slide.
2. Flame the loop and let cool briefly.
3. Touch the bacterial growth with the loop.
4. Briefly touch the loop to the drop, and move it back and forth one time.
5. Flame the loop.
6. Spread the cells to an area the size of a penny.
7. Reflame the loop.
8. Allow the slide to air dry.
9. Heat-fix the slide by passing the slide (smear side up) back and forth through the flame slowly three times.
10. Allow the slide to cool to the touch.

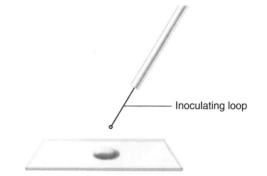

Inoculating loop

Gram Staining

1. To the fixed smear, apply several drops of crystal violet for 1–2 minutes.
2. Wash off excess stain under a stream of water.
3. Flood the slide with Gram's iodine, and let stand at least 1 minute.
4. Rinse with water, and shake off excess water.
5. Holding the slide at an angle, add decolorizing solution dropwise. When color is no longer removed (about 10 seconds), rinse with water.
6. Flood the slide with safranin, and let stand 2–3 minutes.
7. Drain excess and wash slide.
8. Carefully blot dry.
9. Examine with oil immersion lens. Gram-positive organisms will retain the crystal violet and appear purple. Gram-negative organisms will only be stained by safranin and will appear reddish-pink.
10. Complete Table 14.1.

Table 14.1 Gram Staining		
Organism	Color Results	Gram Reaction

Conclusions: Gram Stain

- What do you know about the cell wall of Gram-positive organisms?

_____

- What do you know about the cell wall of Gram-negative organisms?

_____

Colony Morphology

On a nutrient material called agar, bacteria grow as colonies (Fig. 14.3). A **colony** contains cells descended from one original cell. Sometimes, it is possible to identify the type of bacterium by the appearance of the colony.

Figure 14.3 Colony morphology.
Colonies of bacteria on agar plates. Note the variation that helps researchers identify the type of bacterium.

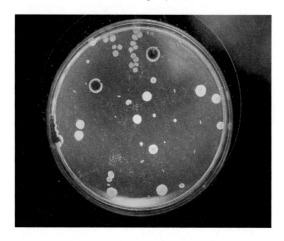 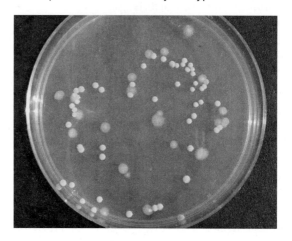

Observation: Colony Morphology

1. View agar plates that have been inoculated with bacteria and then incubated. Notice the "colonies" of bacteria growing on the plates.
2. Compare the colonies' color, surface, and margin, and note your observations in Table 14.2. It is not necessary to identify the type of bacterium.

Table 14.2 Agar Plates	
Plate Number	**Description of Colonies**

3. If available, obtain a sterile agar plate, and inoculate the plate with your thumbprint, or use a swab and inoculate the plate with material from around your teeth or inside your nose. Put your name on the plate, and place it where directed by your instructor. Remember to view the plate next laboratory period. Describe your plate:

4. If available, obtain a sterile agar plate, and expose it briefly (at most for ten minutes) anywhere you choose, such as in the library, your room, or your car. No matter where the plate is exposed, it subsequently will show bacterial colonies. Describe your plate:

Shape of Bacterial Cell

Most bacteria are found in three basic shapes: **spirillum** (spiral or helical), **bacillus** (rod), and **coccus** (round or spherical) (Fig. 14.4). Bacilli may form long filaments, and cocci may form clusters or chains. Some bacteria form endospores. An **endospore** contains a copy of the genetic material encased by heavy protective spore coats. Spores survive unfavorable conditions and germinate to form vegetative cells when conditions improve.

Observation: Shape of Bacterial Cell

1. View the microscope slides of bacteria on display. What magnification is required to view bacteria?

   _____

2. Using Figure 14.4 as a guide, identify the three different shapes of bacteria.

3. Do any of the slides on display show bacterial cells with endospores? _____ What is

   an endospore, and why does it have survival value? _____

Figure 14.4   Diversity of bacteria.
a. Spirillum, a spiral-shaped bacterium. **b.** Bacilli, rod-shaped bacteria. **c.** Cocci, round bacteria.

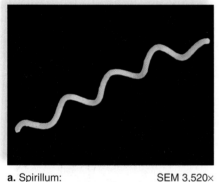 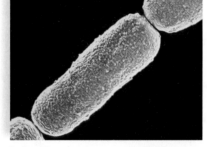 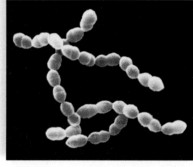

a. Spirillum: SEM 3,520×
    Spirillum volutans

b. Bacilli: SEM 35,000×
    Bacillus anthracis

c. Cocci: SEM 6,250×
    Streptococcus thermophilus

Observation: Yersinia pestis

Commonly known as the Black Death or bubonic plague, *Yersinia pestis* is a Gram-negative, rod-shaped bacterium. It causes plague, normally a **zoonotic** (found in animals) disease of rodents (e.g., rats, mice, ground squirrels). Fleas that live on the rodents can sometimes pass the bacteria to human beings, who then suffer from the **bubonic** form of the plague (the presence of inflamed, tender swelling of lymph nodes, especially in the area of the armpit or groin). *Yersinia pestis* is a potential biological warfare bacterium if infected fleas were released or the bacterium itself can be released as an aerosol.

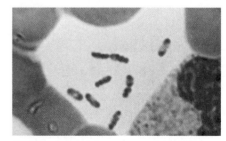

1. Examine the plague bacteria *Yersinia pestis* in Figure 14.5.
2. Try to identify the common "clothespin" shape of the rod-shaped cells.
3. Research the type(s) of antibiotics used to treat patients with bubonic or pneumonic plague in the United States.

Figure 14.5   Plague bacteria in blood smear.

Virtual Lab Classifying Bacteria Using Biotechnology   A Virtual Laboratory called Classifying Bacteria Using Biotechnology is available on the *Biology* website **www.mhhe.com/maderbiology11**. Traditionally, morphological (structural) data have been used to classify organisms; but increasingly, researchers depend directly on molecular data. After all, genetic mutations (permanent DNA changes) are the original source of differences between organisms. After opening this laboratory, follow only the directions below to collect both morphological and molecular data for different species of bacteria.

Collect data on three species of bacteria:

1. *Morphological data:* For the first unknown, click on the dropper of the Gram stain bottle and drag it to slide #1; then drag slide #1 to the microscope. The monitor will show the shape (round, rod, or curved) and arrangement (single, paired, chains, or clusters) of the bacterium and also whether the bacterium is Gram-positive (purple) or Gram-negative (pink).* Record your data in Table 1 under morphological data; use + or − for the Gram stain.

2. *Molecular data:* Determine the percentage of G+C bases in DNA by dragging the #1 DNA red tube to the GC Content Measuring Apparatus (at far right). Record your data in Table 1 under molecular data.

3. *Name of organism:* Drag the #2 rRNA red tube to the Ribosomal RNA gene sequencing electrophoretic unit** (at far left), which analyzes the gene for rRNA in this organism. The name of the bacterium appears on the monitor screen. Record the name in Table 1.

4. Now click on the reset button and collect the same data for the second unknown (use #2 slide and blue tubes) and the third unknown (use #3 slide and green tubes).

5. When you click on the reset button after the third unknown you can collect data for three more unknowns. It's up to you to remember whether you are working with slide #1 (red tubes), or slide #2 (blue tubes), or slide #3 (green tubes).

6. Continue to press the reset button until you have collected data on at least six different bacteria.

Classifying the Organisms

Using the data in Table 1, we will consider how to determine the relationship and classification of bacteria.

1. What staining procedure allows bacteria to be divided into two main groups? _____

What are the two groups? _____

Write these two main groups on the two initial lines provided in this evolutionary tree.

---

*During the Gram staining procedure, a purple stain is applied to the slide and then the slide is washed before a secondary pink stain is applied. Thick-walled bacteria retain the purple stain and are Gram-positive. Thin-walled bacteria lose the purple and show only the pink stain; they are Gram-negative.

**The electrophoretic unit sequences the bases of the rRNA gene. Each organism has its own particular sequence of bases.

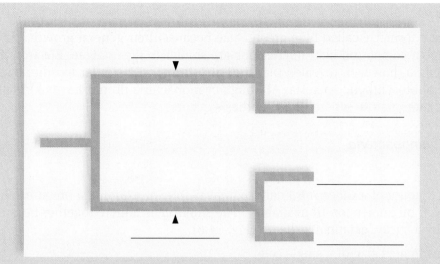

2. On the basis of G-C content each of the groups from question 1 can be divided into bacteria with less than 50% G-C content, called *low G-C bacteria*, and those with more than 50% G-C content, called *high G-C bacteria*. Using these designations, complete the evolutionary tree.

Organisms that share specific characteristics are more closely related; for example, Gram-positive, low G-C bacteria are thought to be more closely related with one another than they are with Gram-positive, high G-C bacteria.

Table 1	Classifying Bacteria			
	Morphological Data Appearance	Gram stain	**Molecular Data** G-C percentage in DNA	**Name of Organism** Based on rRNA sequencing
First Unknown				
Second Unknown				
Third Unknown				
Fourth Unknown				
Fifth Unknown				
Sixth Unknown				

Cyanobacteria

Cyanobacteria were formerly called blue-green algae because their general growth habit and appearance through a compound light microscope are similar to green algae. Electron microscopic study of cyanobacteria, however, revealed that they are structurally similar to other bacteria, particularly other photosynthetic bacteria. Although cyanobacteria do not have chloroplasts, they do have thylakoid membranes, where photosynthesis occurs.

Observation: Cyanobacteria

Gloeocapsa

1. Prepare a wet mount of a *Gloeocapsa* culture, if available, or examine a prepared slide, using high power (45×) or oil immersion (if available). The single cells adhere together because each is surrounded by a sticky, gelatinous sheath (Fig. 14.6).

2. What is the estimated size of a single cell? _____

**Figure 14.6
Gloeocapsa.**

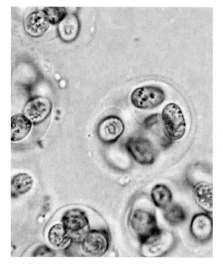

a. Micrograph at low magnification

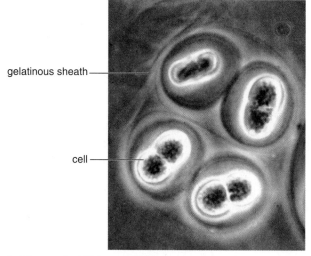

gelatinous sheath ⎯

cell ⎯

b. Micrograph at high magnification

Oscillatoria

1. Prepare a wet mount of an *Oscillatoria* culture, if available, or examine a prepared slide, using high power (45×) or oil immersion (if available). This is a filamentous cyanobacterium with individual cells that resemble a stack of pennies (Fig. 14.7).

2. *Oscillatoria* takes its name from the characteristic oscillations that you may be able to see if your sample is alive. If you have a living culture, are oscillations visible?

_____

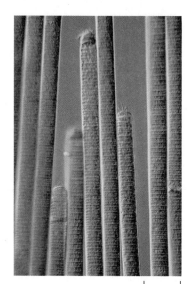

Figure 14.7   *Oscillatoria.*

500 μm

Anabaena

1. Prepare a wet mount of an *Anabaena* culture, if available, or examine a prepared slide, using high power (45×) or oil immersion (if available). This is also a filamentous cyanobacterium, although its individual cells are barrel-shaped (Fig. 14.8).
2. Note the thin nature of this strand. If you have a living culture, what is its color? _____

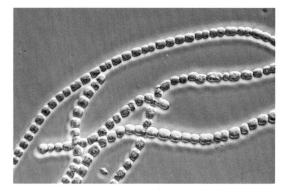

Figure 14.8 *Anabaena.*

14.2 Protists

Protists were the first eukaryotes to evolve. Their diversity and complexity make it difficult to categorize them, but on the basis of molecular data, they are now placed in supergroups, which also include the animals, fungi, and land plants. See the accompanying table.

Protist Diversity

Supergroup	Members	Distinguishing Features	Protist Example
Archaeplastids	Green algae, red algae, land plants, charophytes	Plastids; unicellular, colonial, and multicellular	*Chlamydomonas*, red alga, *Closterium*, a green alga
Chromalveolates	Stramenopiles: brown algae, diatoms, golden brown algae, water molds Alveolates: ciliates, apicomplexans, dinoflagellates	Most with plastids; unicellular and multicellular Alveoli support plasma membrane; unicellular	*Blepharisma*, a ciliate with visible vacuoles; Assorted fossilized diatoms; *Ceratium*, an armored dinoflagellate
Excavates	Euglenids, kinetoplastids, parabasalids, diplomonads	Feeding groove; unique flagella; unicellular	*Giardia*, a single-celled flagellated diplomonad; *Euglena*
Amoebozoans	Amoeboids, plasmodial and cellular slime molds	Pseudopods; unicellular	*Dictyostelium*, a slime mold with fruiting bodies; *Amoeba proteus*, a protozoan
Rhizarians	Foraminiferans, radiolarians	Thin pseudopods; some with tests; unicellular	*Nonionina*, a foraminiferan; Radiolarians (assorted) produce a calcium carbonate shell
Opisthokonts	Choanoflagellates, animals, nucleariids, fungi	Some with flagella; unicellular and colonial	Choanoflagellate (unicellular), animal-like protist

Traditionally, protists were discussed according to their mode of nutrition, and this section will continue to do so while still noting the supergroup to which each type protist belongs. The photosynthetic algae photosynthesize in the same manner as land plants and certain of the green algae share a common ancestor with land plants, as noted again in Laboratory 16. The heterotrophic (must acquire food from an external source) protists commonly called protozoans either ingest their food, as do animals, or they absorb it, as do fungi. Certain of the protozoans may be ancestral to animals, as will be noted again in Laboratory 22. The relationship of any of the protists to fungi is unknown at this time.

Photosynthetic Protists

Algae is a term that is used for aquatic organisms that photosynthesize in the same manner as land plants. All photosynthetic protists contain green chlorophyll, but they also may contain other pigments that mask the chlorophyll color, and this accounts for their common names. Commonly, we speak of the green algae, red algae, brown algae, and golden brown algae.

Observation: Green Algae

The green algae (supergroup Archaeplastids) are ancestral to the first land plants. Both groups possess chlorophylls *a* and *b*, both store reserve food as starch, and both have cell walls that contain cellulose.

If available, view a film loop showing the many forms of green algae. Notice that green algae can be single cells, filaments, colonies, or multicellular sheets. You will examine a filamentous form *(Spirogyra)* and a colonial form *(Volvox)*. A **colony** is a loose association of cells.

Spirogyra

Spirogyra is a filamentous alga, lives in fresh water, and often is seen as a green scum on the surface of ponds and lakes. The most prominent feature of the cells is the spiral, ribbonlike chloroplast (Fig. 14.9).

How do you think *Spirogyra* got its name? _____

_____

Figure 14.9 *Spirogyra*.
a. *Spirogyra* is a filamentous green alga, in which each cell has a ribbonlike chloroplast. **b.** During conjugation, the cell contents of one filament enter the cells of another filament. Zygote formation follows.

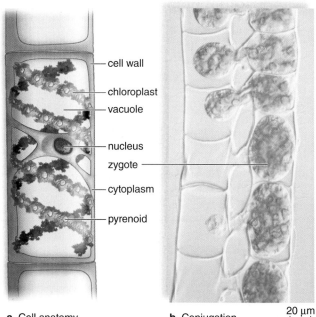

cell wall
chloroplast
vacuole
nucleus
zygote
cytoplasm
pyrenoid

a. Cell anatomy   **b.** Conjugation   20 μm

Spirogyra's chloroplast contains a number of circular bodies, the **pyrenoids,** centers of starch production. The nucleus is in the center of the cell, anchored by cytoplasmic strands. Your slide may show **conjugation,** a sexual means of reproduction illustrated in Figure 14.9*b*. If it does not, obtain a slide that does show this process. Conjugation tubes form between two adjacent filaments, and the contents of one set of cells enter the other set. As the nuclei fuse, a zygote is formed. The zygote overwinters, and in the spring, meiosis and, subsequently, germination occur.

Make a wet mount of live *Spirogyra,* or observe a prepared slide.

Volvox

Volvox is a green algal colony. It is motile (capable of locomotion) because the thousands of cells that make up the colony have flagella. These cells are connected by delicate cytoplasmic extensions (Fig. 14.10).

Volvox is capable of both asexual and sexual reproduction. Certain cells of the adult colony can divide to produce **daughter colonies** (Fig. 14.10) that reside for a time within the parental colony. A daughter colony escapes the parental colony by releasing an enzyme that dissolves away a portion of the matrix of the parental colony. During sexual reproduction, some colonies of *Volvox* have cells that produce sperm, and others have cells that produce eggs.

Using a depression slide, make a wet mount of live *Volvox,* or study a prepared slide.

Figure 14.10   *Volvox.*
Volvox is a colonial green alga. The adult *Volvox* colony often contains daughter colonies, asexually produced by special cells.

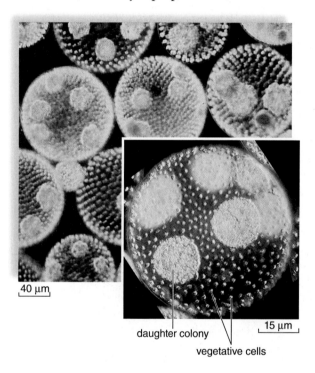

40 µm

15 µm

daughter colony

vegetative cells

Green Algae Diversity

In Table 14.3, list the genus names of each green algae specimen available. Give a brief description of each specimen.

Table 14.3 Green Algae Diversity	
Genus Name	**Description**

Brown algae (supergroup Chromalveolates) are commonly called *seaweed,* along with the multicellular green and red algae. Brown algae contain brown pigments that mask chlorophyll's green color. These algae are large and have specialized parts.

Fucus is called rockweed because it is seen attached to rocks at the seashore when the tide is out (Fig. 14.11). If available, view a preserved specimen. Note the dichotomously branched body plan, so called because the **thallus** (body) repeatedly divides into two branches (Fig. 14.11). Note also the **holdfast** by which the alga anchors itself to the rock; the **air vesicles,** or bladders, that help hold the thallus erect in the water; and the **receptacles,** or swollen tips. The receptacles are covered by small raised areas, each with a hole in the center. These areas are cavities in which the sex organs are located, with the gametes escaping to the outside through the holes. *Fucus* is unique among algae in that it is diploid (2n) and always reproduces sexually.

If available, study preserved specimens of other brown algae (Fig. 14.11). *Laminaria* algae are called **kelps.**

Figure 14.11 Brown algae.
Laminaria and *Fucus* are seaweeds known as kelps. They live along rocky coasts of the north temperate zone. The other brown algae featured, *Macrocystis* and *Nereocystis,* form spectacular underwater "forests" at sea.

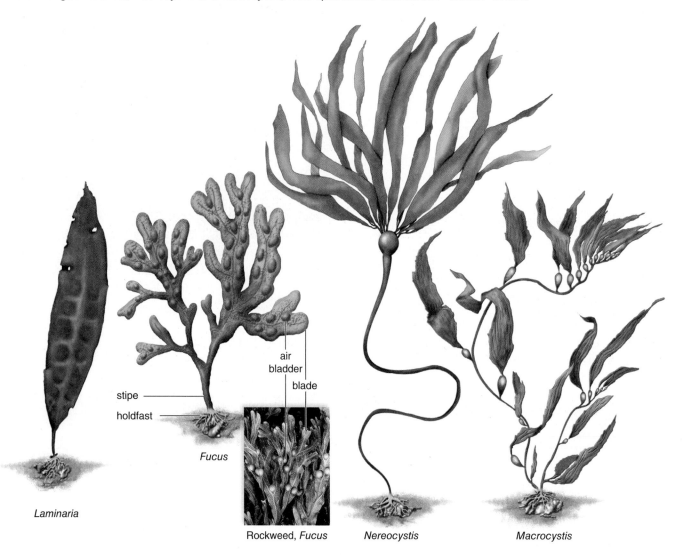

stipe

holdfast

Fucus

air bladder

blade

Laminaria

Rockweed, *Fucus*

Nereocystis

Macrocystis

In Table 14.4, list the genus names of each of the brown algae specimens available, and identify and indicate (using Fig. 14.11) if the following structures are present on your specimens: *holdfast, air bladder,* and *stipe.*

Table 14.4 Brown Algae

Genus Name	Description	Holdfast	Air Bladder	Stipe

Observation: Red Algae

Like most brown algae, the red algae (supergroup Archaeplastids) (Fig. 14.12) are multicellular, but they occur chiefly in warmer seawater, growing both in shallow waters and as deep as light penetrates. Some forms of red algae are filamentous, but more often, they are complexly branched with a feathery, flat, and expanded or ribbonlike appearance. Coralline algae are red algae that have cell walls impregnated with calcium carbonate ($CaCO_3$).

In Table 14.5, list the genus names of each of the red algae specimens available, and give a brief description.

Figure 14.12 Red algae.
Generally, red algae are smaller and more delicate than brown algae. **a.** *Rhodoglossum* has a pronounced filamentous structure. **b.** Calcium carbonate is deposited in the walls of the red alga, *Corallina.*

a.

b.

Table 14.5 Red Algae

Genus Name	Description

Observation: Diatoms

The golden brown algae have chloroplasts that contain a yellow-brown pigment, in addition to chlorophyll. The cell wall of **diatoms** (supergroup Chromalveolates) is in two sections, with the larger one fitting over the smaller as a lid fits over a box. Since the cell wall is impregnated with silica, diatoms are said to "live in glass houses." The glass cell walls of diatoms do not decompose, so they accumulate in thick layers subsequently mined as diatomaceous earth and used in filters and as a natural insecticide. Diatoms, being photosynthetic and extremely abundant, are important food sources for the small heterotrophs (organisms that must acquire food from external sources) in both marine and freshwater environments.

Make a wet mount of live diatoms, or view a prepared slide (Fig. 14.13). Describe what you see:

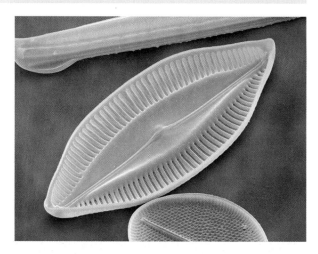

Figure 14.13 Diatoms.
Diatoms are photosynthetic protists of both freshwater and salt water.

Observation: Dinoflagellates

Dinoflagellates (supergroup Chromalveolates) have chloroplasts that are apparently derived from algae. They also have two flagella; one is free, but the other is located in a transverse groove that encircles the animal. The beating of these flagella causes the organism to spin like a top. The cell wall, when present, is frequently divided into closely joined polygonal plates of cellulose. At times there are so many of these organisms in the ocean that they cause a condition called "red tide." The toxins given off in these red tides cause widespread fish kills and can cause paralysis in humans who eat shellfishes that have fed on the dinoflagellates.

Make a wet mount of live dinoflagellates or view a prepared slide (Fig. 14.14). Describe what you see:

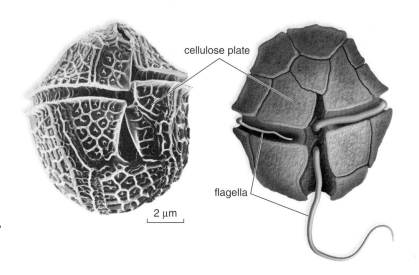

cellulose plate

flagella

2 μm

Figure 14.14 Dinoflagellates.
Dinoflagellates such as *Gonyaulax* have cellulose plates.

Heterotrophic Protists

The term *protozoan* refers to unicellular eukaryotes and its use is often restricted to heterotrophic organisms that ingest food by forming **food vacuoles.** Other vacuoles, such as **contractile vacuoles** that rid the cell of excess water, are also typical. Usually protozoans have some form of locomotion and, as shown in Figure 14.15, some use **pseudopodia,** some move by **cilia,** and some use **flagella**.

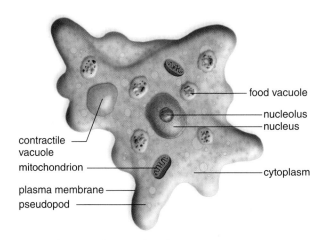

a. *Amoeba* moves by pseudopods

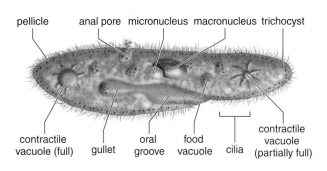

b. *Paramecium* moves by cilia

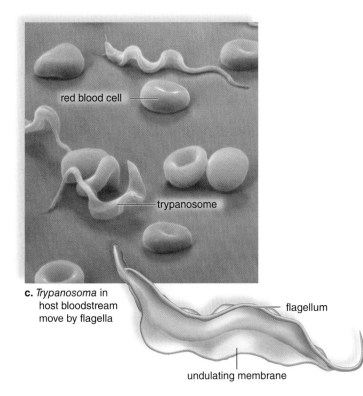

c. *Trypanosoma* in host bloodstream move by flagella

d. *Plasmodium* bursts forth as spores from infected red blood cells.

Figure 14.15   Protozoan diversity.
Many protozoans are motile. **(a)** *Amoeba* (supergroup Amoebozoans) move by pseudopodia. **(b)** *Paramecium* (supergroup Chromalveolates) move by cilia. **(c)** *Trypanosoma* (kinetoplastid in supergroup Excavates) move by flagella. A blood infection is the cause of African sleeping sickness. **(d)** *Plasmodium,* a cause of malarias, (apicomplexan in supergroup Chromalveolates) exist as nonmotile spores inside the red blood cells of humans.

Plasmodium vivax, which causes a common form of malaria is an apicomplexan. Apicomplexans have special organelles; one is an apicomplast at the base of their flagella and the other contains organic chemicals that assist in penetrating host tissues. *Plasmodium* spends a portion of its life cycle in mosquitoes (sexual phase) and the other part in human hosts (asexual phase). During the asexual phase of their life cycle, all apicomplexans exist as particulate spores, which accounts for why they are also called sporozoans. In general, how do sporozoans differ from the other protozoans shown in

Figure 14.15? _____

Plasmodium spores are the cause of malaria—they multiply in and rupture red blood cells.

Observation: Heterotrophic Protists

Individual Protozoans

You may already have had the opportunity to observe a protozoan such as *Euglena* or *Paramecium* in Laboratory 2. However, your instructor may want you to observe these organisms again. *Euglena* have flagella but many also have chloroplasts (see Fig. 2.12) and therefore they do not match our definition of a protozoan. But, then, some *Euglena* do not have chloroplasts. Watch a video if available, and note the various forms of protozoans. Prepare wet mounts or examine prepared slides of protozoans as directed by your instructor. Complete Table 14.6, listing the structures for locomotion in the types of protozoans you have observed.

Table 14.6 Heterotrophic Protists		
Genus Name	**Structures for Locomotion**	**Observations**
Amoeba		
Paramecium		
Trypanosoma		

Pond Water

Pond water typically contains various examples of the protists studied in this laboratory. Your instructor may have an illustrated manual that will help you to identify the ones unfamiliar to you. Prepare a wet mount of a sample of pond water. Be sure to select some of the sediment on the bottom and a few strands of filamentous algae. Identify and classify them as fully as possible.

1. Make a wet mount of pond water by taking a drop from the bottom of a container of pond water.
2. Scan the slide for organisms: Start at the upper left-hand corner, and move the slide forward and back as you work across the slide from left to right.
3. Experiment by using all available objective lenses, by focusing up and down with the fine-adjustment knob, and by adjusting the light so that it is not too bright.
4. Identify the organisms you see by consulting Figure 2.12 on page 25, and use pictorial guides provided by your instructor.

Slime Molds

Slime molds (supergroup Amoebozoans) were once classified as fungi, but unlike fungi they have flagellated cells at one time during their life cycle. Also, unlike fungi, which absorb their food, slime molds phagocytize their food like amoeboids.

There are two types of slime molds: cellular slime molds and plasmodial slime molds. **Cellular slime molds** usually exist as individual amoeboid cells, which aggregate on occasion to form a pseudoplasmodium. **Plasmodial slime molds** usually exist as a **plasmodium,** a fan-shaped, multinucleated mass of cytoplasm.

The plasmodium of a plasmodial slime mold creeps along, phagocytizing decaying plant material in a forest or agricultural field. During times unfavorable for growth, such as a drought, the plasmodium develops many sporangia. A **sporangium** is a reproductive structure that we will see again among fungi which produces wind-blown spores by meiosis. In some plasmodial slime molds, the spores become flagellated cells, and in others, they are amoeboid. In any case, they fuse to form a zygote that develops into a plasmodium (Fig. 14.16).

Plasmodium, *Physarum*

Spongaria, *Hemitrichia* ⌞1 mm⌟

Observation: Plasmodial Slime Molds

1. Obtain a plate of *Physarum* growing on agar. Carefully examine the plate under the stereomicroscope.
2. Describe what you see.

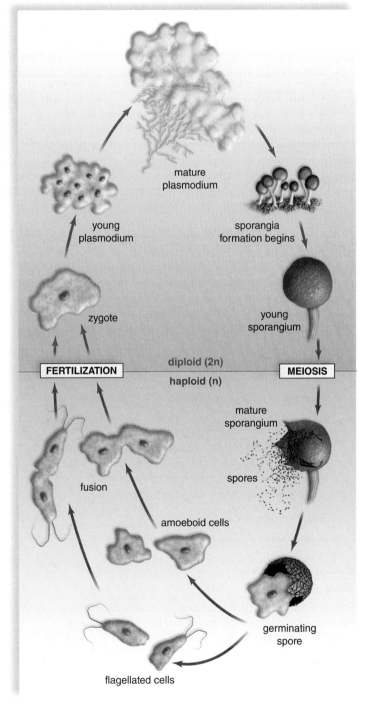

Figure 14.16 Plasmodial slime mold.
The diploid adult forms sporangia during sexual reproduction, when conditions are unfavorable to growth. Wind-blown haploid spores germinate, releasing haploid amoeboid or flagellated cells that fuse.

1. What are two major differences between prokaryotic and eukaryotic cell structures? _____
   _____

2. What chemical associated with the cell wall gives Gram-positive bacteria their final color?
   _____

3. **a.** In what ways are cyanobacteria like algae and land plants? _____
   b. How are they different? _____

4. How can cyanobacteria, in contrast to saprotrophic bacteria, live in an environment that lacks organic
   nutrients?
   _____

5. List the two major types of obtaining organic nutrients and give an example of a protist with each type
   of nutrition.

Nutrition	Examples
a. _____	_____
b. _____	_____

6. List the three different structures for locomotion found among protozoans, and name an organism that
   utilizes each structure.

Structure	Organism Name
a. _____	_____
b. _____	_____
c. _____	_____

7. Name a protist that has both chloroplasts and flagella. _____ Which makes this protist
   algal like? _____ Which makes it protozoan like? _____

8. Name two characteristics of slime molds by explaining how they get their nutrients and what they do to
   survive conditions unfavorable for growth. _____
   _____

9. Complete the following sentence: Plasmodial slime molds usually exist as a _____
   _____.

10. Could an ecosystem that contains only protists remain in continued existence? _____
    Why or why not? _____

15
Fungi

Learning Outcomes

15.1 Zygospore Fungi
- Identify the structures typical of black bread mold, and describe both the sexual and asexual life cycles.   194–96

15.2 Sac Fungi
- Identify types of sac fungi and both sexual and asexual reproductive structures.   196–99

15.3 Club Fungi
- Identify the parts of a mushroom and its sexual life cycle.   199–201

15.4 Fungal Diversity
- Explain the phylum names: Zygomycota, Ascomycota, Basidiomycota.   201

15.5 Fungi as Symbionts
- Explain the structure of a lichen and mycorrhizae.   202–3

Introduction

Fungi are multicellular heterotrophs that send out digestive juices into the environment and then absorb the resulting nutrients. In this way, fungi can be contrasted with animals that consume whole foods and then digest it. The free-living fungal groups you will study in this laboratory (the zygospore fungi, the sac fungi, and the club fungi) are terrestrial, and they produce windblown spores during both asexual and sexual reproduction. Still, the fungi must have evolved in the water as did the plants and animals. The **chytrids** are aquatic fungi with spores that are flagellated. Molecular data suggest that chytrids are the closest living relatives to the common ancestor of all the fungal groups, as shown in the evolutionary tree for fungi in Figure 15.1.

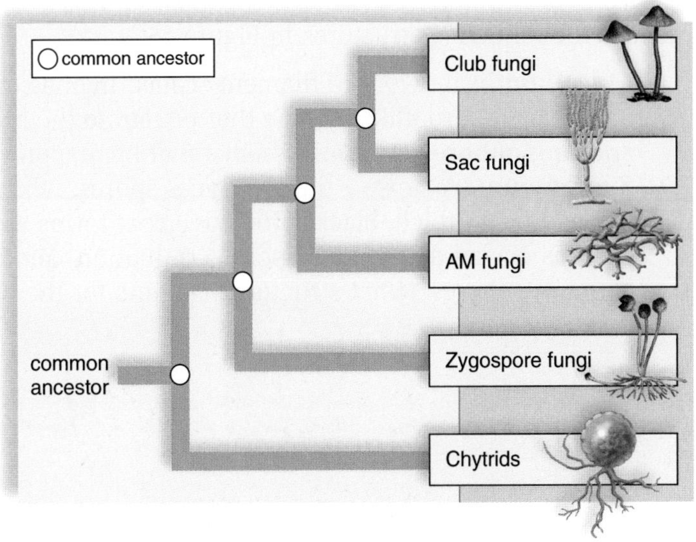

Figure 15.1 Evolutionary relationships among the fungi.
The chytrids are living fungi most closely related to the common ancestor of all fungi. The other groups evolved in the sequence shown.

Typically, the spores of a fungi develop directly into a **mycelium,** which is composed of hyphae. **Hyphae** are filaments of cells joined end to end. During sexual reproduction, hyphae from two different strains (called plus and minus) fuse and give rise to diploid nuclei. In zygospore fungi, the diploid nucleus undergoes meiosis and spore production occurs in a **sporangium.** Sac fungi and club fungi produce **fruiting bodies,** where many instances of meiosis are occurring. In sac fungi, the diploid nucleus undergoes meiosis and spore production occurs in asci, structures that are sac-shaped. In club fungi, the equivalent structures are club shaped and called basidia. As shown in the accompanying table, classification of fungi is based on means of sexual reproduction.

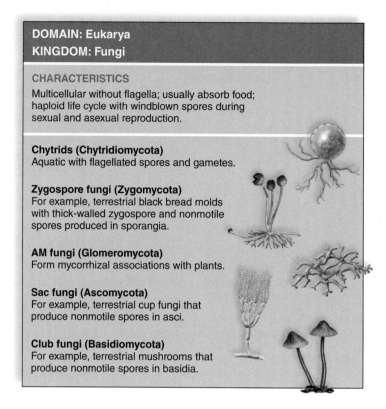

DOMAIN: Eukarya
KINGDOM: Fungi

CHARACTERISTICS
Multicellular without flagella; usually absorb food; haploid life cycle with windblown spores during sexual and asexual reproduction.

Chytrids (Chytridiomycota)
Aquatic with flagellated spores and gametes.

Zygospore fungi (Zygomycota)
For example, terrestrial black bread molds with thick-walled zygospore and nonmotile spores produced in sporangia.

AM fungi (Glomeromycota)
Form mycorrhizal associations with plants.

Sac fungi (Ascomycota)
For example, terrestrial cup fungi that produce nonmotile spores in asci.

Club fungi (Basidiomycota)
For example, terrestrial mushrooms that produce nonmotile spores in basidia.

15.1 Zygospore Fungi

Zygospore fungi, such as black bread mold, commonly grow on bread or any bakery goods kept at room temperature. Identify the sexual and the asexual portion of the life cycle of *Rhizopus,* the black mold shown in Figure 15.2.

Are the nuclei in the mycelium of *Rhizopus* haploid or diploid? _____ Where does meiosis occur in the *Rhizopus* life cycle? _____ Where are spores produced? _____

In zygosphore fungi, the hyphae are nonseptate (without cross walls); therefore, the hyphae are multinucleate.

Black Bread Mold

Identify the following structures in Figure 15.2:

1. **Mycelium:** A network of filaments called hyphae, of which there are three types: **Rhizoids** are somewhat rootlike hyphae that penetrate the bread, **stolons** are horizontal hyphae, and **sporangiophores** are hyphal stalks that bear sporangia.
2. **Sporangium:** A capsule that produces spores, which are black.
3. **Zygospore:** A thick, black, protective coat forms around the zygotes during sexual reproduction. Meiosis occurs during zygospore germination, and asexual sporangia are produced on sporangiophores. What structure accounts for the phylum name Zygomycota (zygospore fungi)?

_____

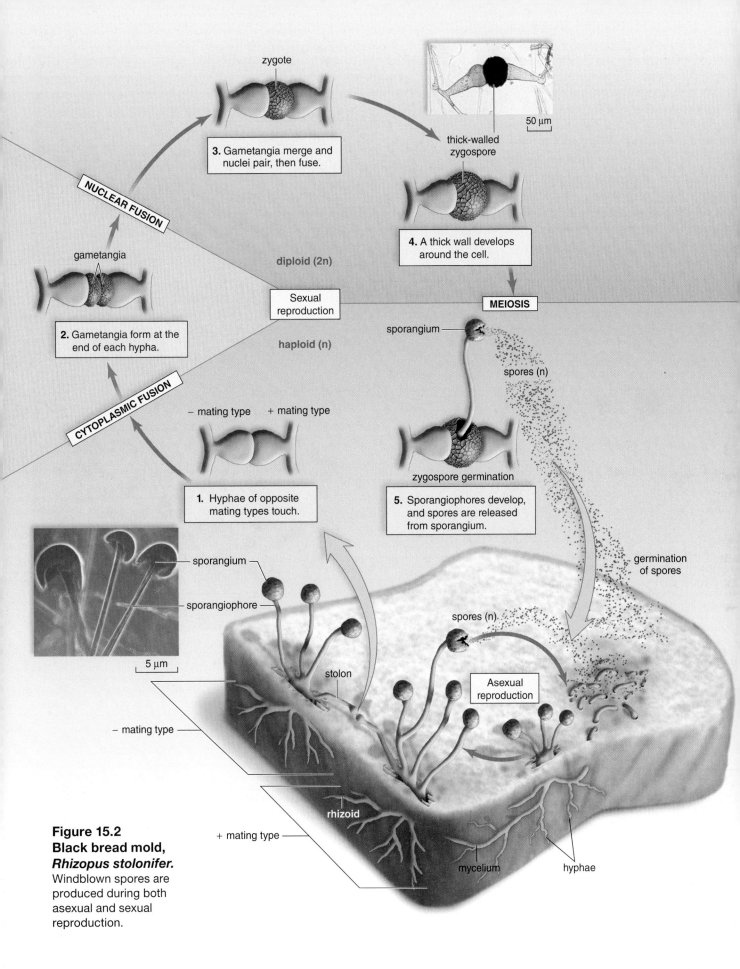

zygote

3. Gametangia merge and nuclei pair, then fuse.

thick-walled zygospore

50 µm

NUCLEAR FUSION

gametangia

4. A thick wall develops around the cell.

diploid (2n)

Sexual reproduction

MEIOSIS

2. Gametangia form at the end of each hypha.

haploid (n)

sporangium

spores (n)

CYTOPLASMIC FUSION

− mating type   + mating type

1. Hyphae of opposite mating types touch.

zygospore germination

5. Sporangiophores develop, and spores are released from sporangium.

germination of spores

sporangium

sporangiophore

spores (n)

Asexual reproduction

5 µm

stolon

− mating type

rhizoid

+ mating type

mycelium

hyphae

Figure 15.2
Black bread mold,
Rhizopus stolonifer.
Windblown spores are produced during both asexual and sexual reproduction.

1. If available, examine bread that has become moldy. Do you recognize black bread mold on the bread? _____ Describe the mold you see. _____

   _____

2. Obtain a petri dish that contains living black bread mold. Observe with a stereomicroscope. Identify the three types of hyphae and the sporangia (black dots)._____

3. View a prepared slide of *Rhizopus,* using both a stereomicroscope and the low-power setting of a light microscope. The absence of cross walls in the hyphae is an identifying feature of Zygomycota. List the structures you can identify. _____

   _____

4. In the micrograph on the left, label structures seen during asexual reproduction: sporangiophore, rhizoid, and sporangium. In the micrograph on the right, label structures seen during sexual reproduction: stolon, gametangium, and zygospore.

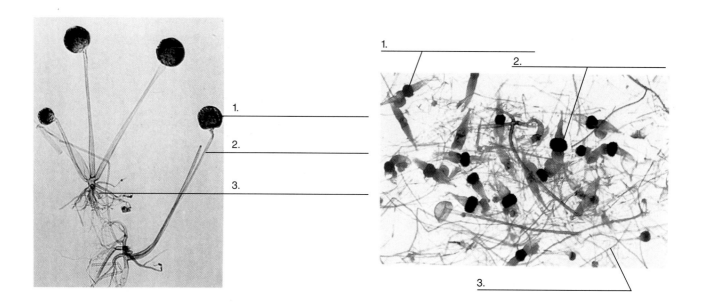

15.2 Sac Fungi

Sac fungi are composed of septate hyphae. The phylum name refers to the **ascus,** a fingerlike sac that develops during sexual reproduction. What structure accounts for the phylum name Ascomycota (sac fungi)? _____

Within each ascus, meiosis followed by mitosis results in eight haploid nuclei, and these become eight ascospores. The asci are usually surrounded and protected by sterile hyphae within a fruiting body. A **fruiting body** is a reproductive structure where spores are produced and released.

Various molds, including red bread mold *(Neurospora),* are sac fungi, as are cup fungi, morels, and truffles. Several molds (e.g., *Penicillium* and *Aspergillus*) are now classified as sac fungi. In 1928, when examining a petri dish, Alexander Fleming observed an area around *Penicillium* that was free of bacteria because the mold produced an antibacterial substance he later called penicillin. This discovery changed medicine and has saved countless lives. Other sac fungi have great medical importance also.

For example, *Candida albicans* is a yeast normally present on the skin and mucous membranes of the mouth, vagina, or rectum. When it overgrows in the mouth, the medical condition is called thrush.

Some sac fungi, such as Dutch elm disease, ergot, and powdery mildews, cause serious plant diseases. Ergot contains LSD, and this hallucinogenic substance sometimes has been ingested accidentally.

Yeasts

Yeasts differ from most other fungi because they are not composed of hyphae. Even though yeasts are unicellular, they form an ascus during sexual reproduction. Usually, however, yeasts reproduce asexually, either by mitosis and cell division or by **budding.** Budding involves an unequal distribution of cytoplasm during cytokinesis.

Yeasts, such as *Saccharomyces,* are used during the production of wine and beer because of the alcohol they produce when fermenting. Yeasts are also helpful in that the carbon dioxide they give off makes bread rise.

Observation: Yeast

1. Obtain a sample of *Saccharomyces* (yeast) culture, and make a wet mount.
2. Observe, using all three objectives of your microscope. Add a drop of methylene blue stain if you cannot see the cells well. See methylene blue caution on page 199.
3. Label the vegetative cell, bud, ascus, and ascospore in the following diagram of *Saccharomyces:*

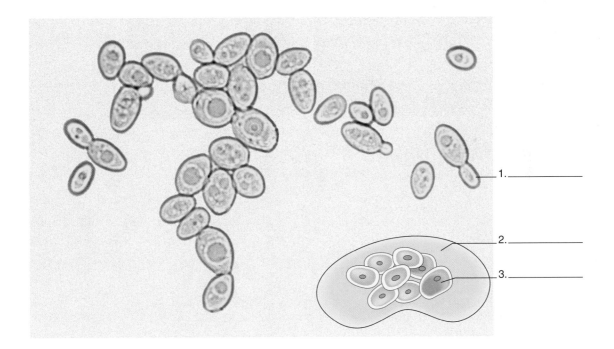

1. _____

2. _____

3. _____

4. View a prepared slide of *Schizosaccharomyces.* Identify an ascus and ascospores. How many ascospores are in each ascus? _____

Cup Fungi

Cup fungi are representative of sac fungi because they produce an **ascocarp** (fruiting body) in which saclike asci develop (Fig. 15.3*a*). The fruiting body contains a mass of sterile hyphae intertwined with reproductive hyphae. The sterile hyphae protect the asci. A diploid (2n) nucleus is produced in each ascus when two haploid nuclei unite. This zygote nucleus undergoes meiosis and then divides mitotically to produce **ascospores**. In some cup fungi, the ascocarp is cup-shaped (Fig. 15.3*a*). In morels, the ascocarp is stalked and crowned by bell-shaped, convoluted tissue that bears the asci (Fig. 15.3*b*).

Figure 15.3 Sac fungi.
a. Cup fungus such as *Sarcoscypha* has a fruiting body shaped like a cup. **b.** A morel such as *Morchella* has a bell-shaped fruiting body.

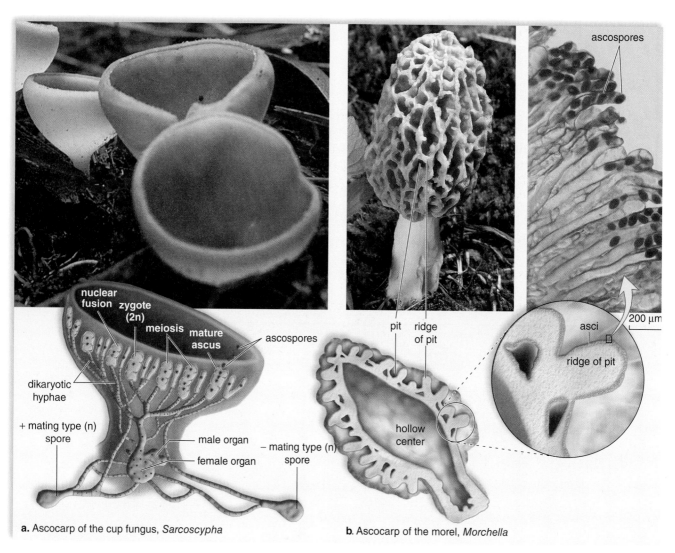

a. Ascocarp of the cup fungus, *Sarcoscypha*

b. Ascocarp of the morel, *Morchella*

Observation: Cup Fungi

View a prepared slide of *Peziza*, and identify the ascocarp, asci, ascospores, and sterile hyphae. Are the ascospores inside or outside the asci? _____

If available, obtain and view samples of morels and other cup fungi. The "pits" of a morel are lined with _____.

Conidiospores

When sac fungi reproduce asexually, they produce **conidiospores** (conidia) on upright hyphae known as conidiophores.

Observation: Conidiophores and Conidiospores

1. Obtain a petri dish of an *Aspergillus* culture. Use a dropper bottle of *methylene blue* to make a streak of stain 2.5 cm long on a microscope slide. Place the center of a 4 cm piece of clear tape over the *Aspergillus* culture so that a sample of the mold sticks to the tape. Put this mold sample on the microscope slide on which methylene blue stain has been placed. Observe with all three objectives of your microscope. What do you see?

> ⚠ **Methylene blue** Avoid ingestion, inhalation, and contact with skin, eyes, and mucous membranes. Exercise care in using this chemical. If any should spill on your skin, wash the area with mild soap and water. Methylene blue will also stain clothing. Follow your instructor's directions for its disposal.

2. Obtain a prepared slide of *Penicillium.* Locate the periphery of the mass under low power, and then switch to high power. You should now be able to see conidiophores.
3. View a prepared slide of *Aspergillus,* and observe how the conidiophore arrangement differs from that of *Penicillium.*
4. Label the following diagram of *Aspergillus:*

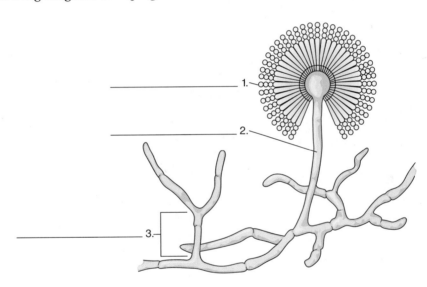

1. _____

2. _____

3. _____

15.3 Club Fungi

Club fungi are composed of septate hyphae. The phylum name refers to the **basidium,** a club-shaped structure that develops during sexual reproduction. What structure accounts for the phylum name

Basidiomycota (club fungi)? _____

Within each basidium, meiosis results in four basidiospores, which project from the basidium.

Study Figure 15.4, and note that when monokaryotic (n) hyphae from two different strains meet, they fuse to produce a dikaryotic (n + n) mycelium that, in turn, sometimes produces a fruiting body called a **basidiocarp.** A mushroom is a commonly known basidiocarp. In gill mushrooms, a button-shaped structure expands to become a full-sized mushroom, which consists of a stalk or stipe, and a terminal cap or pileus with gills on the underside. The gills are thin plates (lamellae) on which the basidia develop. Each basidium produces four basidiospores. Pore mushrooms have pores (tubes) instead of gills (Fig. 15.4).

Figure 15.4 Club fungi.

Life cycle of a mushroom. Sexual reproduction is the norm. After hyphae from two opposite mating types fuse, the dikaryotic mycelium is long-lasting. On the gills of a basidiocarp, nuclear fusion results in a diploid nucleus within each basidium. Meiosis and production of basidiospores follow. Germination of a spore results in a haploid mycelium.

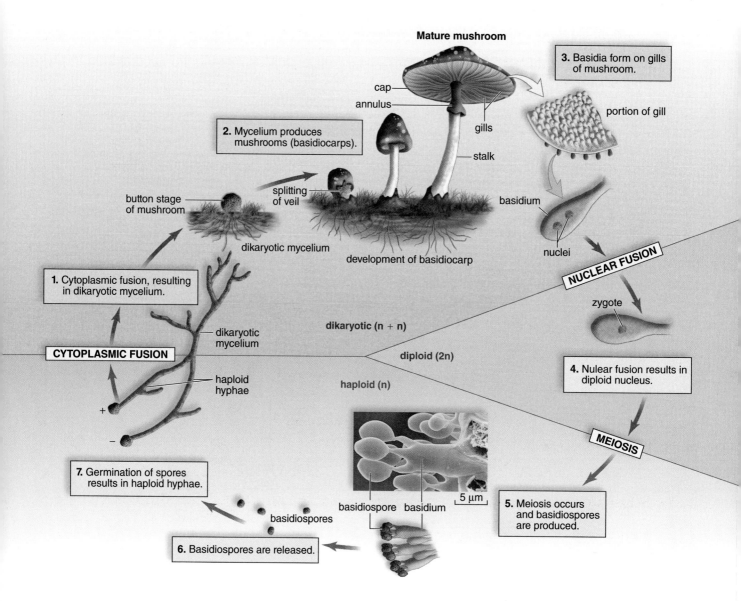

Observation: Mushrooms

1. Obtain an edible mushroom—for example, *Agaricus*—and identify as many of the following structures as possible:

   a. **Stalk:** the upright portion that supports the cap

   b. **Annulus:** a membrane surrounding the stalk where the immature (button-shaped) mushroom was attached

   c. **Cap:** the umbrella-shaped portion of the mushroom

   d. **Gills:** on the underside of the cap, radiating lamellae on which the basidia are located

   e. **Basidia:** on the gills, club-shaped structures where basidiospores are produced

   f. **Basidiospores:** spores produced by basidia

2. View a prepared slide of cross section of *Coprinus.* Using all three microscope objectives, look for the gills, basidia, and basidiospores.

3. Can you see individual hyphae in the gills? _____ Sketch what you see in the space below.

4. Are the basidiospores inside or outside of the basidia? _____

5. Can you suggest a reason for some of the basidia having fewer than four basidiospores? _____
   _____

6. What type of nuclear division(s) took place just before the basidiospores were produced? _____
   _____

7. What happens to the basidiospores after they are released? _____
   _____

15.4 Fungal Diversity

1. Table 15.1 may be helpful when comparing zygospore fungi, sac fungi, and club fungi.

Table 15.1 Comparisons of Fungi		
	Sexual Reproduction	**Asexual Reproduction**
Zygospore fungi	Zygospores	Sporangia
Sac fungi	Asci	Conidiospores
Club fungi	Basidia	Conidiospores

Notice that the formal name refers to the sexual reproductive structure. Why is each group so named?

Zygomycota _____

Ascomycota _____

Basidiomycota _____

2. After observing various fungi on display, complete Table 15.2 to identify and describe various specimens.

Table 15.2 Fungal Diversity			
Type	**Common Name**	**Scientific Name**	**Description**

15.5 Fungi as Symbionts

When two different organisms live and function together, they have a special relationship to each other called **symbiosis.** In **mutualism,** both symbionts benefit. In **parasitism,** one symbiont benefits from the relationship while the other is harmed. Two symbiotic relationships are studied: lichens and mycorrhizas.

Lichens

Lichens are an association between a fungus and a cyanobacterium or a green alga. They can be found growing on trees, rocks, dry soil, and most any harsh environment. The body of a lichen has three layers. The fungus forms a thin, tough upper layer and a loosely packed lower layer. These upper and lower fungal layers shield the photosynthetic cells in the middle layer. The exact relationship between the fungus and the photosynthetic cell is uncertain, but getting the fungus to grow alone is difficult. One possibility is that the fungus benefits from the relationship by utilizing substances produced by the algae and, in turn, the fungus provides its partner with water and a protected environment.

Does it seem as if the fungus is dependent on the photosynthetic cell? _____

What does this suggest? _____

Would you describe the relationship as mutualistic or parasitic? _____ Explain. _____

Lichens can be compact and crustose, shrublike and fruticose, or leaflike and foliose (Fig. 15.5).

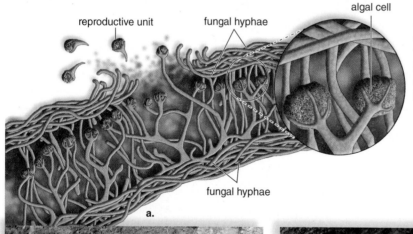

reproductive unit   fungal hyphae   algal cell

fungal hyphae

a.

Figure 15.5 Lichen morphology.
a. A section of a lichen shows the placement of the algal cells and the fungal hyphae, which encircle and penetrate the algal cells. **b.** Crustose lichens are compact. **c.** Fruticose lichens are shrublike. **d.** Foliose lichens are leaflike.

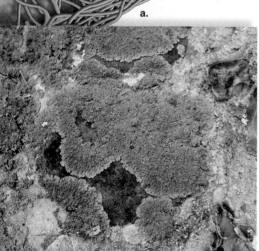

b. Crustose lichen, *Xanthoria*

sac fungi reproductive cups

c. Fruticose lichen, *Cladonia*

d. Foliose lichen, *Xanthoparmelia*

1. Observe numbered samples of lichens, and identify each as crustose, fruticose, or foliose. Check your answers with your instructor.

    Sample 1 _____

    Sample 2 _____

    Sample 3 _____

2. View a prepared slide of a lichen that shows the placement of the photosynthetic cells and

    fungal hyphae. Describe the placement. _____

    _____

Mycorrhizae

My**corrhizae** are mutualistic relationships between a fungus and the roots of most land plants. The fungus partner makes inorganic nutrients available to the land plant, and the plant, in turn, passes organic nutrients to the fungus. However, the relationship greatly aids the growth of plants as shown in Figure 15.6.

There are two types of mycorrhizae: ectomycorrhizae and endomycorrhizae. In **ectomycorrhizae,** the hyphae grow between cells, while in **endomycorrhizae,** the hyphae penetrate cells, where they may form treelike arbuscles or swellings (vesicles). Endomycorrhizae are far more numerous than ectomycorrhizae and the category arbuscular mycorrhizae fungi or AM fungi (Glomeromycota) is now used to include them in the classification of fungi (see page 194).

Observation: Mycorrhizae

View a prepared microscope slide of a mycorrhiza and observe the penetration of the root by the fungal

hyphae. What type of mycorrhiza are you observing? _____

Figure 15.6 Mycorrhizae.
Experimental results show that rough lemon plants with mycorrhizae *(right)* grow much better than plants without mycorrhizae *(left).*

1. Both fungi and animals are heterotrophs although they are markedly different in how they acquire

   nutrients. Explain. _____

   _____

2. Give the name of the sexual spore produced within or on each of the following structures, and cite the
   group of fungi.

	Spore Name	Group Name
a. Sporangium	_____	_____
b. Ascus	_____	_____
c. Basidium	_____	_____

3. The timing of meiosis during sexual reproduction of a fungus ensures that the adult is haploid. Explain.

   _____

4. Ascomycota produce what type of spore during *asexual* reproduction? _____

   _____

5. Which type of fungus studied is not composed of hyphae as are other fungi? _____ Explain.

   _____

6. A student is observing fruiting bodies. (a) If the spores are projecting from a club-shaped structure, the

   fungus is a _____.

   (b) If the spores are produced within a saclike structure, the fungus is a _____.

7. Of what benefit is it to a typical mushroom if the cap opens up when mature? _____

   _____

8. Mycorrhizae are sometimes referred to as "fungus roots." Explain. _____

   _____

9. Of what benefit are the mycorrhizae to the plant? _____

   _____

10. Explain the composition of lichens and why this composition is usually considered a mutualistic

    relationship. _____

    _____

16

Nonvascular Plants and Seedless Vascular Plants

Learning Outcomes

16.1 Evolution and Diversity of Land Plants
- Understand the relationship of charophytes to land plants and the main events in the evolution of plants as they became adapted to life on land.   206–7
- Describe the plant life cycle and the concept of the dominant generation.   208
- Contrast the role of meiosis in the plant and animal life cycles.   208
- Explain why you would expect the sporophyte generation to be dominant in a plant adapted to a land existence.   208

16.2 Nonvascular Plants
- Describe, in general, the nonvascular plants and how they are adapted to reproducing on land.   209–12
- Describe the life cycle of a moss, including the appearance of its two generations.   209–11
- Describe the liverworts, another group of nonvascular plants.   212

16.3 Seedless Vascular Plants
- Describe, in general, the vascular plants and how they are adapted to reproducing on land.   212–19
- Describe the lycophytes and pteridophytes.   213–15
- Describe the life cycle of a fern and the appearance of its two generations.   215–19

Introduction

Plants are multicellar photosynthetic eukaryotes; their evolution is marked by adaptations to a land existence. Among aquatic green algae, the **charophytes** are most closely related to the plants that now live on land. The charophytes have several features that would have promoted the evolution of land plants, including retention of and care of the zygote. The most successful land plants are those that protect all phases of reproduction (sperm, egg, zygote, and embryo) from drying out and have an efficient means of dispersing offspring on land.

In this laboratory, you will have the opportunity to examine various adaptations of plants to living on land. Although this lab will concentrate on land plant reproduction, you will see that much of a land plant's body is covered by a **waxy cuticle** that prevents water loss, and that most land plants are able to oppose the force of gravity and to lift their leaves up toward the sun. **Vascular tissue** offers support and transports water to, and nutrients from, the leaves.

Also, you will examine the adaptations of nonvascular plants and the seedless vascular plants to a land existence, and you will see that reproduction in these plants requires an outside source of water. In Laboratory 17, we will note that reproduction in seed plants is not dependent on external water at all.

16.1 Evolution and Diversity of Land Plants

Figure 16.1 shows the evolution of land plants. Circle the evolutionary events that led to adaptation of plants to a land existence.

Algal Ancestor of Land Plants

In the evolutionary tree, the charophytes are green algae that share a common ancestor with land plants. This common ancestor may have resembled a Charales, such as *Chara*, which can live in warm, shallow ponds that occasionally dry up. Adaptation to such periodic desiccation may have facilitated the ability of certain members of the common ancestor population to invade land, and with time, become the first land plants.

Observation: Chara

Examine a living *Chara* (Fig. 16.2). How does it superficially resemble a land plant?

_____

_____

Chara is a filamentous green algae that consists of a primary branch and many side branches (Fig. 16.2). Each branch has a series of very long cells. Note where one cell begins and the other begins. Measure the length of one

cell. _____ Gently pick up and handle *Chara* while you are examining it. What does it feel like? _____

_____ Its cell walls are covered with calcium carbonate deposit.

Conclusion: Chara

- What characteristics cause *Chara* to resemble land plants? _____

  _____

- Why are *Chara* called stoneworts? _____

  _____

DOMAIN: Eukarya
KINGDOM: Plants

CHARACTERISTICS
Multicellular, usually with specialized tissues; photosynthesizers that became adapted to living on land; most have alternation of generations life cycle.

Charophytes
Live in water; haploid life cycle; share certain traits with the land plants

LAND PLANTS (embryophytes)
Alternation of generation life cycle; protect a multicellular sporophyte embryo; gametangia produce gametes; apical tissue produces complex tissues; waxy cuticle prevents water loss.

Bryophytes (liverworts, hornworts, mosses)
Low-lying, nonvascular plants that prefer moist locations: Dominant gametophyte produces flagellated sperm; unbranched, dependent sporophyte produces windblown spores.

VASCULAR PLANTS (lycophytes, ferns and their allies, seed plants)
Dominant, branched sporophyte has vascular tissue: Lignified xylem transports water, and phloem transports organic nutrients; typically has roots, stems, and leaves; and gametophyte is eventually dependent on sporophyte.

Lycophytes (club mosses)
Leaves are microphylls with a single, unbranched vein; sporangia borne on sides of leaves produce windblown spores; independent and separate gametophyte produces flagellated sperm.

Ferns and Allies (pteridophytes)
Leaves are megaphylls with branched veins; dominant sporophyte produces windblown spores in sporangia borne on leaves; and independent and separate gametophyte produces flagellated sperm.

SEED PLANTS (gymnosperms and angiosperms)
Leaves are megaphylls; dominant sporophyte produces heterospores that become dependent male and female gametophytes. Male gametophyte is pollen grain and female gametophyte occurs within ovule, which becomes a seed.

Gymnosperms (cycads, ginkgoes, conifers, gnetophytes)
Usually large; cone-bearing; existing as trees in forests. Sporophyte bears pollen cones, which produce windblown pollen (male gametophyte), and seed cones, which produce seeds.

Angiosperms (flowering plants)
Diverse; live in all habitats. Sporophyte bears flowers, which produce pollen grains, and bear ovules within ovary. Following double fertilization, ovules become seeds that enclose a sporophyte embryo and endosperm (nutrient tissue). Fruit develops from ovary.

Figure 16.1 Evolutionary relationships among the plants.

The evolution of plants involves five significant innovations (embryo protection, vascular tissue, megaphylls, seeds, and flowers). Protection of a multicellular embryo was seen in the first plants to live on land.

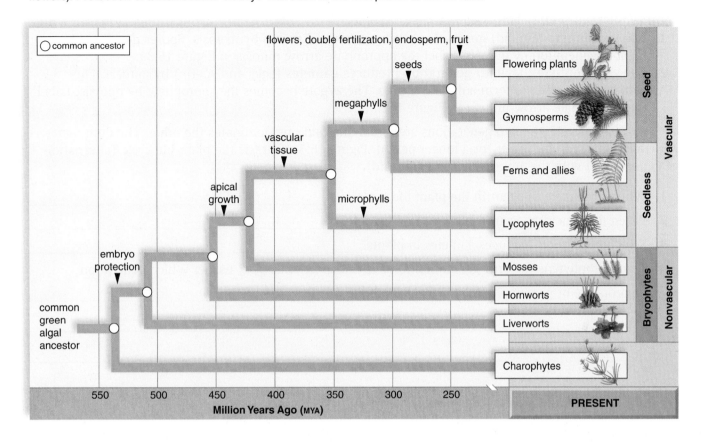

Figure 16.2 *Chara.*

Chara is an example of a stonewort, the type of green alga believed to be most closely related to the land plants.

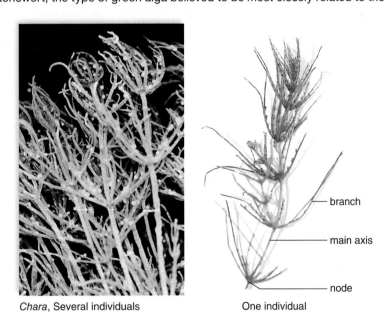

Chara, Several individuals   One individual

Alternation of Generations

Land plants have a two-generation life cycle called **alternation of generations** that involves sporic meiosis:

1. The **sporophyte** (diploid) **generation** produces haploid spores by meiosis. Spores develop into a haploid generation by mitosis. Label the appropriate arrow mitosis in Figure 16.3a.
2. The **gametophyte** (haploid) **generation** produces **gametes** (eggs and sperm) by mitosis. The gametes then unite to form a diploid zygote. The zygote becomes the sporophyte by mitosis. Label the appropriate arrow mitosis in Figure 16.3a.

In this life cycle, the two generations are dissimilar, and one dominates the other. The dominant generation is larger and exists for a longer period. Figure 16.3 contrasts the plant life cycle (alternation of generations) with the animal life cycle **(diploid).**

1. When does meiosis occur in the plant life cycle? _____

2. When does meiosis occur in the animal life cycle? _____

3. Which generation produces gametes in plants? _____

4. The sporophyte is the only generation to have transport (vascular) tissue. Which generation (sporophyte or gametophyte) is better adapted to a land environment? _____

   Explain. _____

Figure 16.3   Plant and animal life cycles.

a. The alternation of generations life cycle is typical of plants. Mitosis occurs as a spore becomes a gametophyte and as the zygote becomes a sporophyte. **b.** The diploid life cycle is typical of animals. Add mitosis to this diagram in one location only.

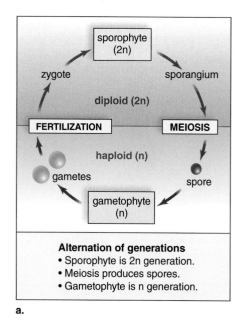
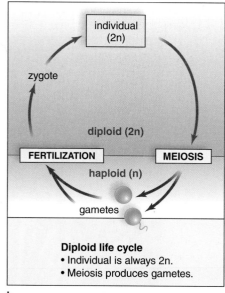

a.   b.

16.2 Nonvascular Plants

The nonvascular plants include the **mosses**, **liverworts**, and **hornworts**. The nonvascular plants are informally grouped together and called **bryophytes** (see Fig. 16.1). The gametophyte is dominant in all bryophtyes. The gametophyte produces eggs within archegonia and flagellated sperm in **antheridia.** The sperm swim to the egg, and the embryo develops within the **archegonium.** The nonvascular sporophyte grows out of the archegonium. The sporophyte is dependent on the female gametophyte (Fig. 16.4).

Figure 16.4 Moss life cycle.
In mosses, the haploid generation (gametophyte) is dominant.

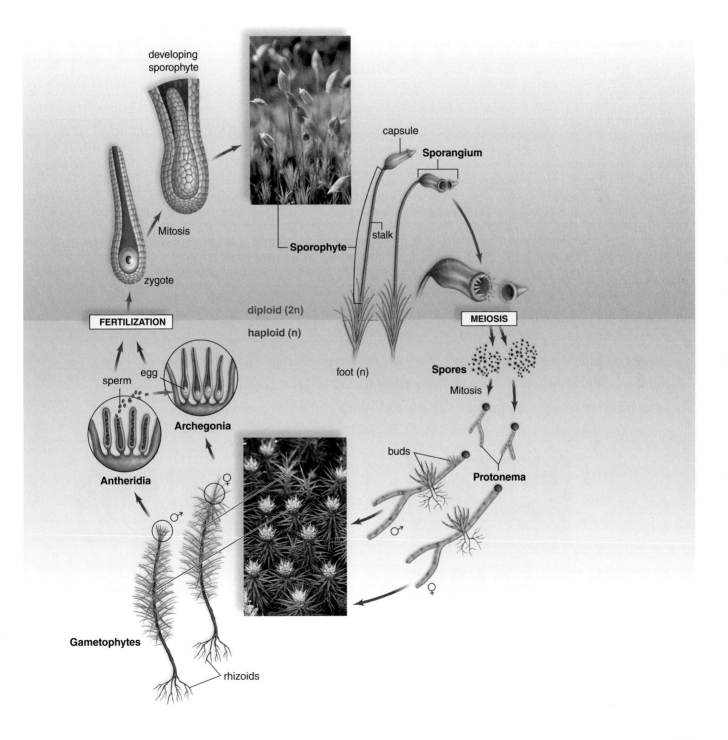

1. Put a check mark beside the phrases that describe nonvascular plants:

I	II
_____ No vascular tissue to transport water	_____ Vascular tissue to transport water
_____ Flagellated sperm that swim to egg	_____ Sperm protected from drying out
_____ Dominant gametophyte	_____ Dominant sporophyte

2. Which two of the features you checked cause nonvascular plants to be low-lying? _____

_____

3. Which listing of features (**I** or **II**) would you expect to find in a plant fully adapted to a land environment? _____ Explain. _____

4. In nonvascular plants, windblown spores are dispersal agents, and some species forcefully expel their spores. How are windblown spores an adaptation to reproduction on land? _____

_____

Observation: Moss Gametophyte

Living or Plastomount

Obtain a living moss gametophyte or a plastomount of this generation. Describe its appearance. _____

_____

   The leafy green shoots of a moss are said to lack true roots, stems, and leaves because, by definition, roots, stems, and leaves are structures that contain vascular tissue.

Microscope Slide

1. Study a slide of the top of a male moss shoot that contains antheridia, the reproductive structures where sperm are produced (Fig. 16.5). What is the chromosome number (choose 2n or n) of the sperm (see Fig. 16.4)? _____ Are the surrounding cells haploid or diploid? _____

2. Study a slide of the top of a female moss shoot that contains archegonia, the reproductive structures where eggs are produced (Fig. 16.6). What is the chromosome number of the egg? _____

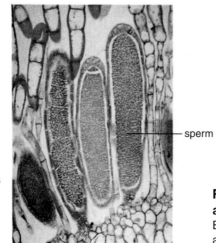

Figure 16.5 Moss antheridia.
Flagellated sperm are produced in antheridia.

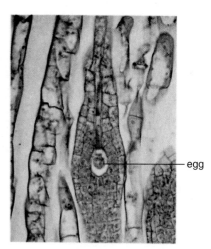

Figure 16.6 Moss archegonia.
Eggs are produced in archegonia.

Are the surrounding cells haploid or diploid? _____ When sperm swim from the antheridia to the archegonia, a zygote results. The zygote develops into the sporophyte. Is the sporophyte haploid or diploid? _____

Observation: Moss Sporophyte

Living Sporophyte

1. Examine the living sporophyte of a moss in a minimarsh, or obtain a plastomount of a female shoot with the sporophyte attached. Identify the **sporangium**, a capsule where spores are produced and released through a lid.
2. Bracket and label the gametophyte and sporophyte in Figure 16.7*a*. Place an n beside the gametophyte and a 2n beside the sporophyte.

Figure 16.7 Moss sporophyte.

a. The moss sporophyte is dependent on the female gametophyte. **b**. The sporophyte produces spores by meiosis.

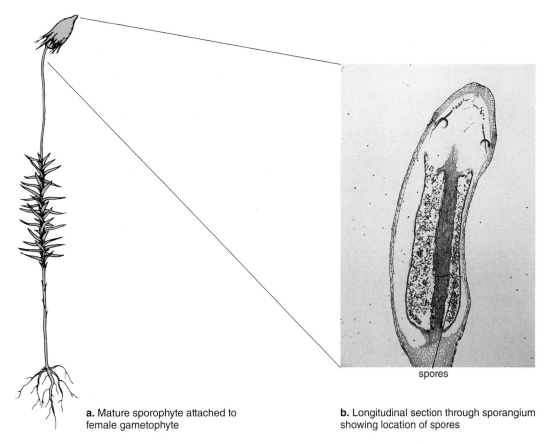

spores

a. Mature sporophyte attached to female gametophyte

b. Longitudinal section through sporangium showing location of spores

Microscope Slide

Examine a slide of a longitudinal section through a moss sporophyte (Fig. 16.7*b*). Identify the stalk and the sporangium, where spores are being produced. By what process are the spores being produced?

_____

When spores germinate, what generation begins to develop? _____

Why is it proper to say that spores are dispersal agents? _____

_____

Obtain a living sample of the liverwort *Marchantia* (Fig. 16.8) thallus, body of the plant. Examine it under the stereomicroscope. What generation is this sample? _____ Identify the following:

1. The gametophyte consisting of lobes. Each lobe is about a centimeter or so in length; the upper surface is smooth, and the lower surface bears numerous **rhizoids** (rootlike hairs).
2. **Gemma cups** on the upper surface of the thallus. These contain groups of cells called **gemmae** that can asexually start a new plant.
3. Disk-headed stalks bear antheridia, where flagellated sperm are produced.
4. Umbrella-headed stalks bear archegonia, where eggs are produced. Following fertilization, tiny sporophytes arise from the archegonia.

Figure 16.8 Liverwort, *Marchantia*.
a. Gemmae can detach and start a new plant. **b.** Antheridia are present in disk-shaped structures, and **c.** archegonia are present in umbrella-shaped structures.

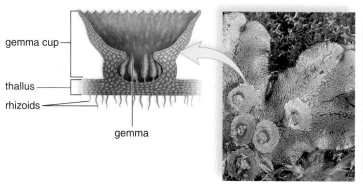

a. Thallus with gemma cups   b. Male gametophytes bear   c. Female gametophytes
                                        antheridia   bear archegonia

(labels: gemma cup, thallus, rhizoids, gemma)

16.3 Seedless Vascular Plants

Seedless vascular plants include the lycophytes (club mosses) and pteridophytes—whisk ferns *(Psilotum)*, horsetails *(Equisetum)*, and ferns (Filicales). These plants were prevalent and quite large during the Carboniferous period. At that time, the coal deposits still used today were formed.

The sporophyte is dominant in the seedless vascular plants. The dominant sporophyte has adaptations for living on land; it has vascular tissue and produces windblown spores. The spores develop into a separate gametophyte generation that is very small (less than 1 cm). The gametophyte generation lacks vascular tissue and produces flagellated sperm.

1. Place a check mark beside the phrases that describe seedless vascular plants:

I	II
_____ Independent gametophyte	_____ Gametophyte dependent on sporophyte, which has vascular tissue
_____ Flagellated sperm	_____ Sperm protected from drying out

2. Which listing (**I** or **II**) would you expect to find in a plant fully adapted to a land environment? _____ Explain. _____

_____

Are seedless vascular plants fully adapted to living on land? _____

Lycophytes are commonly called club mosses. Lycophytes are representative of the first vascular plants to have stems, roots, and leaves. The leaves are called **microphylls** because they have only one strand of vascular tissue. Among club mosses, you will examine the ground pines (*Lycopodium*) and the spike mosses (*Selaginella*).

Ground Pines

1. Examine a living or preserved specimen of *Lycopodium* (Fig. 16.9).
2. Note the shape and the size of the microphylls and the branches of the aerial (erect) stems.
3. Note the terminal clusters of leaves, called strobili, that are club-shaped and bear sporangia. The yellow color of the spore wall is caused by a pigment called sporopollenin. Sporopollenin protects against UV radiation and desiccation.
4. Label strobili, leaves, aerial stem, and root in Figure 16.9. Aerial stems are erect; rhizomes are horizontal stems.
5. Examine a prepared slide of a *Lycopodium* that shows the sporangia with spores inside. *Lycopodium* is homosporous, which means they produce one type of spore. The spore develops into a tiny microscopic gametophyte that remains in the soil.

Figure 16.9 *Lycopodium.*
In the club moss *Lycopodium,* green photosynthetic stems are covered by scalelike leaves, and spore-bearing leaves are clustered in strobili.

Spike Mosses

One type of spike moss that lives in the desert of western Texas is called the resurrection plant because it dries out completely but will become green again within hours of receiving plentiful water. Even though nonseed plants can live in dry environments, they require an outside source of moisture to reproduce because they produce flagellated (swimming) sperm.

1. Examine living or preserved specimens of *Selaginella* (Fig. 16.10).
2. Examine a prepared slide of a *Selaginella* strobilus. Note that *Selaginella* produces two types of spores and is, therefore, called heterosporous. One spore is a microspore that develops into a male gametophyte and the other type of spore is a megaspore that develops into a female gametophyte.

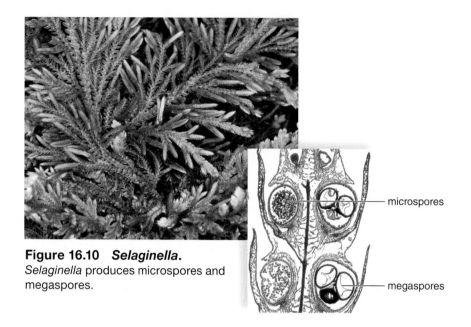

Figure 16.10 *Selaginella*.
Selaginella produces microspores and megaspores.

— microspores

— megaspores

Observation: Pteridophytes

Molecular (DNA) studies tell us that the whisk ferns, horsetails, and ferns are closely related.

Whisk Ferns

Psilotum (phylum Psilotophyta) is representative of whisk ferns, named for their resemblance to whisk brooms.

1. Examine a preserved specimen of *Psilotum*, and note that it has no leaves. The underground stem, called a **rhizome,** gives off upright, aerial stems with a dichotomous branching pattern, where bulbous sporangia are located (Fig. 16.11).

   What generation are you examining? _____

2. Label the aerial stem, the rhizome, and the sporangia in Figure 16.11*b*.

Figure 16.11   Whisk fern, *Psilotum.*
This whisk fern has no roots or leaves—the branches carry on photosynthesis. The sporangia are yellow.

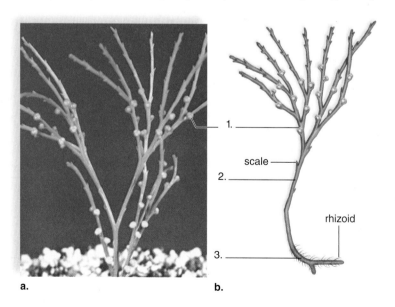

1. _____

scale —

2. _____

rhizoid

3. _____

a.   b.

Horsetails

In horsetails, a rhizome produces aerial stems that stand about 1.3 meters.

1. Examine *Equisetum*, a horsetail, and note the minute, scalelike leaves (Fig. 16.12).
2. Feel the stem. *Equisetum* contains a large amount of silica in its stem. For this reason, these plants are sometimes called scouring rushes and may be used by campers for scouring pots.
3. Strobili appear at the tips of the stems, or else special buff-colored stems bear the strobili. Sporangia are in the strobili.
4. Label strobilus, branches, leaves, rhizomes, and root in Figure 16.12.

Figure 16.12 Horsetail.
In the horsetail *Equisetum,* whorls of branches or tiny leaves appear in the joints of the stem. The sporangia are borne in strobili.

1. _____
2. _____
3. _____
4. _____
5. _____

Ferns

Ferns are quite diverse and range in size from those that are low growing and resemble mosses to those that are as tall as trees. The rhizome grows horizontally, which allows ferns to spread without sexual reproduction (Fig. 16.13). The gametophyte (called a **prothallus**) is small (about 0.5 cm) and usually heart-shaped. The prothallus contains both archegonia and antheridia. Ferns are largely restricted to moist, shady habitats because sexual reproduction requires adequate moisture. Why in particular does sexual reproduction require moisture? _____

_____

Figure 16.13 Fern life cycle.

The sporophyte is the frond, and the gametophyte is the heart-shaped prothallus.

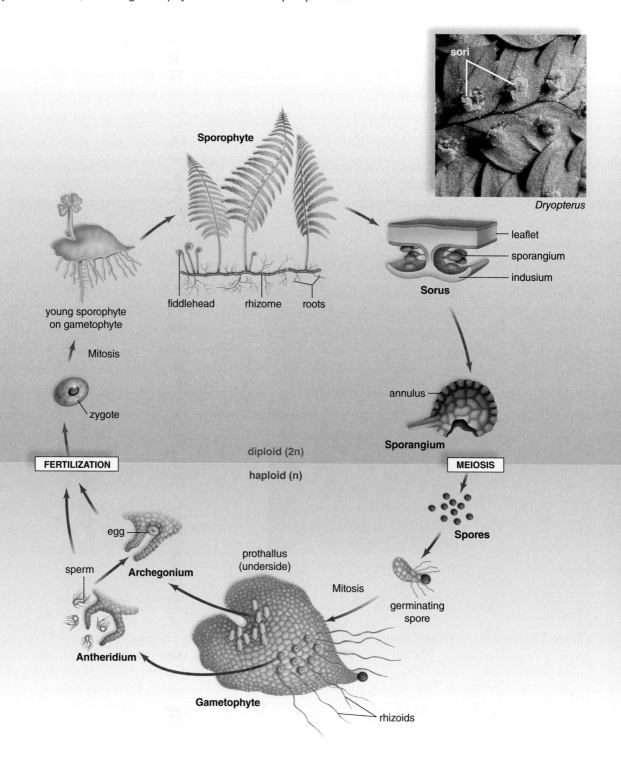

Dryopterus

Sporophyte

young sporophyte on gametophyte

fiddlehead rhizome roots

leaflet
sporangium
indusium

Sorus

Mitosis

zygote

annulus

Sporangium

diploid (2n)

FERTILIZATION

haploid (n)

MEIOSIS

Spores

egg

Archegonium

sperm

prothallus (underside)

Mitosis

germinating spore

Antheridium

Gametophyte

rhizoids

Fern spores are produced by meiosis in structures called sporangia, which in many species, occur on the underside of large leaves called **fronds.** These leaves are called **megaphylls** because they are broad leaves with several strands of vascular tissue. The spores are released when special cells of the **annulus** (a line or ring of thickened cells on the outside of the sporangium) dry out, and the sporangium opens. How do ferns disperse offspring? _____

Observe the ferns on display, and then complete Table 16.1.

Table 16.1 Fern Diversity	
Type of Fern	Description of Frond

Observation: Fern Sporophyte

Study the life cycle of the fern (Fig. 16.13), and find the sporophyte generation. This large, complexly divided leaf is known as a frond. Fronds arise from an underground stem called a rhizome.

Living or Preserved Frond

Examine a living or preserved specimen of a frond, and on the underside, notice a brownish clump called a **sorus** (pl., sori), each a cluster of many sporangia (Fig. 16.14).

What is being produced in the sporangia? _____

Given that this is the generation called the fern, what generation is dominant in ferns? _____

Figure 16.14 Underside of frond leaflets.
Sori occur on the underside of frond leaflets.

sorus

Microscope Slide of Sorus

1. Examine a prepared slide of a cross section of a frond leaflet. Using Figure 16.15 as a guide, locate the fern leaf above and the sorus below.

2. Within the sorus, find the sporangia and spores. Look for an **indusium** (not present in all species), a shelflike structure that protects the sporangia until they are mature. Does this fern have an indusium? _____

Figure 16.15 Cross section of a frond leaflet.
Micrograph of the internal anatomy of a sorus depicts many sporangia, where spores are produced.

leaflet

sporagium

spore

Observation: Fern Gametophyte

Plastomount

1. Examine a plastomount showing the fern life cycle.

2. Notice the **prothallus**, a small, heart-shaped structure. Can you find this structure in your fern minimarsh (if available)? _____ The prothallus is the gametophyte generation of the fern. Most persons do not realize that this structure exists as a part of the fern life cycle. What is the function of this structure? _____

_____

Microscope Slide of Prothallus

1. Examine whole-mount slides labeled fern prothallium-archegonia and fern prothallium-antheridia (Fig. 16.16).

2. If you focus up and down very carefully on an archegonium, you may be able to see an egg inside. What is being produced inside an antheridium? _____ When sperm produced in an antheridium swim to the archegonium in a film of water, what results? _____ This structure develops into what generation? _____

Conclusions: Ferns

- How are ferns dispersed from one area to another? _____

- Is either generation in the fern dependent for any length of time on the other generation? _____ Explain. _____

Figure 16.16 Fern prothallus.

The underside of the heart-shaped fern prothallus contains archegonia where eggs are produced and antheridia where flagellated sperm are produced.

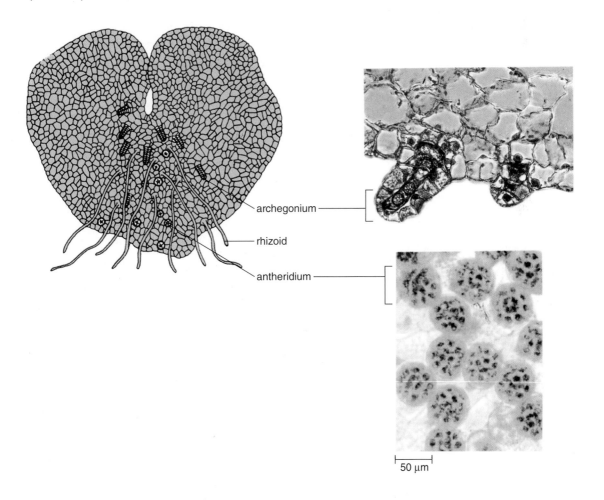

archegonium

rhizoid

antheridium

50 μm

1. Name and describe the life cycle of plants. _____

   _____

   _____

2. Contrast the life cycle of plants to that of animals (e.g., human beings).

	Plants	Animals
Typical life cycle		
Resulting structure from meiosis		
Occurrence of mitosis		

Compare mosses and ferns by completing the following table:

	Dominant Generation	Vascular Tissue	Flagellated Sperm	Method of Offspring Dispersal
3. Moss				
4. Fern				

5. Why do some biologists call the nonvascular plants the "amphibians of the plant kingdom"? _____

   _____

6. Are ferns better adapted to a land environment than mosses? _____

   Why or why not? _____

7. How do both mosses and ferns disperse offspring? _____

   _____

8. How is the gametophyte in ferns similar to that of mosses? _____

   _____

   How is it different (aside from appearance)? _____

   _____

9. Why is it beneficial to have the sporophyte generation dominant in vascular plants? _____

   _____

10. Compare and contrast moss and fern sporophytes. _____

   _____

Biology Website

Enhance your study of the text and laboratory manual with study tools, practice tests, and virtual labs. Also ask your instructor about the resources available through ConnectPlus, including the media-rich eBook, interactive learning tools, and animations.

www.mhhe.com/maderbiology11

McGraw-Hill Access Science Website

An Encyclopedia of Science and Technology Online which provides more information including videos that can enhance the laboratory experience.

www.accessscience.com

LABORATORY

17

Seed Plants

Learning Outcomes

17.1 Life Cycle of Seed Plants
- Compare the life cycle of nonseed and seed plants.   222
- List significant innovations of seed plants.   222

17.2 Gymnosperms
- Name the four groups of gymnosperms and give the distinguishing features of each group.   223–25
- Distinguish between pollination and fertilization in the life cycle of a pine tree.   226–27
- Identify the structure and function of pollen cones and seed cones, and tell where you would find the male and female gametophytes and seeds.   227–29

17.3 Angiosperms
- Identify the parts of a flower and compare eudicot flowers and monocot flowers.   230-31
- Identify which part of a flower contains pollen sacs and which contains ovules, and tell where you would find the male and female gametophytes and seeds.   231–34

17.4 Comparison of Gymnosperms and Angiosperms
- Compare the adaptations of flowering plants to those of conifers.   235
- Contrast the life cycle of gymnosperms and angiosperms.   235

> **Planning Ahead** Your instructor may suggest that you prepare the pollen grain slide (page 229) at the start of the laboratory session.

Introduction

Among plants, **gymnosperms** and **angiosperms** are seed plants. Review the plant evolutionary tree (see Fig. 16.1) and note that the gymnosperms and angiosperms share a common ancestor, which produced **seeds,** a structure that contains the next sporophyte generation. We shall see that the use of seeds to disperse the next generation requires a major overhaul of the alternation of generations life cycle. Nonseed plants disperse the gametophyte, the haploid (n) generation, by production of spores. Seed plants disperse the sporophyte, the diploid (2n) generation. Which generation—gametophyte or sporophyte—is better

adapted to a land environment because it contains vascular tissue? _____

Nonseed plants utilize an archegonium to protect the egg; and following fertilization, the sporophyte develops immediately within the archegonium. Seed plants protect the entire female gametophyte within an **ovule**. Following fertilization the ovule develops into a sporophyte-containing seed. You will see that the formation of the ovule and also the **pollen grains** (male gametophytes) are radical innovations in the life cycle of seed plants. In gymnosperms, pollen grains are windblown, but many flowering plants have a mutualistic relationship with animals, particularly flying insects, which disperse their pollen. Various animals also help disperse the seeds of many flowering plants. Relationships with animals help explain why flowering plants are so diverse and widespread today.

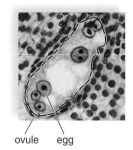

ovule   egg

Female gametophyte of angiosperms

17.1 Life Cycle of Seed Plants

Figure 17.1 shows alternation of generations as it occurs in nonseed plants, and Figure 17.2 shows alternation of generations as it occurs in seed plants. Note which structures are haploid and which are diploid in these diagrams.

1. In which life cycle, nonseed or seed, do you note pollen sacs (microsporangia) and ovules

   (megasporangia)? _____ In which life cycle, nonseed or seed, do you note two types of

   spores, microspores, and megaspores? _____ The formation of **heterospores** (unlike spores)

   is an innovation that leads to the production of pollen grains and the formation of ovules in seed
   plants. *Label heterospores where appropriate in Figure 17.2.* In which life cycle do you note male

   gametophyte (pollen grain) and female gametophyte (embryo sac in ovule)? _____

2. In nonseed plants, flagellated sperm must swim in external water to the egg. In seed plants,
   pollination (the transport of male gametophytes by wind or pollen carrier to the vicinity of
   female gametophytes) does not require external water. This is an innovation in seed plants. *Label
   pollination where appropriate in Figure 17.2.*

3. In which life cycle does a **seed** appear between the zygote and the sporophyte? _____

   What generation is present in a seed? _____ Formation of a seed from an ovule

   following fertilization is an innovation in seed plants.

Figure 17.2   Alternation of generations in flowering plants, which are seed plants.

Figure 17.1   Alternation of generations in nonseed plants.

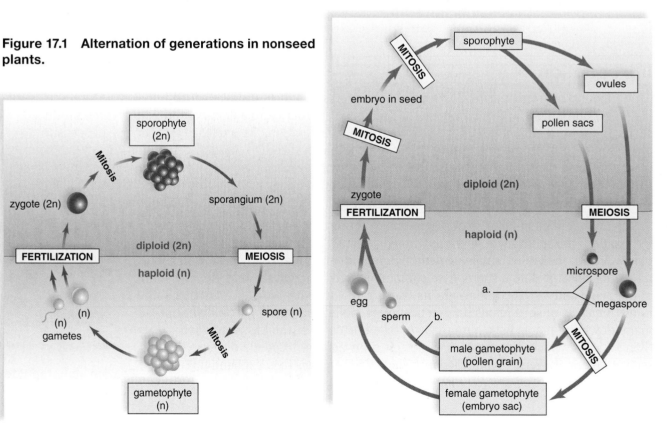

17.2 Gymnosperms

The term *gymnosperm* means "naked seed." The ovules and seeds of gymnosperm are exposed on a cone scale or some other comparable structure. Most gymnosperms do produce cones. The four groups of gymnosperms are:

Cycadophyta: cycads   Gnetophyta: gnetophytes
Ginkgophyta: ginkgoes   Coniferophyta: conifers

Cycads

Cycads are cone-bearing, palmlike plants found today mainly in tropical and subtropical regions. Cycads flourished during the Mesozoic era. They were so numerous that this era is sometimes referred to as the Age of Cycads and Dinosaurs. Herbivorous dinosaurs most likely fed on cycad seeds and foliage.

Observation: Cycads

Among cycads, male and female plants are separate. If available, examine cycad specimens, including *Zamia* and *Cycas*. If you are examining *Zamia*, note the male and female cones. If you are examining *Cycas*, note that only the male plant bears cones; the female reproductive structure is much more leaflike.

After examining cycads and the following photograph, give three characteristics you could use to recognize a cycad.

1. _____
2. _____
3. _____

Ginkgoes

Only one species of **ginkgo**—the maidenhair tree—survives today. The maidenhair tree was largely restricted to ornamental gardens in China until it was found to do quite well in polluted areas. Since the female trees produce rather smelly seeds, usually only male trees are used as ornamentals, such as those planted in city parks.

Observation: The Maidenhair Tree

The maidenhair tree can be quite tall (12–25 meters). A few of the lower branches may be pendulous. The fan-shaped leaves turn a vivid yellow in the fall. The tree is deciduous. The male tree bears cones, and the female tree produces the seeds later surrounded by a fleshy covering.

After examining ginkgo leaves and the following photograph, give three characteristics you could use to recognize a male ginkgo tree.

1. _____

2. _____

3. _____

cone

Cones of male tree

Mature maidenhair tree

Leaves of tree

Gnetophytes

Gnetophytes are seed-bearing plants that can grow as shrubs, trees, or vines and share similarities with both gymnosperms and angiosperms. Three genera survive today: *Ephedra* is a shrub found in arid regions of the southwestern United States. It has small, scalelike leaves. *Gnetum* is a vine or tree found in tropical rainforests of Asia, South America, and Africa. Its large, leathery leaves resemble those of angiosperms. *Welwitschia* is found only in South Africa. A woody central disk has only two enormous straplike leaves that split lengthwise with age. Cone-bearing branches arise from the margin of the disk. Most of the plant is buried in sandy soil. Molecular data supports the long-held belief that these plants are related.

If available, examine gnetophyte specimens and then label the following photographs on the lines provided.

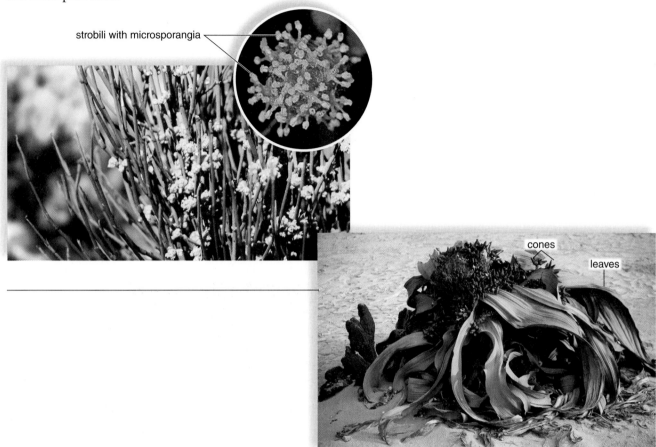

strobili with microsporangia

cones

leaves

Conifers

The **conifers** are by far the largest group of gymnosperms. Pines, hemlocks, and spruces are evergreen conifers because their leaves remain on the tree through all seasons. The cypress tree and larch (tamarack) are examples of conifers that are not evergreen.

The gymnosperms have the first real development of **wood,** which is derived from dead transport tissue. Pines in particular are grown to serve as sources for wood and paper.

seed cones

pollen cones

seed cones

a. A northern coniferous forest of evergreen trees   **b.** Cones of lodgepole pine, *Pinus contorta*   **c.** Fleshy seed cones of juniper, *Juniperus*

Figure 17.3 Pine life cycle.

The sporophyte is the tree. The male gametophytes are windblown pollen grains shed by pollen cones. The female gametophytes are retained within ovules on seed cones. The ovules develop into windblown seeds.

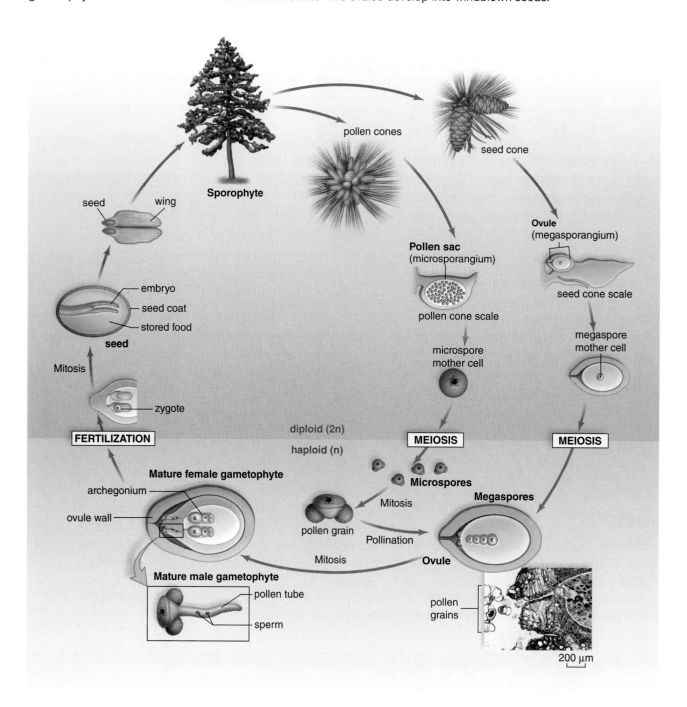

Pine Trees

The pine tree is the dominant sporophyte generation (Fig. 17.3) in the life cycle of a pine. Vascular tissue extends from the roots, through the stem, to the leaves. Pine tree leaves are needlelike; leathery; and covered with a waxy, resinous cuticle. The **stomata,** openings in the leaves for gas exchange, are sunken. The structure of the leaf and the leaf's internal anatomy are adaptive to a drier climate.

In pine trees, the **pollen sacs** and **ovules** are located in cones. Pollen sacs on pollen cone scales contain microspore mother cells which undergo meiosis to produce **microspores.** Microspores (n) become sperm-bearing **male gametophytes** (pollen grains). Seed cone scales bear ovules where megaspore mother cells undergo meiosis to produce **megaspores.** A megaspore (n) becomes an egg-bearing **female gametophyte.**

Pollination occurs when pollen grains are windblown to the seed cones. After **fertilization,** the egg becomes a sporophyte (2n) embryo enclosed within the ovule, which develops a seed coat. The seeds are winged and are dispersed by the wind.

1. Which part of the pine life cycle is an adult sporophyte? _____

2. Which part of the pine life cycle is the male gametophyte? The female gametophyte?

   _____

   _____

3. Where does fertilization occur and what structure becomes a seed? _____

   _____

Observation: Pine Leaf

Obtain a cluster of pine leaves (needles). A very short woody stem is at its base. Each type of pine has a typical number of leaves in a cluster (Fig. 17.4). How many leaves are in the cluster you are examining?

_____ What is the common name of your specimen? _____

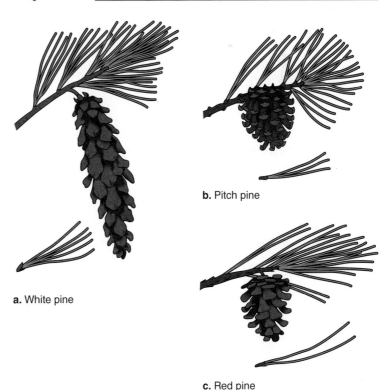

a. White pine

b. Pitch pine

c. Red pine

Figure 17.4   Pine leaves (needles).
a. The needles of white pines are in clusters of five. **b.** The needles of pitch pines are in clusters of three. **c.** The needles of red pines are in clusters of two.

Observation: Pine Cones

Preserved Cones

1. Compare a pine pollen cone to a pine seed cone. _____

   _____

   _____

   Note the size and texture of the pollen cones relative to the larger, woody, seed cones.

2. Pollen cones. Remove a single scale (sporophyll) from the pollen cone and examine with a stereomicroscope. Note the two pollen sacs on the lower surface of each scale (Fig. 17.5*a*). What do the pollen sacs produce? ____

   _____

   _____

3. Seed cones.

   a. Seed cones of three distinct ages are present. First-year cones are about 1 cm wide. Second-year cones are about 10 cm long. They are green and the scales are tightly closed together. In third-year cones the scales have opened, revealing the mature seeds at the base of the cone scale.

Figure 17.5   Pine cones.
a. The scales of pollen cones bear pollen sacs where microspores become pollen grains. **b.** The scales of seed cones bear ovules that develop into winged seeds.

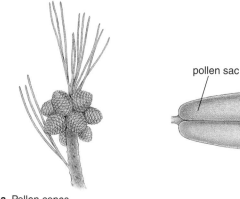

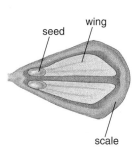

a. Pollen cones

b. Seed cone

   b. If available, examine a first-year cone cut lengthwise. The ovules are visible as small, milky-white domes at the base of the cone scale. Ovules will hold the female gametophyte generation, and then they become seeds.

   c. Examine mature seed cones. See if any seeds are present (Fig. 17.5*b*). Where are they located?

   _____ Note the hardness of the seed coat.

   What is the function of the seed coat? _____

   Are the seeds covered by tissue donated by the original sporophyte? _____

   What does gymnosperm mean? _____

   If instructed to do so, use tweezers to pull out a seed and note the wing. What is the wing for?

   _____

   Replace the seed in the cone when you are finished.

4. Pine seeds. If available, examine a pine seed (called a pine nut) that has the seed coat removed. These seeds can be used for cooking foods such as pesto. Carefully cut the seed lengthwise and examine. Can you find an embryo inside? _____

Microscope Slides of Pine Cones

1. Examine a prepared slide of a longitudinal section through a mature pine pollen cone. Label a pollen sac in Figure 17.6*a* and a pollen grain in Figure 17.6*b*. A pollen grain (male gametophyte) has a central body and two attached hollow bladders. How do these help in the dispersal of pine pollen? _____

_____

The central body has two cells. One cell will divide to become two nonflagellated sperm, one of which fertilizes the egg after pollination. The other cell forms the **pollen tube** through which a sperm travels to the egg.

2. Examine a prepared slide of a longitudinal section through an immature pine seed cone. As mentioned, seed cone scales bear ovules. The ovule contains a megaspore mother cell, which undergoes meiosis to produce four megaspores, three of which disintegrate. This megaspore (n) becomes an egg-bearing female gametophyte. Label the ovule and the megaspore mother cell in Figure 17.7. Also, label the pollen grains that you can see just outside the ovule.

Figure 17.6 Pine pollen cone.

Pollen cones bear **a.** pollen sacs (microsporangia) in which microspores develop into pollen grains. **b.** Enlargement of pollen grains (male gametophytes).

1. _____

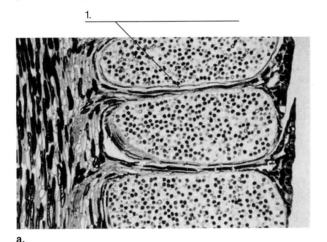

a.

2. _____

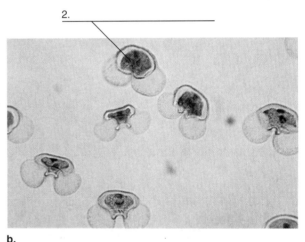

b.

Figure 17.7 Seed cone.
Seed cones bear ovules, each of which will contain a female gametophyte. Note pollen grains near the entrance.

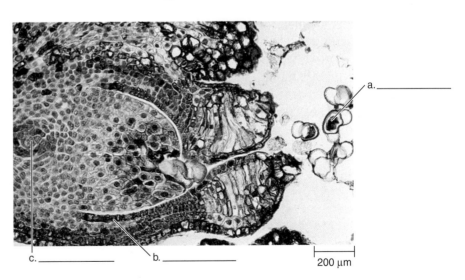

a. _____

c. _____ b. _____

200 μm

17.3 Angiosperms

Flowering plants are the dominant plants today. They occur as trees, shrubs, vines, and garden plants. At some point in their life cycle, all flowering plants bear flowers (Fig. 17.8).

Figure 17.8 Generalized flower.
A flower has four main kinds of parts: sepals, petals, stamens, and a carpel. A stamen has an anther and filament. A carpel has a stigma, style, and ovary. An ovary contains ovules.

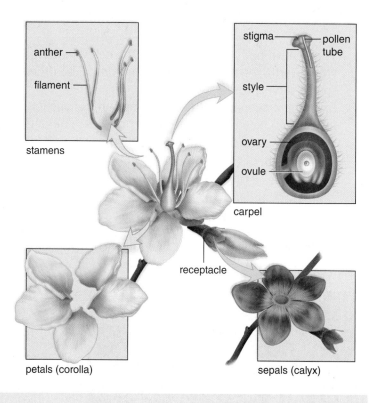

Observation: A Flower

1. With the help of Figure 17.8, identify the following structures on a model of a flower:
   a. **Receptacle:** The portion of a stalk to which the flower parts are attached.
   b. **Sepals:** An outermost whorl of modified leaves, collectively called the **calyx.** Sepals are green in most flowers. They protect a bud before it opens.
   c. **Petals:** Usually colored leaves that collectively constitute the **corolla.**
   d. **Stamen:** A swollen terminal **anther** and the slender supporting **filament.** The anther contains two pollen sacs, where microspores develop into male gametophytes (pollen grains).
   e. **Carpel:** A modified sporophyll consisting of a swollen basal ovary; a long, slender **style** (stalk); and a terminal **stigma** (sticky knob).
   f. **Ovary:** The enlarged part of the carpel that develops into a fruit.
   g. **Ovule:** The structure within the ovary where a megaspore develops into a female gametophyte (embryo sac). The ovule becomes a seed.

2. Carefully inspect a fresh flower. What is the common name of your flower? _____

3. Remove the sepals and petals by breaking them off at the base. How many sepals and petals are there? _____

4. Are the stamens taller than the carpel? _____

5. Remove a stamen, and touch the anther to a drop of water on a slide. If nothing comes off in the water, crush the anther a little to release some of its contents. Place a coverslip on the drop, and observe with low- and high-power magnification. What are you observing? _____

6. Remove the carpel by cutting it free just below the base. Make a series of thin cross sections through the ovary. The ovary is hollow, and you can see nearly spherical bodies inside. What are these bodies? _____

   _____

7. Flowering plants are divided into two groups, called **monocots** and **eudicots**, as discussed more fully in Laboratory 18. Table 17.1 lists significant differences between the two classes of plants. Is your flower a monocot or eudicot? _____

   _____

Table 17.1 Monocots and Eudicots	
Monocots	**Eudicots**
One cotyledon	Two cotyledons
Flower parts in threes or multiples of three	Flower parts in fours or fives or multiples of four or five
Usually herbaceous	Woody or herbaceous
Usually parallel venation	Usually net venation
Scattered bundles in stem	Vascular bundles in a ring
Never woody	Can be woody

Life Cycle of Flowering Plants

The life cycle of a flowering plant (Fig. 17.9) is like that of the pine tree except for these innovations:

- In angiosperms, the often brightly colored flower contains the pollen sacs and ovules. Locate these structures in Figure 17.9, and trace the life cycle of flowering plants from the sporophyte generation (the tree) through the various stages to the sporophyte generation once again.

- **Pollination** in flowering plants—when pollen is delivered to the stigma of the carpel—is sometimes accomplished by wind but more likely by the assistance of an animal pollinator. The pollinator acquires nutrients (e.g., nectar) from the flower and inadvertantly collects pollen, which it takes to the next flower.

- Notice that flowering plants, unlike gymnosperms, practice **double fertilization.** A mature pollen grain contains two sperm; one fertilizes the egg, and the other joins with the two polar nuclei to form **endosperm** (3n), which serves as food for the developing embryo.

- Also, flowering plants produce seeds the same as gymnosperm, but their seeds are enclosed within fruits. **Fruits** protect the seeds and aid in seed dispersal. Sometimes, animals eat the fruits, and after the digestion process, the seeds are deposited far away from the parent plant. The term *angiosperm* means "covered seeds." The seeds of angiosperm are found in fruits, which develop from parts of the flower, particularly the ovary.

Figure 17.9 Flowering plant life cycle.

The parts of the flower involved in reproduction are the anthers of stamens and the ovules in the ovary of a carpel.

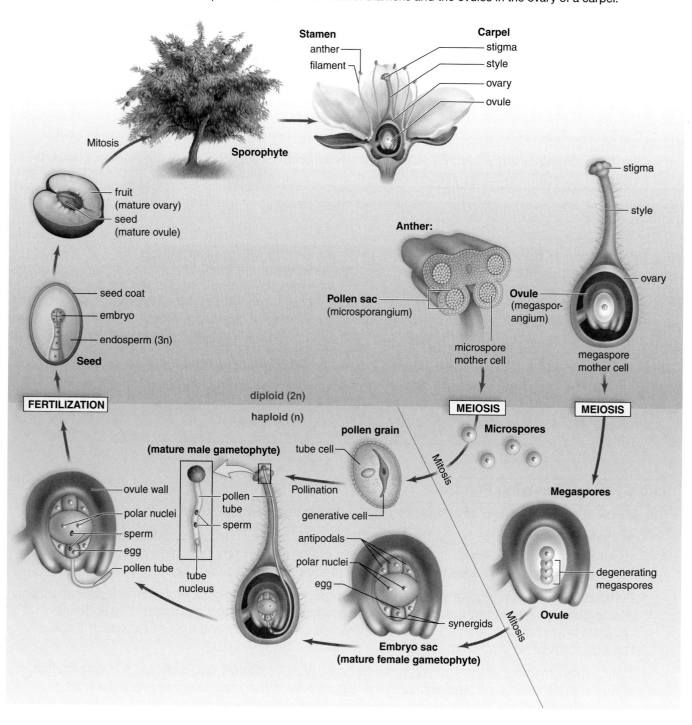

The Male Gametophyte

Notice that in the life cycle of flowering plants (Fig. 17.9), meiosis in pollen sacs produces four microspores, each of which will become a two-celled pollen grain. In flowering plants, _____ is the transfer of the pollen grain from the anther to the stigma, where the pollen grain germinates and becomes the mature male gametophyte. By what means is pollination accomplished in the flowering plant life cycle? _____

Observation: Pollen Grain Slide

1. Pollen grain slide before germination. In this observation, you will observe nongerminated pollen grains and germinated pollen grains. A pollen grain newly released from the anther has two cells. The larger of the two cells is the **tube cell,** and the smaller is the **generative cell.** Obtain and examine a prepared slide of pollen grains. Identify the tube cell and the generative cell. Sketch your observation here.

2. Pollen grain slide following germination. During germination, the pollen grain's tube cell gives rise to the pollen tube. As it grows, the pollen tube passes through the stigma and grows through the style of the carpel and into the ovary. Two sperm cells produced by division of the pollen grain's generative cell migrate through the pollen tube into the embryo sac. Obtain and examine a prepared slide of germinated pollen grains with pollen tubes. You should be able to see the tube nucleus and two sperm cells. What portion of a germinated pollen grain tells you that it is the mature male gametophyte?

   _____

*Experimental Procedure: Pollen Grains

Inoculate an agar-coated microscope slide with pollen grains. Invert the slide, and place it onto a pair of wooden supports in a covered petri dish. Leave the inverted slide in the covered petri dish for one hour. Then remove the slide, and examine it with the compound light microscope. Have any of the pollen grains germinated? _____

   If so, describe. _____

_____

*Experimental Procedure from Richard Carter and Wayne R. Faircloth, *General Laboratory Studies,* Lab 17.2, 1991. Used with permission of Kendall/Hunt Publishing Company.

The Female Gametophyte

In the ovule, the surviving megaspore undergoes three mitotic divisions to produce a seven-celled (eight-nuclei) structure called an **embryo sac** (Fig. 17.10). One of these cells is an egg cell. The largest cell contains two **polar nuclei.** The embryo sac is the mature female gametophyte.

What portion of the embryo sac tells you that it is the female gametophyte? _____

_____

Observation: Embryo Sac Slide

1. Examine the demonstration slide of the mature embryo sac of *Lilium.* Identify the egg labeled in Figure 17.10.
2. When the pollen tube delivers sperm to the embryo sac, double fertilization, as defined previously, occurs. Due to the first fertilization, what happens to the egg?_____

   _____

   Due to the second fertilization, what happens to the polar nuclei? _____

   _____

3. Following fertilization, the ovules develop into seeds, and the ovary develops into the fruit (see Laboratory 21, pages 284–86, for fruits). What are the three parts of a seed? _____

   _____

Figure 17.10   Embryo sac in a lily ovule.

An embryo sac is the female gametophyte of flowering plants. It contains seven cells, one of which is the egg.

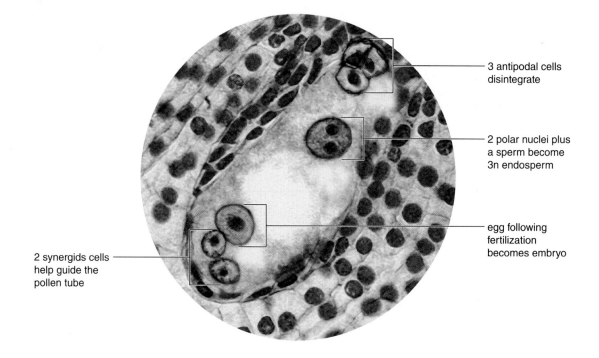

3 antipodal cells disintegrate

2 polar nuclei plus a sperm become 3n endosperm

egg following fertilization becomes embryo

2 synergids cells help guide the pollen tube

17.4 Comparison of Gymnosperms and Angiosperms

1. Beneath the photos, list ways to tell a gymnosperm from an angiosperm.

a.

b.

_____   _____

_____   _____

_____   _____

2. Complete Table 17.2 using "yes" or "no" to compare gymnosperms to angiosperms.

Table 17.2 Comparison of Gymnosperms and Angiosperms					
	Heterospores	Pollen grains/ Ovule	Cones	Flower	Fruit
Gymnosperms					
Angiosperms					

3. What structure in gymnosperms and angiosperms delivers sperm from pollen sacs to the vicinity of the egg? _____

   Does delivery require external water? _____

4. What structure in gymnosperms and angiosperms becomes a seed? _____

5. Is a sporophyte or gametophyte embryo in a seed? _____

6. What innovation in angiosperms led to the seeds being covered by fruit? _____

1. Name two differences between the life cycles of seedless vascular plants and seed plants.

_____

2. Name the structures that precede the ones listed in the life cycle of a seed plant.

   a. _____   Female gametophyte

   b. _____   Male gametophyte (pollen grain)

   c. _____   Sporophyte

3. The pine tree, unlike a fern, is able to reproduce sexually in a dry environment. Explain. _____

_____

Compare conifers to flowering plants by completing the following table:

	Dominant Generation	Vascular Tissue (Present or Absent)	Dispersal of Offspring	Fruit (Present or Absent)
4. Conifer				
5. Flowering plant				

6. a. Compare the location of the pollen sacs in the pine and the flowering plant. _____

_____

   b. Compare the location of the ovule in the pine and the flowering plant. _____

_____

7. Suppose you wanted to show a friend the female gametophyte of the pine. What would you do?

_____

8. Name two innovations (aside from fruit) of the flowering plant life cycle not found in the pine tree life cycle. _____

_____

9. What is the difference between pollination and fertilization? _____

_____

10. What is the difference between monocot and eudicot flowers? _____

_____

18

Organization of Flowering Plants

Introduction

Despite their great diversity in size and shape, all flowering plants have three vegetative organs that have nothing to do with reproduction*: the root, the stem, and the leaf (Fig. 18.1). Roots anchor a plant and absorb water and minerals from the soil. A stem usually supports the leaves so that they are exposed to sunlight. Leaves carry on photosynthesis, and thereby produce the nutrients that sustain a plant and allow it to grow.

Each of these organs contains various tissues, arranged differently depending on whether a flowering plant is a monocot or a eudicot. The arrangement of tissues is distinctive enough that you should be able to identify the plant as a monocot or eudicot when examining a slide of a root, stem, or leaf.

Flowering plants can also be grouped according to whether they are herbaceous or woody. All flowering trees are woody; their stems contain wood. Many flowering garden plants and all grasses are herbaceous (nonwoody). Herbaceous plants have only primary growth, which increases their height. Woody plants have both primary and secondary growth. Secondary growth increases the girth of a tree.

*See Laboratory 21 for the flower and fruits.

18.1 External Anatomy of a Flowering Plant

Figure 18.1 shows that a plant has a root system and a shoot system. The **root system** consists of the roots. The **shoot system** consists of the stem and the leaves.

Observation: A Living Plant

Shoot System

What structures are in the shoot system?

_____

_____

The Leaves

Leaves carry on photosynthesis. Which part of a leaf (blade or petiole) is the most expansive part?

_____

_____

The Stem

1. In the **stem**, locate a **node** and an **internode**.
2. Measure the length of the internode in the middle of the stem. Does the internode get larger or smaller toward the apex of the stem? _____ Toward the roots? _____ Based on the fact that a stem elongates as it grows, explain your observation. _____

   _____

   _____

3. Where is the **terminal bud** (also called the shoot tip) of a stem?_____

   _____

   Where are axillary buds? _____

   _____

Root System

Observe the root system of a living plant if the root system is exposed. Does this plant have a strong primary root or many roots of the same size? _____

What structures are in the root system?

_____

_____

Where is the root tip? _____

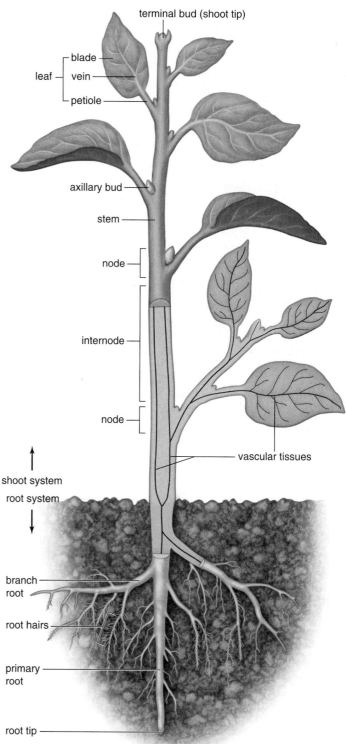

Figure 18.1   Organization of a plant.
Roots, stems, and leaves are vegetative organs. The flower and fruit are reproductive structures studied in Laboratory 21.

terminal bud (shoot tip)

blade
leaf — vein
petiole

axillary bud

stem

node

internode

node

vascular tissues

shoot system
root system

branch root

root hairs

primary root

root tip

18.2 Major Tissues of Roots, Stems, and Leaves

Unlike humans, flowering plants grow in size their entire life because they have immature **meristematic tissue** composed of cells that divide. **Apical meristem** is located at the terminal end of the stem, the branches, and at the root tip and the root branches. When apical meristem cells divide, some of the cells differentiate into the mature tissues of a plant:

Dermal tissue: Forms the outer protective covering of a plant organ.

Ground tissue: Fills the interior of a plant organ; photosynthesizes and stores the products of photosynthesis.

Vascular tissue: Transports water and sugar, the product of photosynthesis, in a plant and provides support.

Note in Table 18.1, roots, stems, and leaves have all three tissues but they are given different specific names.

Table 18.1 Mature Tissues of Vegetative Organs

Tissue Type	Roots	Stems	Leaves
1. Dermal tissue (epidermis)	Protect inner tissues. Root hairs absorb water and minerals.	Protect inner tissues exchange.	Protect inner tissues. Cuticle prevents H_2O loss. Stomata carry on gas
2. Ground tissue	Cortex: Store products of photosynthesis Pith: Store products of photosynthesis	Cortex: Carry on photosynthesis, if green Pith: Store products of photosynthesis	Mesophyll: Photosynthesis
3. Vascular tissue (xylem and phloem)	Vascular cylinder: Transports water and nutrients.	Vascular bundle: Transports water and nutrients.	Leaf vein: Transports water and nutrients.

Observation: Tissues of Roots, Stems, and Leaves

Meristematic tissue. A slide showing the apical meristem in a shoot tip and another showing the apical meristem in a root tip are on demonstration. Meristematic cells are spherical and stain well when they are dense and have thin cell walls (Fig. 18.2). What is the function of immature meristematic tissue?

_____

Figure 18.2 Apical meristem.
A shoot tip and a root tip contain meristem tissue, which allows them to grow longer the entire life of a plant.

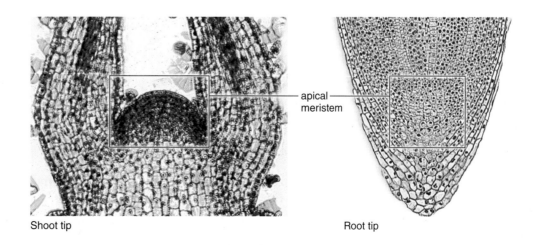

Shoot tip   apical meristem   Root tip

Mature Tissues

1. Dermal tissue. In a cross section of a leaf (Fig. 18.3), for example, focus only on the upper or lower epidermis (Fig. 18.3). Epidermal cells tend to be square or rectangular in shape. In a leaf, the epidermis is interrupted by openings called **stomata** (sing., stoma). Later you will have an opportunity to see the epidermis in roots, stems, and leaves. What is a function of epidermis in all three organs (see Table 18.1)?

   _____

   _____

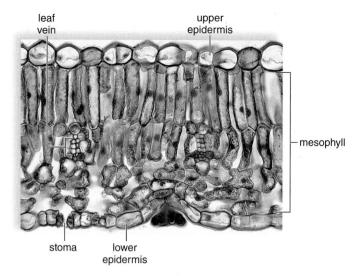

Figure 18.3 Microscopic leaf structure.
Like the stem and root, a leaf contains epidermal tissue, vascular tissue (leaf vein), and ground tissue (mesophyll).

2. Ground tissue. The ground tissue fills the space between epidermis in roots, stems, and leaves. In leaves, for example, ground tissue is called mesophyll (see Fig. 18.3). Ground tissue largely contains parenchyma cells and sclerenchyma cells. **Paranchyma cells** can be of different sizes from fairly circular to oval. Those that contain chloroplasts carry on photosynthesis. Those that contain leucoplasts store starches and oils. **Sclerenchyma cells** are usually elongate in shape and have thick walls impregnated with lignin. These dead cells appear hollow and the presence of lignin means that they stain a red color. Sclerenchyma cells are strong and provide support. Which type cell (parenchyma or sclerenchyma) carries on photosynthesis or stores the products

   of photosynthesis in a leaf? _____ Which one lends strength to ground

   tissue in roots and stems? _____

3. Vascular tissue. In a leaf, strands of vascular tissue are called leaf veins (see Fig. 18.3). There are two types of vascular tissue, called xylem and phloem. **Xylem** contains hollow dead cells that transport water. The presence of lignin makes the cell walls strong, stains red, and makes xylem easy to spot. **Phloem** contains thin-walled, smaller living cells that transport sugars in a plant.

   Phloem is harder to locate than xylem, but it is always found in association with xylem. Which

   type tissue (xylem or phloem) transports sugars in a plant? _____

   Which type tissue transports water? _____

Monocots Versus Eudicots

Flowering plants are classified into two groups: **monocots** and **eudicots**. In this laboratory, you will be studying the differences between monocots and eudicots with regard to the roots, stems, and leaves as noted in Figure 18.4.

Experimental Procedure: Monocot Versus Eudicot

1. Examine the live plant again (see Fig. 18.1). The leaf vein pattern—that is, whether the veins run parallel to one another or whether the veins spread out from a central location (called the

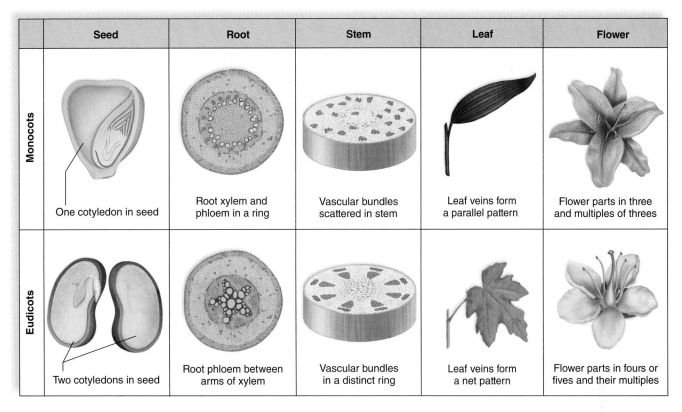

	Seed	Root	Stem	Leaf	Flower
Monocots	One cotyledon in seed	Root xylem and phloem in a ring	Vascular bundles scattered in stem	Leaf veins form a parallel pattern	Flower parts in three and multiples of threes
Eudicots	Two cotyledons in seed	Root phloem between arms of xylem	Vascular bundles in a distinct ring	Leaf veins form a net pattern	Flower parts in fours or fives and their multiples

Figure 18.4   Monocots versus eudicots.
The five features illustrated here are used to distinguish monocots from eudicots.

net pattern)—indicates that a plant is either a monocot or a eudicot. Is the plant in Figure 18.1 a monocot or eudicot? _____ The leaf pattern in Figure 18.3 appears to be parallel. Is this the leaf of a monocot or eudicot? _____

2. Observe any other available plants, and note in Table 18.2 the name of the plant and whether it is a monocot or a eudicot based on leaf vein pattern.

3. Aside from leaf vein pattern, other external features indicate whether a plant is a monocot or a eudicot. For example, open a peanut if available. The two halves you see are cotyledons. Is the plant that produced the peanut a monocot or eudicot? _____

If available, examine a flower. If a flower has three petals or six petals, or any multiple of three, is the plant that produced this flower a monocot or a eudicot? _____

In this laboratory you will have the opportunity to examine the cross sections of roots and stems microscopically; the arrangement of vascular tissue roots and stems also indicates whether a plant is a monocot or a eudicot.

Table 18.2 Monocots Versus Eudicots		
Name of Plant	**Organization of Leaf Veins**	**Monocot or Eudicot?**

18.3 Root System

The **root system** anchors the plant in the soil, absorbs water and minerals from the soil, and stores the products of photosynthesis received from the leaves.

Anatomy of a Root Tip

Primary growth of a plant increases its length. Note the location of the root apical meristem in Figure 18.5. As primary growth occurs, root cells enter zones that correspond to various stages of differentiation and specialization.

Observation: Anatomy of Root Tip

1. Examine a model and/or a slide of a root tip (Fig. 18.5).
2. Identify the **root cap** (dead cells at the tip of a plant that provide protection as the root grows).
3. Locate the **zone of cell division.** Apical meristem is found in this zone. As mentioned previously, meristematic tissue is composed of embryonic cells that continually divide, providing new cells for root growth.
4. Find the **zone of elongation.** In this zone, rows of newly produced cells elongate as they begin to grow larger.
5. Identify the **zone of maturation.** In this zone, the cells become differentiated into particular cell types. When epidermal cells differentiate, they produce **root hairs,**[*] small extensions that absorb water and minerals. Also noticeable are the cells that make up the xylem and phloem of vascular tissue.

Figure 18.5 Eudicot root tip.
In longitudinal section, the root cap is followed by the zone of cell division, zone of elongation, and zone of maturation.

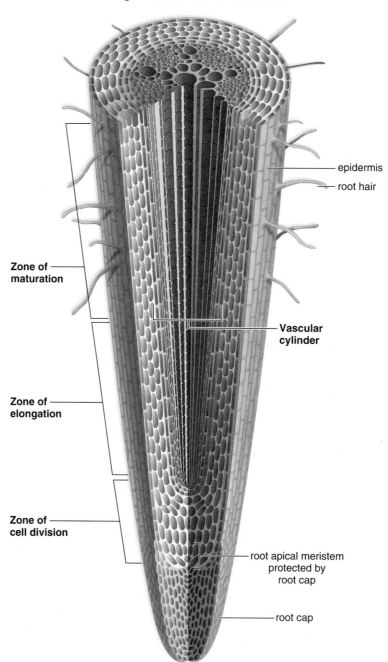

epidermis

root hair

Zone of maturation

Vascular cylinder

Zone of elongation

Zone of cell division

root apical meristem protected by root cap

root cap

*See Laboratory 19 for more on root hairs.

Anatomy of Eudicot and Monocot Roots

Eudicot and monocot roots differ in the arrangement of their vascular tissue.

Observation: Cross-Section Anatomy of Eudicot and Monocot Roots

Eudicot Root

1. Obtain a prepared cross-section slide of a buttercup *(Ranunculus)* root. Use both low power and high power to identify the **epidermis** (the outermost layer of small cells that gives rise to root hairs). The epidermis protects inner tissues and absorbs water and minerals.
2. Locate the **cortex**, which consists of several layers of thin-walled cells (Fig. 18.6a, b). In Figure 18.6b, note the many stained starch grains in the cortex cells. The cortex is ground tissue that functions in food storage.
3. Find the **endodermis**, a single layer of cells whose walls are thickened by a layer of waxy material known as the **Casparian strip**. (It is as though these cells are glued together with a waxy glue.) Because of the Casparian strip, the only access to the xylem is through the living endodermal cells. The endodermis regulates what materials that enter a plant through the root? _____

_____

Use this illustration to trace the path of water and minerals from the root hairs to xylem: _____

4. Identify the **pericycle**, a layer one or two cells thick just inside the endodermis. Branch roots originate from this tissue.
5. Locate the **xylem** in the vascular cylinder of the root. Xylem has several "arms" that extend like the spokes of a wheel. This tissue conducts water and minerals from the roots to the stem.
6. Find the **phloem**, located between the arms of the xylem. Phloem conducts organic nutrients from the leaves to the roots and other parts of the plant.

Figure 18.6   Eudicot root cross section.
The vascular cylinder of a dicot root contains the vascular tissue. Xylem is typically star-shaped, and phloem lies between the points of the star. **a.** Drawing. **b.** Micrograph.

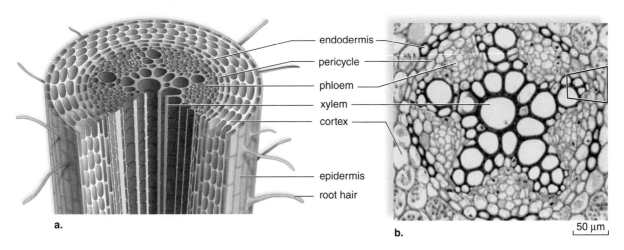

Monocot Root

1. Obtain a prepared cross-section slide of a corn *(Zea mays)* root (Fig. 18.7a, b). Use both low power and high power to identify the six tissues mentioned for the eudicot root.
2. In addition, identify the **pith,** a centrally located ground tissue that functions in food storage.

Figure 18.7   Monocot root cross section.
a. Micrograph of a monocot root cross section and in **b.** an enlarged portion.

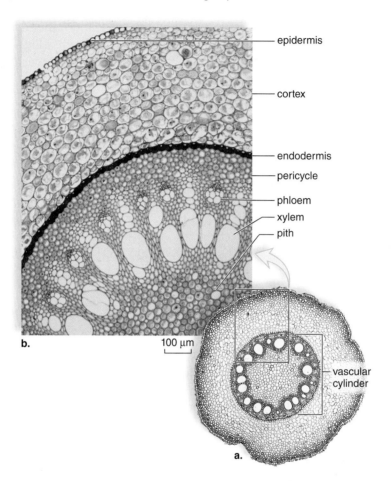

epidermis

cortex

endodermis
pericycle
phloem
xylem
pith

b.   100 μm

vascular cylinder

a.

Comparison

Contrast the arrangement of vascular tissue (xylem and phloem) in the vascular cylinder of monocot roots and eudicot roots by writing "monocot" or "eudicot" on the appropriate line.

_____ Xylem has the appearance of wheel and phloem is between spokes of wheel.

_____ Ring of xylem (inside) and phloem (outside) surrounds pith.

Root Diversity

Roots are quite diverse, and we will take this opportunity to become acquainted with only a few select types.

Observation: Root Diversity

Taproots and fibrous roots. Most plants have either a taproot or they have a fibrous root. Note in Figure 18.8*a* that carrots have a **taproot**. The main root is many times larger than the branch roots. Grasses such as in Figure 18.8*b* have a **fibrous root**: All the roots are approximately the same size.

Examine the taproots on display, and name one or two in which the taproot is enlarged for storage.

_____

The dandelion on display has a fibrous root. Describe. _____

_____

Taproots in particular function in food storage.

Adventitious roots. Some plants have **adventitious** roots. Roots that develop from nonroot tissues, such as nodes of stems, are called adventitious roots. Examples include the prop roots of corn (Fig. 18.8*c*) and the aerial roots of ivy that attach this plant to structures such as stone walls.

Which plants on display have adventitious roots? _____

Other types of roots. Mangroves and other swamp-dwelling trees have roots called pneumatophores that rise above the water line. Pneumatophores have numerous lenticels, which are openings that allow gas exchange to occur.

What root modifications not noted here are on display in your laboratory? _____

_____

Figure 18.8   Root diversity.
a. Carrots have a taproot. **b.** Grass has a fibrous root system. **c.** A corn plant has prop roots, and **d.** black mangroves have pneumatophores to increase their intake of oxygen.

a. Taproot system   **b.** Fibrous root system   **c.** Prop roots, a type of adventitious root   **d.** Pneumatophores of black mangrove trees

18.4 Stems

Stems are usually found aboveground where they provide support for leaves and flowers. Vascular tissue extends from the roots, through the stem and its branches to the leaves. Therefore, what function do

botanists assign to stems in addition to support for branches and leaves? _____

_____ Explain why a branch cannot live if severed from the rest of the plant.

_____

Stems that do not contain wood are called **herbaceous,** or nonwoody, stems. Usually, monocots remain herbaceous throughout their lives. Some eudicots, such as those that live a single season, are also herbaceous. Other eudicots, namely trees, become woody as they mature.

Anatomy of Herbaceous Stems

Herbaceous stems undergo primary growth. **Primary growth** results in an increase in length due to the activity of the apical meristem located in the terminal bud (see Fig. 18.2a) of the shoot system.

Observation: Anatomy of Eudicot and Monocot Herbaceous Stems

Eudicot Herbaceous Stem

1. Examine a prepared slide of a eudicot herbaceous stem (Fig. 18.9), and identify the **epidermis** (the outer protective layer). Label the epidermis in Figure 18.9.
2. Locate the **cortex,** which may photosynthesize or store nutrients.
3. Find a **vascular bundle,** which transports water and organic nutrients. The vascular bundles in a eudicot herbaceous stem occur in a ring pattern. Label the vascular bundle in Figure 18.9a. Which

   vascular tissue (xylem or phloem) is closer to the surface? _____
4. Identify the central **pith,** which stores organic nutrients. Both cortex and pith are composed of

   what type of tissue? _____

Figure 18.9 Eudicot herbaceous stem.

The vascular bundles are in a definite ring in this photomicrograph of a eudicot herbaceous stem. Complete the labeling as directed by the Observation.

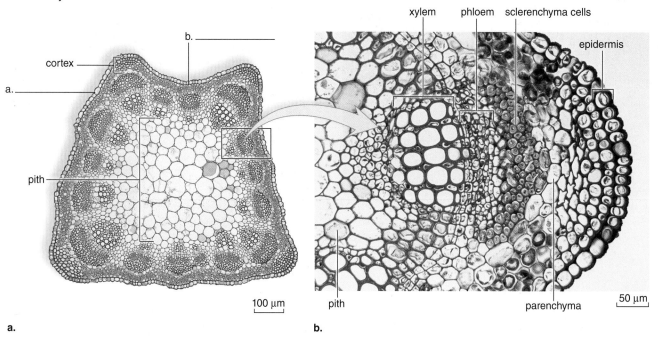

a.

b.

Monocot Herbaceous Stem

1. Examine a prepared slide of a monocot herbaceous stem (Fig. 18.10). Locate the epidermal, ground, and vascular tissues in the stem.

2. The vascular bundles in a monocot herbaceous stem are said to be scattered. Explain. _____

_____

Figure 18.10   Monocot stem.
The vascular bundles, one of which is enlarged, are scattered in this photomicrograph of a monocot herbaceous stem.

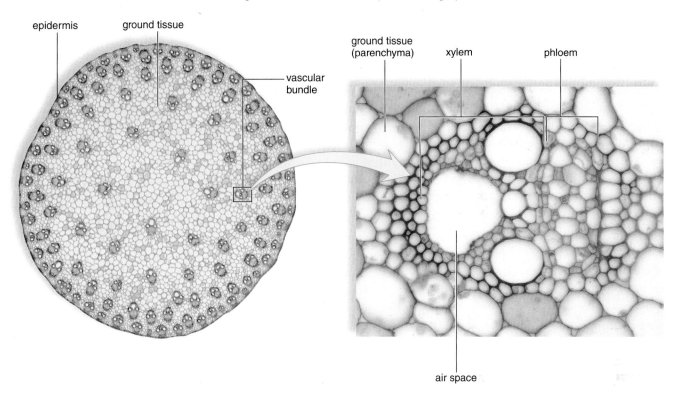

Comparison

1. Compare the arrangement of ground tissue in eudicot and monocot stems. _____

_____

2. Compare the arrangement of vascular bundles in the stems of eudicots and monocots. _____

_____

Stem Diversity

Stems can be quite diverse as well. We take this opportunity to become acquainted with those that allow a plant to accomplish vegetative reproduction and/or function in food storage. Several of these and other types of stems may be on display in the laboratory.

Stolons. The strawberry plant in Figure 18.11*a* has a horizontal aboveground stem called a runner or **stolon**. The stolon produces adventitious roots and new shoots at nodes. Label an adventitious root and a new shoot in Figure 18.11*a*.

    List any other plants on display that spread and produce new shoots by sending out stolons.

_____

Rhizomes. An iris has a belowground horizontal stem called a **rhizome** which functions as a fleshy food storage organ (Fig. 18.11*b*). New plants can grow from a single piece of the rhizome.

    List any other plants on display that have the same belowground horizontal stems (rhizomes) as

the iris. _____

Tubers. A white potato has a belowground rhizome that gives off food storage **tubers** (Fig. 18.11*c*). Each eye is a node that can produce new plants.

    List any other types of plants on display whose belowground stems have tubers. _____

_____

Corms. A gladiolus has a belowground vertical stem called a **corm**, which functions in food storage and has thin, papery leaves (Fig. 18.11*d*).

    List any other types of plants on display that have a vertical stem called a corm. _____

_____

Figure 18.11 Stem diversity.
a. Stolon of a strawberry plant. **b.** Rhizome of an iris. **c.** Tuber of a potato. **d.** Corm of a gladiolus.

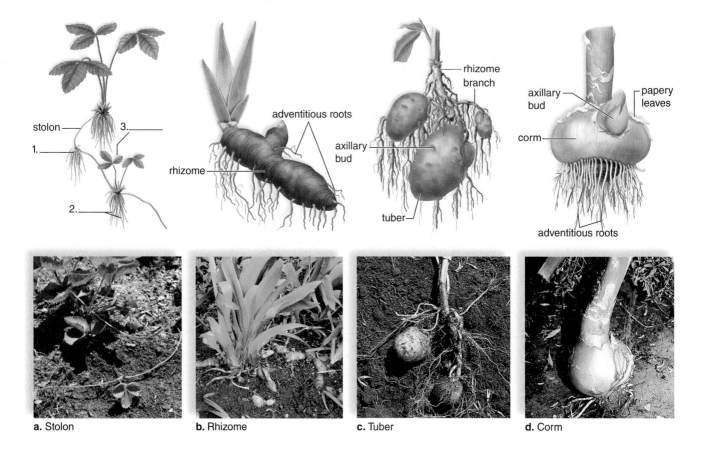

a. Stolon   b. Rhizome   c. Tuber   d. Corm

Anatomy of Woody Stems

Woody stems undergo both primary growth (increase in length) and secondary growth (increase in girth). When *primary growth* occurs, the apical meristem within a terminal bud is active. When *secondary growth* occurs, the vascular cambium is active. **Vascular cambium** is meristem tissue, which produces new xylem and phloem called **secondary xylem** and **phloem** each year. The buildup of secondary xylem year after year is called **wood.** Complete Table 18.3 to distinguish between primary growth and secondary growth of a stem.

Table 18.3 Primary Growth Versus Secondary Growth		
	Primary Growth	**Secondary Growth**
Active meristem		
Result		

Observation: Anatomy of a Winter Twig

1. A winter twig typically shows several years' past primary growth. Examine several examples of winter twigs (Fig. 18.12), and identify the **terminal bud** located at the tip of the twig. This is where new primary growth will originate. During the next growing season, the terminal bud produces new tissues including vascular bundles and leaves.
2. Locate a **terminal bud scar.** These scars encircle the twig and indicate where the terminal bud was located in previous years. The distance between two adjacent terminal bud scars equals one year's primary growth.
3. Find a **leaf scar.** This is where a leaf was attached to the stem.
4. Note the **vascular bundle scars.** Complete this sentence: Vascular bundle scars appear where the vascular tissue _____.
5. Identify a **node.** This is the region where you find leaf scars and bundle scars. The region between nodes is called an _____
6. Locate an **axillary bud.** This is where new branch growth can occur.
7. Note the numerous lenticels, breaks in the outer surface where gas exchange can occur.

Figure 18.12 External structure of a winter twig.
Counting the terminal bud scars tells the age of a particular branch.

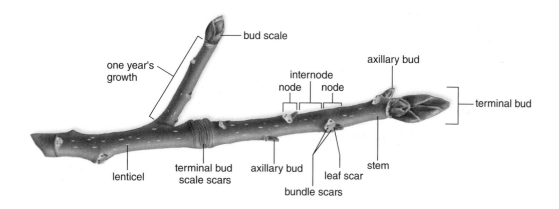

1. Examine a prepared slide of a cross section of a woody stem (Fig. 18.13), and identify the **bark** (the dark outer area), which contains **cork,** a protective outer layer; **cortex,** which stores nutrients; and **phloem,** which transports organic nutrients.

Figure 18.13   Woody eudicot stem cross section.
Because xylem builds up year after year, it is possible to count the annual rings to determine the age of a tree. This tree is three years old.

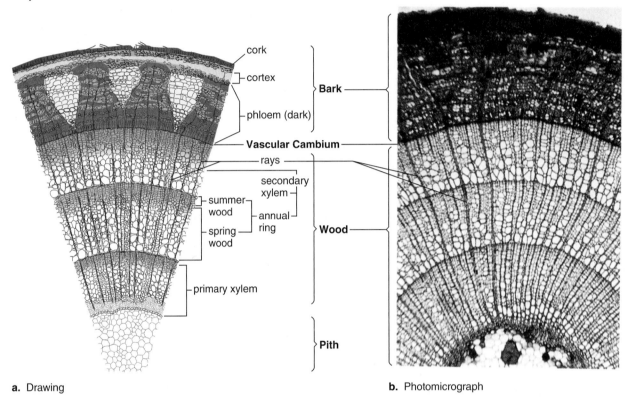

a. Drawing   b. Photomicrograph

2. Locate the **vascular cambium** at the inner edge of the bark, between the bark and the wood. Vascular cambium is meristem tissue whose activity accounts for secondary growth which causes increased girth of a tree. Secondary phloem (disappears) and secondary xylem (builds up) are produced by vascular cambium each growing season.
3. Find the **wood,** which contains annual rings. An **annual ring** is the amount of xylem added to the plant during one growing season. Rings appear to be present because spring wood has large xylem vessels and looks light in color, while summer wood has much smaller vessels and appears much darker. How old is the stem you are observing? _____ Are all the rings the same

   width? _____

4. Identify the **pith,** a ground tissue that stores organic nutrients and may disappear.
5. Locate **rays,** groups of small, almost cuboid cells that extend out from the pith laterally.

18.5 Leaves

A **leaf** is the organ that produces food for the plant by carrying on photosynthesis. Leaves are generally broad and quite thin. An expansive surface facilitates the capture of solar energy and gas exchange. Water and nutrients are transported to the cells of a leaf by leaf veins, extensions of the vascular bundles from the stem.

Anatomy of Leaves

Observation: Anatomy of Leaves

1. Examine a model of a leaf. With the help of Figure 18.14, identify the waxy **cuticle,** the outermost layer that protects the leaf and prevents water loss.
2. Locate the **upper epidermis** and **lower epidermis,** single layers of cells at the upper and lower surfaces. Trichomes are hairs that grow from the upper epidermis and help protect the leaf from insects and water loss.
3. Find the leaf veins in your model. The bundle sheath is the outer boundary of a vein; its cells surround and protect the vascular tissue. If this is a model of a monocot, all the leaf veins will be

   _____. If this is a model of a eudicot, some leaf veins will be circular and

   some will be oval. Why? _____
4. Identify the **palisade mesophyll,** located near the upper epidermis. These cells contain chloroplasts and carry on most of the plant's photosynthesis. Locate the **spongy mesophyll,** located near the lower epidermis. These cells have air spaces that facilitate the exchange of gases across the plasma membrane. Label the layers of mesophyll in Figure 18.14. Collectively, the mesophyll represents

   which of the three types of tissue found in all parts of a plant (see Table 18.1)? _____
5. Label the two layers of epidermis in Figure 18.14. Find the **stoma** (pl., stomata),* openings through which gas exchange occurs. These are more numerous in the lower epidermis. Each has two guard cells that regulate opening and closing of a stoma.

Figure 18.14 Leaf anatomy.

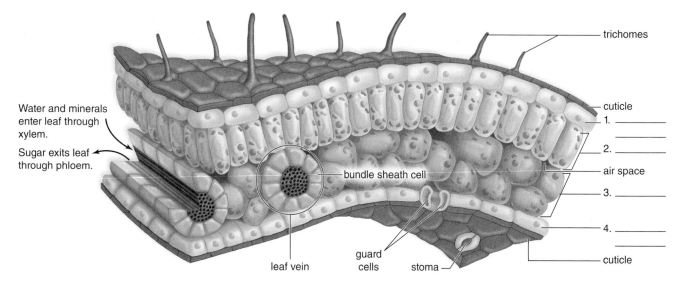

Water and minerals enter leaf through xylem.

Sugar exits leaf through phloem.

bundle sheath cell

leaf vein

guard cells

stoma

trichomes

cuticle

1. _____

2. _____

air space

3. _____

4. _____

cuticle

*See Laboratory 19 for more on stomata.

Leaf Diversity

A eudicot leaf consists of a flat blade and a stalk, called the petiole. An axillary bud appears at the point where the petiole attaches a leaf to the stem. In other words, an axillary bud is a tip-off that you are looking at a single leaf.

Observation: Leaf Diversity

Several types of eudicot leaves will be on display in the laboratory. Examine them using these directions.
- A leaf may be **simple**, in which case it consists of a single blade; or a leaf may be **compound**, meaning its single blade is divided into leaflets. In Figure 18.15*a* write "simple" or "compound" next to the word *leaf* in 1–3. Among the leaves on display find one that is simple and one that is compound.
- A compound leaf can be **palmately** compound, meaning the leaflets are spread out from one point. Which leaf in Figure 18.15*a* is palmately compound? Add "palmately" in front of "compound" where appropriate. See if you can find a palmately compound leaf among those on display.
- A compound leaf can be **pinnately** compound, meaning the leaflets are attached at intervals along the petiole. Which leaf in Figure 18.15*a* is pinnately compound? Add pinnately in front of "compound" where appropriate. See if you can find a pinnately compound leaf among those on display.
- As shown in Figure 18.15*b*, leaves can be in various positions on a stem. On which stem in Figure 18.15*b*, 4–6, are the leaves **opposite** one another? Write the word "opposite" where appropriate. On which stem do the leaves **alternate** along the stem? Write the word "alternate" where appropriate. On which stem do the leaves **whorl** about a node? Add the word "whorl" where appropriate in Figure 18.15*b*. See if you can find different arrangements of leaves among stems on display.

Figure 18.15   Classification of leaves.

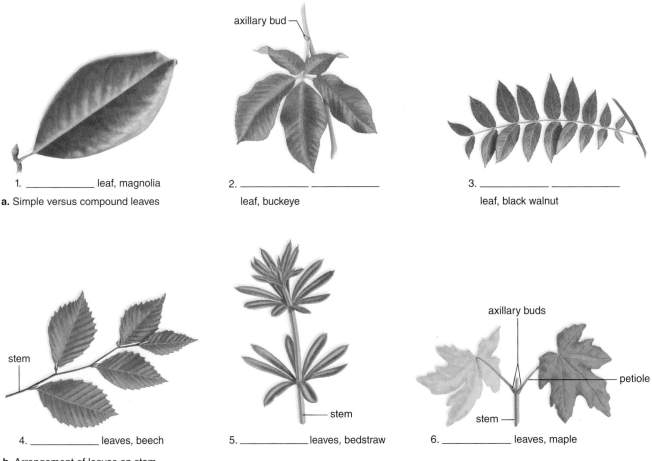

1. _____ leaf, magnolia

a. Simple versus compound leaves

2. _____ _____
   leaf, buckeye

3. _____ _____
   leaf, black walnut

4. _____ leaves, beech

5. _____ leaves, bedstraw

6. _____ leaves, maple

b. Arrangement of leaves on stem

1. How would you distinguish between a monocot and eudicot plant based on leaf vein pattern?

   _____

2. How is meristem different from all other types of plant tissue? _____

3. Show that a root has epidermal, ground, and vascular tissue by using these terms to label this illustration.

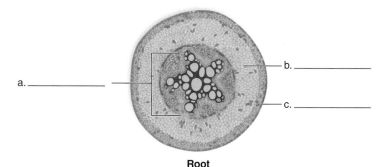

a. _____

b. _____

c. _____

Root

4. In which zone of a eudicot root would you expect to find mature vascular tissue? Explain.

   _____

5. In a eudicot root, what structural feature allows the endodermis to regulate the entrance of water and materials into the vascular cylinder, where xylem and phloem are located? _____

6. What type of root does a carrot have? _____

7. How would you microscopically distinguish a eudicot stem from a monocot stem?

   _____

8. Compare how primary and secondary growth of a woody stem arises.

   _____

9. Contrast how you could determine one year's growth by looking at a winter twig with how you determine one year's growth in a cross section of a woody tree stem. _____

   _____

10. Contrast the manner in which water reaches the inside of a leaf with the manner in which carbon dioxide reaches the inside of a leaf. _____

    _____

19

Water Absorption and Transport in Plants

Learning Outcomes

Introduction

Water from the soil is absorbed into a plant through extensions of epidermal root cells, called **root hairs** (Fig 19.1, *bottom*). It enters **xylem,** the vascular tissue that transports water up through the stem to the leaves. Water forms a continuous column in xylem for two reasons due to the presence of hydrogen bonding between water molecules: First, water molecules exhibit **adhesion** because they momentarily stick to the sides of the xylem vessels. Second,

Water evaporates, pulling the water column from the roots to the leaves.

Water molecules cling together and adhere to sides of vessels in stems.

H_2O

Water enters a plant at root hairs.

Figure 19.1 Water transport in a tree.
When water evaporates from the leaves, the water column in xylem is pulled upward due to the clinging together (cohesion) of water molecules with one another and the sticking (adhesion) of water molecules to the sides of the vessels.

water molecules exhibit **cohesion** because they cling together and do not ordinarily break apart.

The water column moves upward due to **transpiration** (evaporation of water from the leaves at the stomata). As water molecules exit at the leaves, they are replaced by other water molecules from leaf veins. Therefore, transpiration exerts a tension that pulls the water column in the xylem up from the roots to the leaves. This explanation for xylem transport is called the cohesion-tension model of xylem transport.

⏱ **Planning Ahead** Section 19.3 requires students to take readings for 40 minutes; therefore, you may wish to start this procedure early during the laboratory session. Alternately, you may wish to have students do the virtual lab, "Plant Transpiration," also described in Section 19.3.

19.1   Water Absorption by Root Hairs

As we learned in Laboratory 18, the tip of a root consists of a root cap, the zone of cell division, the zone of elongation, and the zone of maturation, where root hairs appear (Fig. 19.2). Root hairs are extensions of epidermal cells. Root hairs increase the surface area for absorption of water that moves through the cortex by way of osmosis and enters the vascular cylinder where xylem is located.

Figure 19.2   Root hairs.
Root epidermis has root hairs to absorb water.

Observation: Root Hairs

1. Obtain a young, germinated *seedling,* and float it in some water in a petri dish while you examine it. The root is rapidly elongating and therefore has grown some distance beyond the seed coat.
2. Using a stereomicroscope, locate the root cap, which covers the root tip, and the region where the root hairs have formed on the root surface.
3. Remove the root from the seedling, and make a wet mount of the root, using *water* and/or a solution of *0.1% neutral red.*
4. Observe your slide under the microscope. Does every epidermal cell have a root hair? _____

   How does the structure of a root hair aid absorption? _____

   _____

Experimental Procedure: Absorption of Water by Osmosis

This Experimental Procedure uses potato cylinders to determine under which osmotic conditions water is likely to enter root hairs. Recall from your previous study of tonicity (Laboratory 4) that **isotonic** refers to equal tonicity, where the concentration of solute in two solutions is the same. **Hypertonic** refers to solutions that contain more solute particles (and less water) than the one to which it is being compared. **Hypotonic** refers to a solution that contains fewer solute particles (more water) than the one to which it is being compared.

1. Obtain five test tubes and a rack or beaker to hold the tubes. The tubes must be large enough to hold a small piece of potato (see step 3).
2. Number the test tubes and fill them with enough solution to cover a piece of potato. Record the contents in Table 19.1.

   Fill tube 1 with a solution of *0.05% sucrose.*   Fill tube 4 with a solution of *0.12% sucrose.*
   Fill tube 2 with a solution of *0.07% sucrose.*   Fill tube 5 with a solution of *0.14% sucrose.*
   Fill tube 3 with a solution of *0.09% sucrose.*

3. Using a cork borer, remove cylinders of tissue from a large *potato*. With a scalpel, cut the cylinders into uniform 2 cm lengths until you have five sections, placing them under a damp paper towel as you cut them. Cut all sections from the *same* potato.
4. Weigh each potato piece before placing it in one of the numbered test tubes. Record the pretreatment weights of each in Table 19.1. (*Note:* You must know which potato piece you put in which tube.)
5. After one hour; remove, dry with a paper towel, and weigh each potato piece and record its post-treatment weight in Table 19.1. (*Note:* You must know which potato piece goes with which tube.)
6. Complete the last two columns in Table 19.1.

Table 19.1 Absorption of Water by Osmosis					
Test Tube	Sucrose %	Pretreatment Weight	Post-treatment Weight	Movement of Water (into or out of potato)	Tonicity of Original Solution Compared to Potato
1					
2					
3					
4					
5					

Conclusions: Absorption of Water by Osmosis

- Since potato cells (modified stems) are believed to behave much as root hairs do, this experiment indicates that only if groundwater is _____ to cytoplasm in root hairs will root hairs be able to absorb water.
- If, by chance, a plant's roots are surrounded by a hypertonic solution, the roots will _____ water.

19.2   The Water Column in Stems

The water column in stems is continuous and moves up the stem for three reasons: (1) Xylem (Fig. 19.3) contains two types of nonliving and hollow conducting cells: vessel elements and tracheids. A **vessel element** has performation plates as end walls. The plate may have a single, large opening or it may have a series of openings. The vessel elements align one on top of the other; and the performation plates allow water and minerals to freely flow from one to the other. **Tracheids** are nonliving and hollow but they are longer and more narrow than a vessel element. Water can move from one tracheid to the other because they have pitted walls. Pits are depressions where the secondary cell wall does not form; making it easier for water to flow from one tracheid to the other. (2) The column of water in xylem is continuous because water molecules are **cohesive** (they cling together) and because water molecules **adhere** to the sides of xylem cells. (3) Water evaporation from leaf surfaces creates a force that pulls the water column upward.

Figure 19.3 Xylem structure.
Photomicrograph of xylem vascular tissue (left) and drawings (right) showing general organization of xylem tissue.

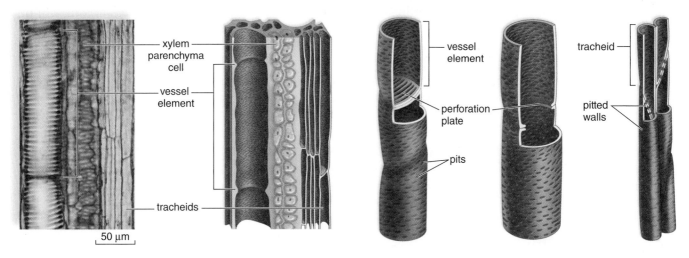

Experimental Procedure: The Water Column in Stems

1. Place a small amount of *red-colored water* in two beakers. Label one beaker "wet" and the other beaker "dry."
2. Transfer a stalk of *celery* (which was cut and then immediately placed in a container of water) into the "wet" beaker so that the large end is in the colored water.
3. Transfer a stalk of *celery* of approximately the same length and width (but that was kept in the air after being cut) into the "dry" beaker so that the large end is in the colored water.
4. With scissors, cut off the top end of each stalk, leaving about 10 cm total length.
5. Time how long it takes for the red-colored water to reach the top of each stalk, and record these data in Table 19.2.

Table 19.2 The Water Column in Stems

Stalk	Speed of Dye (Minutes)	Conclusion
Cut end placed in water prior to experiment		
Cut end kept in air prior to experiment		

Conclusions: The Water Column in Stems

- Exposing a celery stalk to air breaks the water column. Is a continuous water column helpful to the conduction of water in plants? _____ Explain on the basis of your results.

_____

- Conclude why speed of conduction was faster/slower, and write this conclusion in the last column in Table 19.2.

19.3 Transpirational Pull at Leaves

Evaporation of water from leaves is called **transpiration.** As transpiration occurs, the continuous water column is pulled upward—first within the leaf, then from the stem, and finally from the roots.

Experimental Procedure: Transpiration

Assembling a Transpirometer

1. Tightly fit one end of a 4 cm piece of *rubber tubing* over one end of a 15–20 cm glass pipette (Fig. 19.4*a*).
2. Immerse the assembled glass and rubber tube in a large tub of *water* so that it will fill. Check that there are no air bubbles in the tube.
3. Cut off a portion of a *geranium plant* (stem with five to seven leaves), and place the cut end of this stem into the tub of water, but keep the leaves dry (Fig. 19.4*a*). With sharp scissors held under water, cut off a 1.5 cm piece of the stem at an oblique angle. (If this were done in the air, a vacuum would occur in the vascular system.)
4. Keeping the leaves dry and the stem end submerged, fit the cut stem end into the rubber tubing that has been filled with water. Squeeze out all of the air bubbles. If necessary, tightly wind a rubber band or tie a string around the juncture (while still submerged) for added tightness, but do not crush the stem.
5. Remove the assembled apparatus from the water. Hold it so that the geranium stem cutting is upright.
6. Clamp the assembly to a *ring stand* (Fig. 19.4*b*). The water will not run out of the tube due to water's adhesive and cohesive properties.

Determining Transpiration Rate Under Normal Conditions

1. Wait 5 minutes, and then mark with a wax pencil the water level at the lower end of the transpirometer.
2. Then, every 10 minutes for the next 40 minutes, mark the water level, and measure (in mL) the amount the water has moved. Record your data in the first two columns of Table 19.3. These figures indicate the mL of water transpired.

Figure 19.4 Assembling a transpirometer.
Complete directions are given for **a.** assembling the transpirometer apparatus, and **b.** clamping the finished assembly to a ring stand.

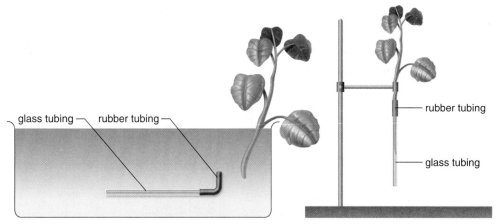

a. Assemble the transpirometer apparatus under water.   b. Clamp the finished assembly to a ring stand.

Determining Transpiration Rates Under Varied Environmental Conditions

1. Your instructor will assign you one of these conditions. Repeat the experiment using the assigned condition.

   a. Focus a light source on the plant (to simulate heat). The plant should be located at least 25 cm away from the light source. Hypothesize how an increase in *temperature* will affect the rate of transpiration.

   _____

   b. Spray the plant and the inside of a plastic bag with water (to simulate a rise in humidity). Put the plastic bag over the plant, and use string to draw it closed around the tubing. Hypothesize how *humidity* will affect the rate of transpiration.

   _____

   c. Use a small fan to gently blow air across the plant (to simulate wind). Hypothesize how *wind* will affect the rate of transpiration.

   _____

2. Remove your previous marks from the glass pipette. Again wait 5 minutes. Then, as before, every 10 minutes for the next 40 minutes mark the water level and measure (in mL) the amount the water has moved. Record your data in Table 19.3.

Table 19.3 Effect of [Temperature, Humidity, Wind]* on Transpiration Rate

| Time | Normal Conditions | | Test Conditions | |
	Reading (in mL)	Total Change (in mL)	Reading (in mL)	Total Change (in mL)
After 10 minutes				
After 20 minutes				
After 30 minutes				
After 40 minutes				

*Circle the condition you tested.

3. Plot the results of your two experiments on the graph provided, using one color to show normal conditions and a different color to show the environmental condition you tested. Calculate the transpiration rate.

   Total mL/40 min = mL/min

   Control: _____

   Condition tested (_____):

   _____

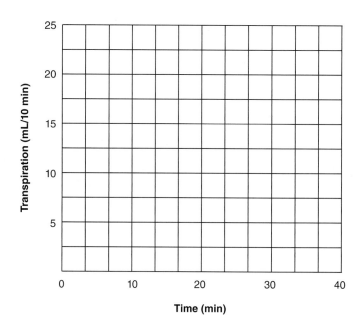

Conclusion: Transpiration

Conclude one of the following according to your data. Conclude the other two according to data collected by other laboratory groups.

1. Effect of temperature on transpiration rate. Did the results support or not support your hypothesis? Explain. _____

    _____

2. Effect of humidity on transpiration rate. Did the results support or not support your hypothesis? Explain. _____

    _____

3. Effect of wind on transpiration rate. Did the results support or not support your hypothesis? Explain. _____

    _____

Virtual Lab Plant Transpiration   A Virtual Laboratory called Plant Transpiration is available on the *Biology* website **www.mhhe.com/maderbiology11**. Your instructor may direct you to do this virtual lab in place of the exercise in your laboratory manual. After opening, you will follow the directions provided in the virtual lab (see Procedure in the write-up next to the experimental items). To save time, reduce the number of different plants selected but be sure to test any selected plant in the four available ways. Before you begin the experiment,

Hypothesize which of these—a heater, a fan, or a lamp—will have the greatest effect on the rate of transpiration.

Hypothesis: _____

_____

Results

Record your data in this table.

Amount of Water Transpired in 1 Hour (mL)				
	Normal Conditions (21°C)	**With Fan (21°C)**	**With Heater (27°C)**	**With Lamp (21°C)**
Arrowhead				
English Ivy				
Geranium				
Zebra Plant				

Conclusion

Did your results support or fail to support your hypothesis? _____

_____

Speculate on why the heater and the fan produced approximately the same results. _____

_____

Speculate on why the rate of transpiration with the lamp approximated the rate under normal conditions. _____

_____

Answer the journal questions and record your answers here:

1. _____

2. _____

3. _____

4. _____

5. _____

6. _____

7. _____

19.4   Stomata and Their Role in Transport of Water

In leaves, the lower epidermis in particular has openings called **stomata.** The opening and closing of a stoma is regulated by **guard cells** on either side of a stoma. When stomata are open, gas exchange occurs; also, water evaporates from the leaves.

The more stomata present and the more they are open (degree of aperture) the greater the rate of transpiration.

Observation: Number of Stomata

1. Calculate the area of the high-power microscopic field with the following formula:

$$\text{Area} = \pi r^2$$

where r (radius) of the field = 0.2 mm (or as determined in Laboratory 2) and π = 3.14.

Area = πr^2 = _____ mm^2

2. Obtain a *leaf* from a plant designated by your instructor, and have a slide with a drop of *distilled water* ready.

3. Using the technique shown in Figure 19.5*a*, obtain a strip of *epidermis* from the underside of the leaf, and put it in the drop of water on the slide, outer side up. Add a coverslip, and examine it microscopically, using both low power and high power.

4. Count the number of stomata (Fig. 19.5*b*) you see in the high-power field. _____ stomata

5. Divide the number of stomata by the area of the field calculated in step 1. This will tell you the number of stomata in 1 mm^2. _____ stomata/mm^2

6. If the underside surface area of your leaf were 400 mm^2, how many stomata would be present on its surface? _____ Such a large number accounts for the large amount of water lost by transpiration.

Figure 19.5 Stomata.

a. Method of obtaining a strip of epidermis from the underside of leaf. **b.** False-colored scanning electron micrograph of leaf surface.

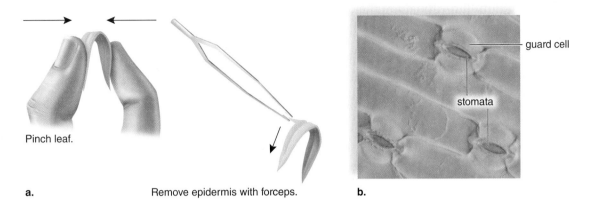

Pinch leaf.

a.

Remove epidermis with forceps.

b.

guard cell

stomata

Experimental Procedure: Open Versus Closed Stomata

Stomata are expected to close when plants are water stressed because closure prevents further water loss. In this Experimental Procedure, we will simulate water stress by placing a strip of epidermis in a salty environment.

1. Prepare two slides, one of which has a drop of distilled water and the other a drop or two of 5% salt solution. Mark the slides *W* (for water) and *S* (for salts).
2. Obtain two strips of epidermis (Fig. 19.5*a*) and place one on each slide.
3. Wait five minutes.
4. Observe each slide in turn. Which slide contains open stomata, and which slide contains closed stomata? _____
5. Explain your results. _____

Summary

Complete Table 19.4 to explain the mechanism of water transport in plants.

Table 19.4	Water Transport in Plants	
Process	**Where**	**Mechanism**
Absorption of water		
Formation of water column		
Transpirational pull		

1. How is the epidermis of a root specialized to absorb water? _____

2. What tonicity promotes water absorption in a root? _____
_____

3. Name and briefly describe the two types of conducting cells in xylem. _____
_____
_____

4. What two characteristics of water cause it to fill the conducting cells of xylem from the roots to the leaves? _____
_____

5. Why is it best to cut flower stems under water? _____

6. Name two reasons the water column is continuous in xylem. _____
_____
_____

7. What is transpiration? _____

8. Name two environmental conditions that can affect transpiration. _____
_____ _____

9. What is the function of stomata? _____

10. How is transpiration prevented when a plant is water stressed? _____
_____

Biology Website

Enhance your study of the text and laboratory manual with study tools, practice tests, and virtual labs. Also ask your instructor about the resources available through ConnectPlus, including the media-rich eBook, interactive learning tools, and animations.

www.mhhe.com/maderbiology11

McGraw-Hill Access Science Website

An Encyclopedia of Science and Technology Online which provides more information including videos that can enhance the laboratory experience.

www.accessscience.com

20.2 Phototropism

Phototropism is the bending of plants in response to unidirectional light. When a plant is exposed to unidirectional light, the hormone *auxin* migrates from the bright side of a stem to the shady side. The cells on that side elongate faster than those on the bright side, causing the stem to curve toward the light (Fig. 20.5).

As you will see, blue light brings about phototropism. Therefore, it is hypothesized that a plant pigment capable of absorbing blue light initiates phototropism.

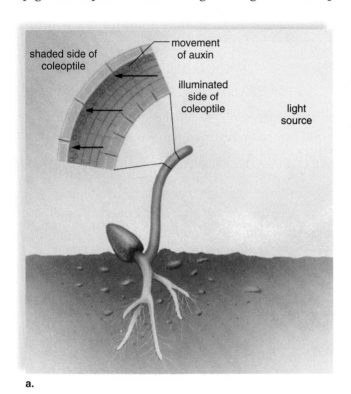

a.

b.

Figure 20.5 Phototropism in a seedling and plant.
a. Experiments with seedlings show that when a seedling bends toward the light, auxin (arrows) migrates from the bright side to the shady side. Cell elongation follows and this causes bending toward the light. **b.** Adult plants also respond to light by bending toward it.

Experimental Procedure: Phototropism

Phototropism chambers can be made with film canisters (Fig. 20.6). Three windows are punched into the sides of the film canister, and three *Brassica rapa* seeds are placed on wet blotting paper. Each window of the canister then is covered with a red, blue, or green acetate square. A control canister has windows covered with clear acetate.

1. Obtain a control canister and a canister with colored windows.
2. Open the canisters and observe the growth of the seedlings placed there by your instructor. Reclose the canisters, being sure that the orientation of the colored windows is the same as before.
3. Record the response of the seedlings in Table 20.3.

Figure 20.6 Color of light and phototropism.
Seeds are exposed to lights of different colors in a phototropism chamber.

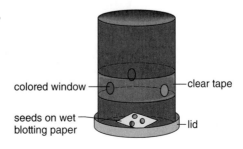

Film canister

Table 20.3 Phototropism

Type of Canister	Response of Seedlings	Conclusion
Clear windows (control)		
Colored window (red)		
Colored window (blue)		
Colored window (green)		

Conclusions: Phototropism

- Record the results of this experiment and your conclusion in Table 20.3.
- Do your results support the model that blue light (but not red and green light) reception is involved in positive phototropism of stems? _____
  Explain how you came to this conclusion. _____
  _____

- Investigators have found that the activation of a plasma membrane photoreceptor for blue light, now called phototropin, leads to the binding of auxin by cells and the bending of stems. Why is it adaptive for plants to have a way to increase the bending of stems in reponse to unidirectional light? _____

20.3 Gibberellins and Stem Elongation

Gibberellins are plant hormones that cause stem elongation (Fig. 20.7). The presence of gibberellin in a cell activates a gene that codes for amylase. Amylase breaks down starch, providing an energy source for wall construction during elongation.

Experimental Procedure: Stem Elongation

Do you hypothesize that gibberellins would cause dwarf plants (called dwarf plants because the stem cells do not usually elongate) to grow taller? _____ Explain. _____

Figure 20.7 Gibberellins cause stem elongation.
a. The *Cyclamen* plant on the right was treated with gibberellins; the plant on the left was not treated. b. The grapes are larger on the right because gibberellins caused an increase in the space between the grapes, allowing them to grow larger.

a.   b.

Normal Plants

1. Observe three normal *Brassica rapa* plants of increasing age. Your instructor will tell you how many days each plant has been growing. Record the ages of the plants in Table 20.4.
2. See the boxed instructions on how to measure an internode. Choose the two longest internodes on each plant. Carefully measure (and record) their lengths, and determine the average internode distance for these two measurements. Record your data in Table 20.4.
3. Return the plants to the location specified by your instructor.

Untreated Dwarf Plants

1. Observe three dwarf plants (a mutant strain of *Brassica rapa*) of increasing age. Your instructor will tell you how many days each plant has been growing. Record the ages of the plants in Table 20.4.
2. See the boxed instructions on how to measure an internode. Choose the two longest internodes on each plant. Carefully measure (and record) their lengths, and determine the average internode distance for these two measurements. Record your data in Table 20.4.
3. Return the plants to the location specified by your instructor.

Treated Dwarf Plants

1. Observe three dwarf plants of increasing age that have been sprayed with *gibberellin* since germination. Your instructor will tell you how many days each plant has been growing. Record the ages of the plants in Table 20.4.
2. See the boxed instructions on how to measure an internode. Choose the two longest internodes on each plant. Carefully measure (and record) their lengths, and determine the average internode distance for these two measurements. Record your data in Table 20.4.
3. Return the plants to the location specified by your instructor.

Measuring an Internode

A **node** is where a leaf is attached to the stem and an **internode** is the region of a stem between nodes. To arrive at the length of an internode, use a metric ruler to measure from the **base of the petiole** (leaf stalk) of the upper leaf to the **leaf axil** of the lower leaf.

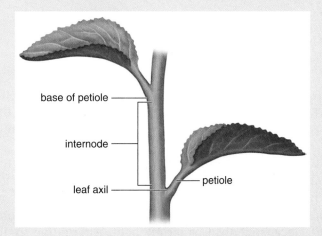

Table 20.4   Effect of Gibberellin

	Internode 1 Measurement	Internode 2 Measurement	Average Internode Distance
Normal Plants			
Age			
Age			
Age			
Untreated Dwarf Plants			
Age			
Age			
Age			
Treated Dwarf Plants			
Age			
Age			
Age			

Conclusions: Stem Elongation

- How do the three groups of plants differ from each other? _____

_____

_____

- Did your data support your hypothesis? _____
- If not, why not? _____

20.4   Etiolation

In a plant grown in the light, the leaves are lifted up above the ground, and chlorophyll is synthesized so that photosynthesis may begin. Therefore, the leaves are green. **Etiolation** is the summation of altered growth patterns that result when a seedling is grown without sunlight. Etiolation may assist the seedling in reaching sunlight.

A blue-green pigment called **phytochrome** is involved in etiolation. Phytochrome is composed of two identical proteins (Fig. 20.8). Each protein has a light-sensitive region that absorbs red light during the day and far-red light at night. When phytochrome absorbs red light it becomes Pfr the active form of phytochrome, which promotes normal growth; the leaves expand and become green. This form of phytochrome is known as Pfr because it absorbs far-red light. After doing so, it becomes Pr, the inactive form of phytochrome, which is so called because it absorbs red light during the day. As long as Pr is present and not Pfr, a plant will undergo etiolation.

Figure 20.8 Phytochrome conversion cycle.

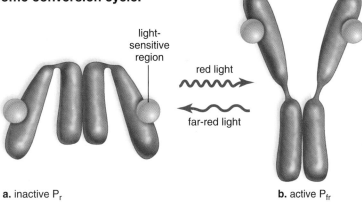

light-sensitive region

red light →

← far-red light

a. inactive P_r

b. active P_{fr}

Experimental Procedure: Bean Seedling Etiolation

1. Compare the appearance of a bean seedling that germinated and grew in the dark with one that germinated and grew in the light (Fig. 20.9). Record your data in Table 20.5.
2. Complete Table 20.5 by explaining why the etiolated plant differs from the normal plant in the characteristics listed.

Figure 20.9 Etiolation.
Bean seedlings on the left are etiolated because they were grown in the dark. Bean seedlings on the right are much shorter with well-formed leaves because they were grown in the light.

Table 20.5 Comparison of Normal and Etiolated Plants

	Normal Plant	Etiolated Plant	Explanation
Color of stem			
Color of leaf			
Length of stem			
Size of leaf			
Stiffness of stem (examine gently)			
Other comments, if any			

1. Contrast the response of a stem and root to gravity. _____

_____

2. Does a stem or root show negative gravitropism? _____

3. Explain why the stem of a plant placed in the dark and one placed in the light both respond to gravity.

_____

4. Contrast the response of a stem to gravity and to unidirectional light. _____

_____

5. Explain why stem responses to gravity and light can both be due to the presence of auxin. _____

_____

_____

_____

6. Which color light is the best stimulus for phototropism seen in stems? Explain. _____

_____

7. What happens to a dwarf plant sprayed with gibberellin? _____

8. Offer an explanation for your answer in question 7. _____

_____

9. What is phytochrome? _____

10. What is etiolation, and how is it related to phytochrome? _____

_____

_____

Biology Website

Enhance your study of the text and laboratory manual with study tools, practice tests, and virtual labs. Also ask your instructor about the resources available through ConnectPlus, including the media-rich eBook, interactive learning tools, and animations.

www.mhhe.com/maderbiology11

McGraw-Hill Access Science Website

An Encyclopedia of Science and Technology Online which provides more information including videos that can enhance the laboratory experience.

www.accessscience.com

21
Reproduction in Flowering Plants

Learning Outcomes

Introduction

The evolution of the flower accounts for the great success of the flowering plants. The flower is the center of sexual reproduction for angiosperms and is involved in the production of male and female gametophytes, gametes, and embryos enclosed within seeds. The flower also gives rise to fruits that cover the seeds.

Flowering plants are stationary but they have evolved a successful mechanism for pollination, a process so necessary to completion of their life cycle. They employ the services of animals, often insects, which are motile! The relationship is mutualistic because before gathering pollen and taking off to the next flower, the insect acquires nectar, a nutrient substance produced by the flower (Fig. 21.1). Animals also help flowering plants disperse their seeds. When animals feed on fruits they inadvertently take the seeds to new locations.

Figure 21.1   Landing platform.
This composite flower contains many individual flowers and the reproductive parts of these flowers are used as a place to land by a bee. The bee will feed on nectar and collect pollen.

21.1 Flowering Plant Life Cycle

Flowering plants, like all land plants, undergo a life cycle called alternation of generations in which the sporophyte alternates with the gametophyte. However, in seed plants the sporophyte produces two types of spores called microspores and megaspores. These spores develop into male gametophytes (pollen grains) and female gametophytes (embryo sacs), respectively. Because pollen grains are either windblown or carried by a pollinator to the female gametophyte, no outside moisture is needed for fertilization to occur. Clearly, flowering plants are well adapted to a land existence.

Observation: Flowering Plant Life Cycle

Use Figure 21.2 as a guide to describe the life cycle of flowering plants. The number of a question matches with a number in the diagram.

1. The parts of the flower involved in reproduction are the _____ and the _____ .

2. The anther at the top of the stamen has _____ , which contain numerous microspore mother cells that undergo meiosis to produce _____.

3. The carpel contains an ovary that encloses the _____. Within an ovule, a megaspore mother cell undergoes meiosis to produce four _____.

4. A microspore undergoes mitosis to produce a _____, which has two cells, the tube cell and the _____ cell.

5. One megaspore undergoes mitosis and develops into a(n) _____, which contains two _____ nuclei and the _____ cell.

6. Pollination is _____

_____.

   After a pollen grain lands on the stigma of a carpel, it develops a pollen tube that passes down the style and takes _____ sperm into the embryo sac.

7. During double fertilization, one sperm from the pollen tube fertilizes the egg within the embryo sac, and the other joins with two _____ nuclei.

8. The fertilized egg becomes a(n) _____ , and the joining of polar nuclei and sperm becomes the triploid (3n) _____ inside a _____ . A seed coat protects the seed. In angiosperms, seeds are enclosed by _____ . A fruit assists the dispersal of seeds (e.g., when animals eat fleshy fruits, they may ingest the seeds and disperse them by defecation sometime later).

9. After germination of the seed, a new plant begins to grow and eventually becomes the adult sporophyte.

Figure 21.2 Flowering plant life cycle.

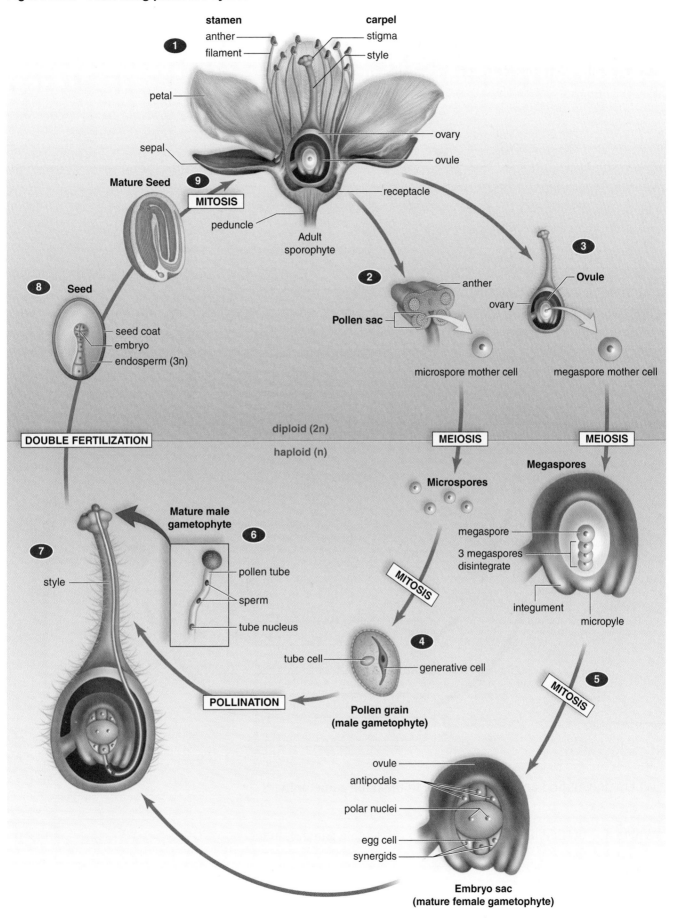

1

stamen
anther
filament

carpel
stigma
style

petal

sepal

Mature Seed

9

MITOSIS

peduncle

ovary
ovule

receptacle

Adult
sporophyte

2

Pollen sac

anther

3

Ovule
ovary

8

Seed

seed coat
embryo
endosperm (3n)

microspore mother cell

megaspore mother cell

diploid (2n)

DOUBLE FERTILIZATION

haploid (n)

MEIOSIS

MEIOSIS

Megaspores

Microspores

megaspore

3 megaspores
disintegrate

integument

micropyle

**Mature male
gametophyte**

6

7

style

pollen tube

sperm

tube nucleus

MITOSIS

4

tube cell

generative cell

5

MITOSIS

POLLINATION

**Pollen grain
(male gametophyte)**

ovule
antipodals

polar nuclei

egg cell
synergids

**Embryo sac
(mature female gametophyte)**

Summary

Complete the following diagram that summarizes the flowering plant life cycle. Name a structure for b, c, e, f, and h. Name a process for a, d, g, and i.

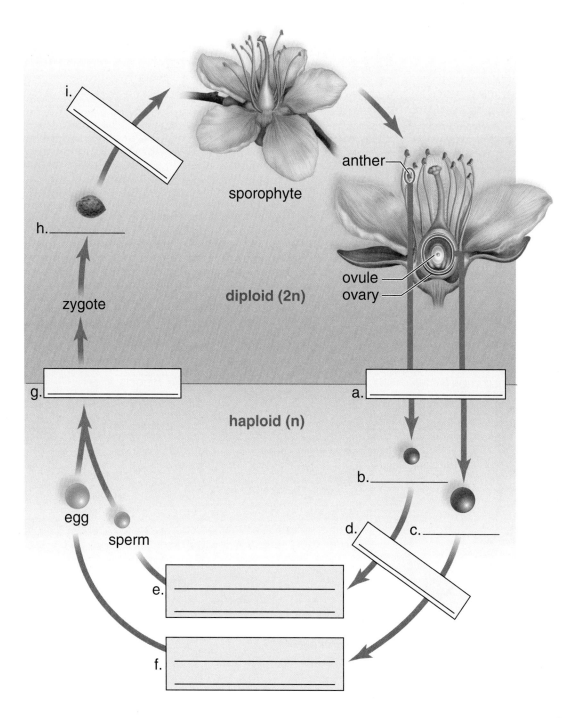

Where do you find and what happens to the male gametophyte? _____

_____

Where do you find and what happens to the female gametophyte? _____

_____

Structure of a Flower

The typical parts of a flower are shown in Figure 21.3 and described below:

Figure 21.3 Flower structure.
See if you can label the flower on page 282 without looking at this one!

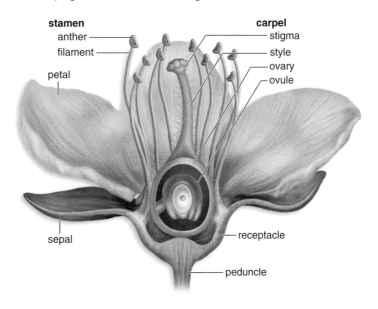

Peduncle	The flower stalk.
Receptacle	The portion of a flower that bears the other parts.
Sepals	Collectively called the **calyx,** the **sepals** protect the flower bud before it opens. The sepals may drop off or may be colored like the petals. Usually, sepals are green and remain attached to the receptacle.
Petals	Collectively called the **corolla,** the **petals** are quite diverse in size, shape, and color. Their color or arrangement can attract a pollinator.
Stamens	Each **stamen** consists of two parts—the **anther,** also called a pollen sac, and the **filament,** a slender stalk. Pollen grains develop from microspores produced in the anther. In what way is the anther analogous to the testes in human males? _____
Carpel	A vaselike structure with three major regions from top to bottom: the **stigma,** an enlarged sticky knob; the **style,** a slender stalk; and the **ovary,** an enlarged base that contains one or more ovules. Inside the **ovule** a megaspore becomes an embryo sac and then develops into a seed, and an ovary becomes a fruit. **Fruit** is instrumental in the distribution of seeds. In what way is the ovary of a carpel analogous to the ovary in human females? _____

Observation: The Structure of a Flower

1. Obtain a live flower, such as that of a lily (Fig. 21.4). What is the scientific and common name of the plant that produced this flower? ______________________________

   Sketch the overall appearance of the whole flower here. Label your drawing using only the main terms on page 279.

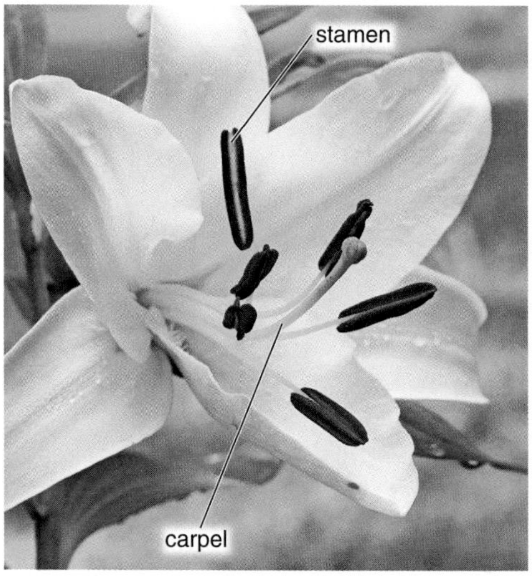

Figure 21.4 Lily (*Lilium longiflorum*).

   a. Compare the structure of your flower to that of Figure 21.4. ______________________________

   b. What is the color of your flower and does it have a scent? ______________________________

   c. Count the number of petals and sepals (see Fig. 18.4). Is this flower a monocot or eudicot? ____ Explain. ______________________________

2. Determine where the petals and sepals are attached and remove them. What color are the sepals? ____________ Do they resemble the petals in any way? ______________________ If yes, how? ______________________________

3. Count the number of stamen and record the number here. ____________ Is this the same as the petal number or a multiple of the petal number? ____________ Remove the stamens and examine one to locate the anther and filament. Place a stamen on a slide and use a scalpel to remove and open the anther to disperse pollen on the slide. Remove all but a few pollen grains from the slide. Add a coverslip and examine microscopically, using high power. Can you find two different nuclei? ______________ The generative cell has a small spindle-shaped nucleus; the tube cell has a larger and more centrally placed nucleus.

4. Examine the carpel and identify the stigma, style, and ovary. Remove the carpel and place it on a slide. Describe. ______________________________

   Use the scalpel to make a cross section of the ovary. Does this ovary contain chambers, each with ovules? ____________

5. If available, examine the fruit produced by this type flower. A fruit develops from what part of the carpel? ____________ The seeds came from what structures in the ovary? ______________

Plants and Their Pollinators

It might seem as if pollinators go to all types of flowers but they don't (Fig. 21.5). For example, bees only visit certain flowers—the ones that provide them with a lot of nectar, a surgery liquid that serves as their food. Bee-pollinated flowers have a shape and appearance preferred by bees, even a particular type of bee. What process causes particular bees and particular flowers to be suited to one another? _____ _____How is this mutualistic relationship advantageous to both the plant and the bee? _____

_____

Observation: Plants and Their Pollinators

1. The description you gave of your flower can help you decide what type of pollinator would be attracted to your flower.

   Bees and moths can smell. But bees like a delicate, sweet smell, while moths prefer a strong smell that can allow them to find flowers in the dark. Moths are nocturnal and feed at night. If the flower has a smell, which of these two pollinators might pollinate your flower? _____

   Bees and butterflies generally need a landing platform. Bees can land on a small petal but butterflies typically walk around on a cluster of flowers. Moths and hummingbirds don't need a landing platform because they hover (flap their wings to stay in one place) (see Fig. 21.5b).

Figure 21.5 Pollinators.
a. The butterfly-pollinated flower is often a brightly colored composite containing many individual flowers. The broad exposure provides room for the butterfly to land. **b.** A moth-pollinated flower is usually light in color; a moth depends on scent to find the flower it prefers in the dark. **c.** Hummingbird-pollinated flowers are curved back, allowing the bird to insert its beak to reach the rich supply of nectar.

butterfly

bright-colored flower with landing platform

a.

proboscis

moths hover

light-colored flower

b.

long, thin beak

hummingbirds hover

floral tube with curved back margins

c.

Moths have a long proboscis and hummingbirds have a long tongue to reach nectar at the bottom of the floral tube. Based on this information, which of these pollinators might pollinate your flower? _____

_____

Moth-pollinated flowers are typically white; bees can't see red but can see yellow; butterflies like brightly colored flowers; and hummingbirds prefer the color red. Which of these pollinators might prefer your flower? _____

2. Justify your answer by completing 1 in Table 21.2. Examine several other available flowers and conclude what type of pollinator would be attracted to different flowers and complete Table 21.1.

Table 21.1   Plants and Their Pollinators

Description of Flower	Possible Pollinator	Explanation
1.		
2.		
3.		

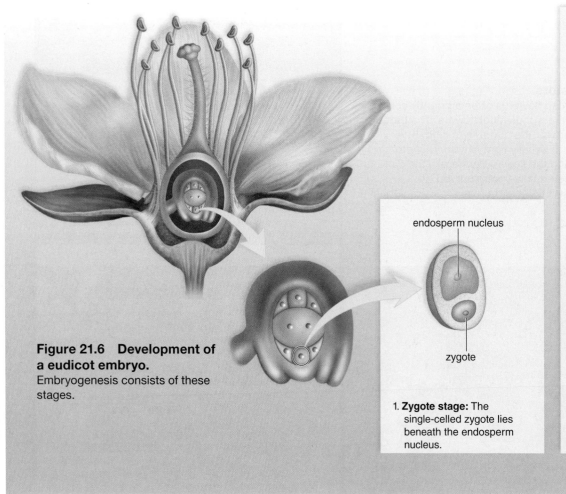

Figure 21.6   Development of a eudicot embryo.
Embryogenesis consists of these stages.

1. Zygote stage: The single-celled zygote lies beneath the endosperm nucleus.

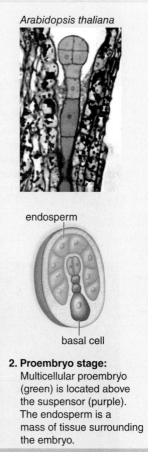

Arabidopsis thaliana

2. Proembryo stage: Multicellular proembryo (green) is located above the suspensor (purple). The endosperm is a mass of tissue surrounding the embryo.

21.2 Development of a Eudicot Embryo

Stages in the development of a eudicot embryo are shown in Figure 21.6. During development, the **suspensor** anchors the embryo and transfers nutrients to it from the mature plant. The **cotyledons** store nutrients that the embryo uses as nourishment. An embryo consists of the **epicotyl,** which becomes the leaves; the **hypocotyl,** which becomes the stem; and the **radicle,** which becomes the roots.

Observation: Development of the Embryo

Study prepared slides and identify the stages described in Figure 21.6. List the stages you were able to identify in the prepared slides. _____

_____

_____

_____

_____

A. thaliana

endosperm

3. Globular stage: As cell division continues, the proembryo (green) becomes globe-shaped. The stalklike suspensor (purple) anchors the embryo.

A. thaliana

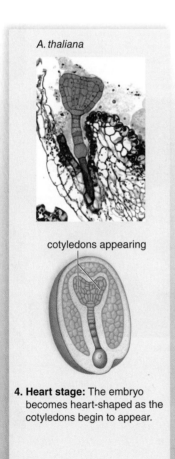

cotyledons appearing

4. Heart stage: The embryo becomes heart-shaped as the cotyledons begin to appear.

Capsella

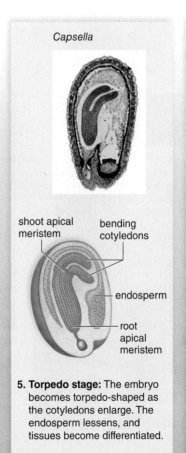

shoot apical meristem — bending cotyledons

endosperm

root apical meristem

5. Torpedo stage: The embryo becomes torpedo-shaped as the cotyledons enlarge. The endosperm lessens, and tissues become differentiated.

Capsella

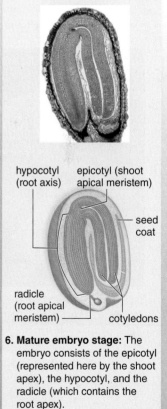

hypocotyl (root axis) — epicotyl (shoot apical meristem)

seed coat

radicle (root apical meristem) — cotyledons

6. Mature embryo stage: The embryo consists of the epicotyl (represented here by the shoot apex), the hypocotyl, and the radicle (which contains the root apex).

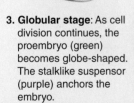

21.3 Fruits

Because fruits are defined as the mature ovary, what people normally call a vegetable can be a fruit to a botanist. For example, aside from a blackberry, an almond, a pea pod, and a tomato are fruits (Fig. 21.7).

As a fruit develops from an ovary, the ovary wall thickens to become the **pericarp**. The pericarp can have as many as three layers: exocarp, mesocarp, and endocarp. The **exocarp** forms the outermost

a. Almond, *Prunus*

seed

b. Tomato, *Lycopersicon*

one chamber of ovary

c. Pea, *Pisum*

developing fruit

seed

d. Blackberry, *Rubus*

many carpels

Figure 21.7   Fruit diversity.
a. The almond fruit is fleshy, with a single seed enclosed by a hard covering. **b.** The tomato is derived from a compound ovary. **c.** Dry fruit of the pea plant develops from a simple ovary. **d.** The blackberry fruit is an aggregate fruit derived from a flower that had many ovaries. Each ovary produces one of the "berries" in the aggregate fruit.

skin of a fruit. The **mesocarp** between the exocarp and endocarp can be fleshy. The receptacle of a flower often contributes to the flesh of a fruit also. The **endocarp** serves as the boundary around the seed(s). The endocarp may be fleshy (as in tomatoes), hard (as in peach pits), or papery (as in apples). Botanists classify fruits in the manner described in the key. This is a simplified key but it does contain some of the most common types of fruits.

Dichotomous Key to Major Types of Fruit

I. Fleshy fruits
  A. Simple fruits (i.e., from a single ovary)
    1. Flesh mostly of ovary tissue particularly the mesocarp.
      a) Endocarp is hard and stony; ovary superior and single-seeded (cherry, olive, coconut): **drupe**
      b) Endocarp is fleshy or slimy; ovary usually many-seeded (tomato, grape, green pepper): **berry**
    2. Flesh mostly of receptacle tissue (apple, pear, quince): **pome**
  B. Complex fruits (i.e., from more than one ovary)
    1. Fruit from many carpels on a single flower (strawberry, raspberry, blackberry): **aggregate fruit**
    2. Fruit from carpels of many flowers fused together (pineapple, mulberry): **multiple fruit**
II. Dry fruits
  A. Fruits that split open at maturity (usually more than one seed)
    1. Split occurs along two seams in the ovary. Seeds borne on one of the halves of the split ovary (pea and bean pods, peanuts): **legume**
    2. Seeds released through pores or multiple seams (poppies, irises, lilies): **capsule**
  B. Fruits that do not split open at maturity (usually one seed)
    1. Pericarp hard and thick, with a cup at its base (acorn, chestnut, hickory): **nut**
    2. Pericarp thin and winged (maple, ash, elm): **samara**
    3. Pericarp thin and not winged (sunflower, buttercup): **achene** (cereal grains): **caryopsis**

Source: Vodopich and Moore: *Biology Laboratory Manual* 8/e, p. 344.

To use this key understand that at each step you have at least two choices and you pick one of these in order to proceed further. Therefore, first determine if the fruit is fleshy or dry. If fleshy, proceed further with I and if dry proceed further with II. Next choose either A or B and so forth until you arrive at the particular type of fruit. The last column in Table 21.2 will be one of the boldface terms.

Observation: Fruits

1. Again examine the fruit of your flower. Use the key to determine the fruit type and list this fruit as an example in Table 21.2.

2. Examine an apple that has been sliced to give a longitudinal and cross-section view of the interior. The flesh of an apple is from the receptacle and only the core of an apple is from the ovary. Locate the outer limit of the pericarp and the limit of the endocarp. An ovary can be simple (have one chamber) or can be compound (have more than one chamber). What type of ovary does an apple have? _____

   How could an animal, such as a deer, help disperse the seeds of an apple? _____

   _____

   Use apple as an example of a fruit in Table 21.2.

3. Examine the pod of a string bean or pea plant. How many seeds (beans or peas) are in the pod? _____ Would it help disperse the seeds of a pea plant if an animal were to eat the peas? _____ Why or why not? _____

Split the pea or bean and look for the embryo. Is this plant a monocot or eudicot plant?_____
How do you know? _____

Use a pea pod as an example of a fruit in Table 21.2.

4. Examine a sunflower fruit and remove the seed. The outercoat of a sunflower seed is actually _____. How can examining the seed tell you that the sunflower plant is a eudicot? _____

Except for the apple, all the fruits you have examined are dry fruits? What does this mean? _____
_____

Add sunflower fruit to Table 21.2.

5. Examine other available fruits and complete Table 21.2.

Table 21.2 Identification of Simple Fruits			
Common Name	Fleshy or Dry	Eaten as a Vegetable, Fruit, Other	Type of Fruit (from Dichotomous Key)
1			
2			
3			
4			
5			
6			
7			
8			
9			
10			

21.4 Seeds

The seeds of flowering plants develop from ovules. As noted previously in Figure 21.6, 6, a seed contains an embryonic plant, stored food, and a seed coat. Monocot seeds have one **cotyledon** (seed leaf); eudicots have two cotyledons.

Observation: Eudicot and Monocot Seeds

Bean Seed

1. Obtain a presoaked bean seed (eudicot). Carefully dissect it, using Figure 21.8 to help you identify:

   a. **Seed coat:** The outer covering. Remove the seed coat with your fingernail.

   b. **Cotyledons:** Food storage organs. The endosperm was absorbed by the cotyledons during development. What is the function of these cotyledons? _____

   _____

   c. **Epicotyl:** The small portion of the embryo located above the attachment of the cotyledons. The first true leaves (**plumules**) develop from the epicotyl.

   d. **Hypocotyl:** The small portion of embryo located below the attachment of the cotyledons. The lower end develops into the embryonic root, or **radicle.**

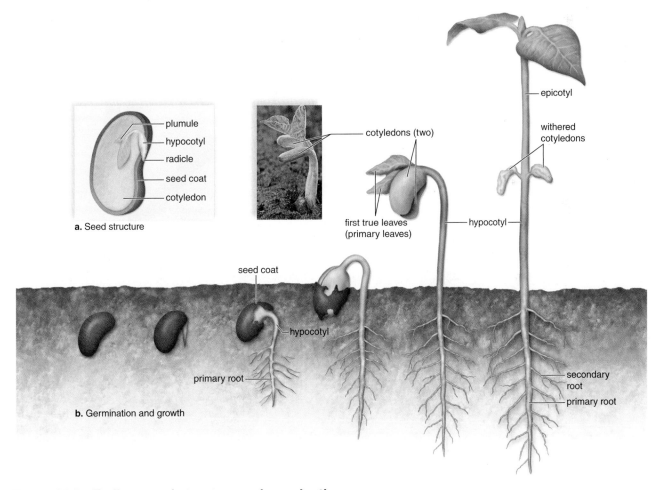

Figure 21.8 Eudicot seed structure and germination.
a. A seed contains an embryo, stored food, and a seed coat as exemplified by a bean seed. A eudicot seed has two cotyledons. Following **b.** germination, a seedling grows to become a mature plant.

2. As observed in Figure 21.8, which organ emerges first from a dicot seed—the plumule or the radicle? _____

Of what advantage is this to the plant? _____

_____

3. The hypocotyl is the first part to emerge from the soil. What is the advantage of the hypocotyl pulling the plumule up out of the ground instead of pushing it up through the ground? _____

_____

4. Do cotyledons stay beneath the ground in monocots? _____

Corn Kernel

1. Obtain a presoaked corn kernel (monocot). Lay the seed flat, and with a razor, carefully slice it in half. A corn kernel is a fruit, and the seed coat is tightly attached to the pericarp (Fig. 21.9).
2. Identify the cotyledon, plumule, and radicle. In addition, identify the:
   a. **Endosperm:** Stored food for the embryo; passes into the cotyledon as the seedling grows.
   b. **Coleoptile:** A sheath that covers the emerging leaves.
3. As observed in Figure 21.9, does the cotyledon of a corn seed, our example of a monocot, stay beneath the ground? _____

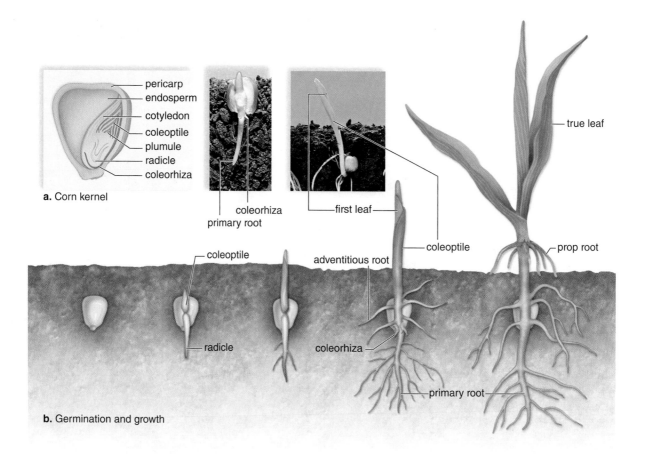

Figure 21.9 Monocot seed structure and germination.
a. A monocot seed has only one cotyledon as exemplified by a corn kernel. A corn kernel is a fruit—the seed is covered by a pericarp. Following **b.** germination, a corn seedling grows to become a mature plant.

Seed Germination

Mature seeds contain an embryo that does not resume growth until after germination, which requires the proper environmental conditions. Mature seeds are dry, and for germination to begin, the dry tissues must take up water in a process called **imbibition.** After water has been imbibed, enzymes break down the food source into small molecules that can provide energy or be used as building blocks until the seedling is ready to photosynthesize.

Experimental Procedure: Seed Germination

Your instructor has placed sunflower seeds in five containers and watered them. The seedlings are in increasing stages of growth.

1. Order the seedlings in a series according to increasing stages of growth. What criteria did you use to order the seedlings? _____

   _____

2. Is this plant a monocot or eudicot? _____ What criteria did you use to decide? _____

   _____

3. Can you see the cotyledons? _____ Explain why the cotyledons of a eudicot seedling shrivel as the seedling grows. _____

4. Use the space below to draw two contrasting stages of growth and add these labels to your drawings: hypocotyl, cotyledons, epicotyl, leaves, stem, terminal bud, node. For terminal bud and node, see Figure 18.1.

1. What parts of a flower might you be tempted to call the "male"part? Why?

   _____

   _____

2. Tell what part of a carpel is involved in (1) germination, (2) growth, and (3) function of a pollen tube.

   _____

   _____

3. How can you tell a monocot flower from a eudicot flower?_____

   _____

4. What process results in pollinators that are specific to a particular type of flower? _____

   _____

5. Explain why a fruit ordinarily contains seeds._____

   _____

6. A string bean, tomato, okra, and cucumber are fruits. Explain._____

   _____

7. How can you tell a monocot seed from a eudicot seed? _____

   _____

8. Name three general parts of a seed. _____

9. Relate the plumule and radicle to parts of an adult plant. _____

   _____

10. Name two growth pattern differences between monocot and eudicot seeds after germination.

    _____

    _____

22

Introduction to Invertebrates

Learning Outcomes

Introduction

Introduction

In our survey of the animal kingdom, we will see that animals are very diverse in structure. Even so, all animals are multicellular and **heterotrophic,** which means their food consists of organic molecules made by other organisms. Consistent with the need to acquire food, animals have some means of locomotion by use of muscle fibers. Animals are always diploid, and during sexual reproduction, the embryo undergoes specific developmental stages.

> **Planning Ahead** To see hydra and planarians feed, have students observe the animals at the start of lab, add food, and then check frequently until food engulfment occurs.

While we tend to think of animals in terms of **vertebrates** (e.g., dogs, fishes, squirrels), which have a backbone, most animal species are those that lack a backbone, commonly known as **invertebrates.** In this laboratory, we will examine those invertebrates that lack a true body cavity, called a **coelom.** A survey of the rest of the animal kingdom follows in Laboratory 23 and Laboratory 24.

22.1 Evolution of Animals

Today, molecular data is used to trace the evolutionary history of animals. These data tell us, as shown in the phylogenetic (evolutionary) tree (Fig. 22.1), that all animals share a common ancestor. This common ancestor was most likely a colonial protist consisting of flagellated cells. All but one of the phyla depicted in the tree consists of only invertebrates—phylum Chordata contains a few invertebrates and also the vertebrates.

Figure 22.1 Evolutionary tree of animals.

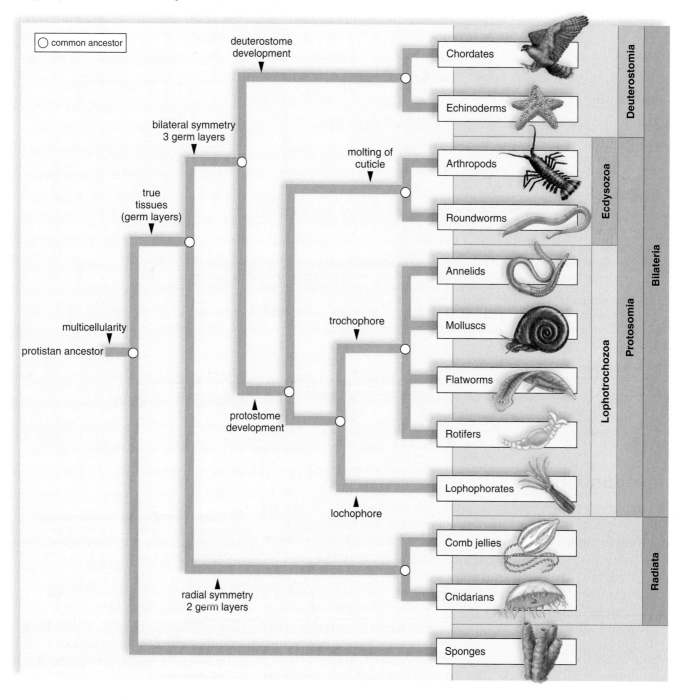

Certain anatomical features of animals can be used to substantiate the tree. The first feature of interest is **symmetry**. **Asymmetry** means the animal has no particular symmetry. The Radiata have **radial symmetry**—the animal is organized circularly and, just as with a wheel, two identical halves are obtained no matter how the animal is longitudinally sliced. Which of the phyla in the tree have radial

symmetry? _____

The other phyla are the Bilateria with **bilateral symmetry,** which means the adult animal has a definite right and left half.

One of the main events during the development of animals is the formation of tissue layers called **germ layers** because all other structures are derived from them. The outer germ layer of an animal is the ectoderm (a tissue layer that becomes the skin and nervous system) and the inner germ layer is the endoderm (a tissue layer that becomes the gut). The middle germ layer is the mesoderm (a tissue layer that becomes the muscles of an animal). Which of the phyla in the tree have only two germ layers (i.e.,

ectoderm and endoderm)? _____

_____

Finally, complex animals are either Protostomia or Deuterostomia. Protostomia have the **protostome** pattern of development in which the first opening of the embryo is the mouth. Deuterostomia have the **deuterostome** pattern of development in which the second opening of the embryo is the mouth. Which pattern of development do the flatworms, rotifers, and roundworms

(animals included in this laboratory) have? _____

22.2   Sponges (Phylum Porifera)

Sponges live in water, mostly marine, attached to rocks, shells, and other solid objects. An individual sponge is typically shaped like a tube, cup, or barrel. Sponges grow singly or in colonies whose overall appearances vary widely. A single sponge can become a colony by asexual budding.

Anatomy of Sponges

Sponges consist of loosely organized cells and have no well-defined tissues. They are asymmetrical or radially symmetrical and **sessile** (immotile). They can reproduce asexually by budding or fragmentation, but they also reproduce sexually by producing eggs and sperm.

Sponges have a few types of specialized cells. Most notably they have flagellated **collar cells** (**choanocytes**). The movement of their flagella keep water moving through the pores into the central cavity and out the osculum of a sponge (Fig. 22.2). Collar cells also take in suspended food particles from the water and digest them for the benefit of all the other cells in a sponge.

Observation: Anatomy of Sponges

Preserved Sponge

1. Examine a preserved sponge (Fig. 22.2a). Note the main excurrent opening **(osculum)** and the multiple incurrent pores. Water is constantly flowing in through the pores and out the osculum. *Label the arrows in the left-hand drawing of Figure 22.2a to indicate the flow of water.* Use the labels *water out* and *water in* through pores.
2. Examine a sponge specimen cut in half. Note the central cavity and the sponge wall. The wall is convoluted in some sponges, and the pores line small canals. Does this particular sponge have

   pore-lined canals? _____

Figure 22.2 Sponge anatomy.

a. Movement of water through pores into the central cavity and out the osculum is noted. Label the water flow. **b.** Collar cells line the central cavity of a sponge and the movement of their flagella keeps the water moving through a sponge. **c.** Draw an enlargement of spicules here.

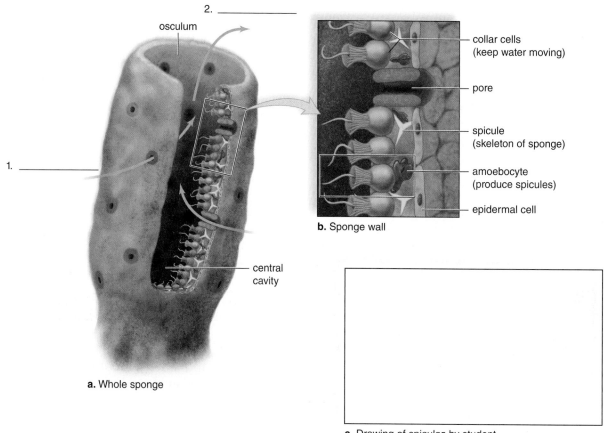

2. _____

osculum

collar cells
(keep water moving)

pore

spicule
(skeleton of sponge)

amoebocyte
(produce spicules)

epidermal cell

b. Sponge wall

1. _____

central
cavity

a. Whole sponge

c. Drawing of spicules by student

3. You may be able to see **spicules,** fine projections over the body and especially encircling the osculum. Does this sponge have spicules? _____

Prepared Slides

1. Examine a prepared slide of *Grantia.*

   a. Find the collar cells that line the interior (Fig. 22.2*b*). A sponge is a **sessile filter feeder.** Collar cells phagocytize (engulf) tiny bits of food that come through the pores along with the water flowing through the sponge. They then digest the food in food vacuoles. Explain the expression *sessile filter feeder.* _____

   _____

   _____

   b. Do you see any spicules? _____ Do they project from the wall of a sponge? _____

   c. Depending on the sponge, spicules are made of either calcium carbonate, silica (glass), or protein. Calcium carbonate and silica produce hard sharp spicules. Name two possible advantages of spicules to a sponge.

   _____

   _____

2. Examine a prepared slide of sponge spicules. What do you see? _____

_____

Draw a sketch of four spicules, each having a different appearance in the space provided in Figure 22.2c.

Diversity of Sponges

Sponges are very diverse and come in many shapes and sizes. Some sponges live in fresh water although most live in the sea and are a prominent part of coral reefs, areas of abundant sea life discussed in the next section. Zoologists have described over 5,000 species of sponges, which are grouped according to the type spicule (Fig. 22.3).

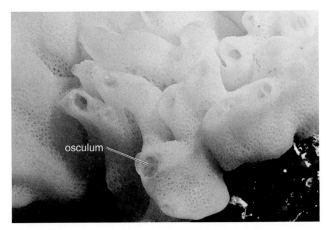

a. Calcareous sponge, *Clathrina canariensis*

b. Bath sponge, *Xestospongia testudinaria*

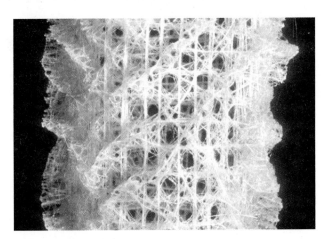

Figure 22.3   Diversity of sponges.
a. Calcareous (chalk) sponges have spicules of calcium carbonate. **b.** Bath sponges have a skeleton of spongin. **c.** Glass sponges have glassy spicules.

c. Glass sponge, *Euplectella aspergillum*

Conclusions: Anatomy of Sponges

- The anatomy and behavior of a sponge aid its survival and its ability to reproduce. How does a sponge:

    a. Protect itself from predators? _____

    b. Acquire and digest food? _____

    c. Reproduce asexually and sexually? _____

22.3 Cnidarians (Phylum Cnidaria)

Cnidarians are tubular or bell-shaped animals that live in shallow coastal waters, except for the oceanic jellyfishes. Two basic body forms are seen among cnidarians. The mouth of a **polyp** is directed upward, while the mouth of a jellyfish, or **medusa,** is directed downward. At one time, both body forms may have been a part of the life cycle of all cnidarians. When both are present, as in *Obelia*, the sessile polyp stage produces medusae, and this motile stage produces egg and sperm (Fig. 22.4). Today in some cnidarians, one stage is dominant and the other is reduced; in other species, one form is absent altogether. How can a life cycle that involves two forms, called polymorphism, be of benefit to an

animal, especially if one stage is sessile (stationary)? _____

_____

Anatomy of Cnidarians

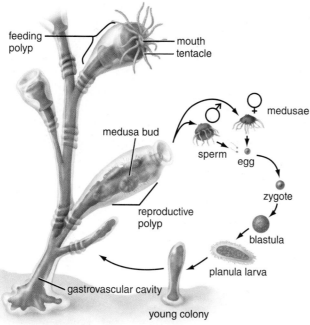

Figure 22.4   Life cycle of *Obelia*.
The stationary *Obelia* colony has both feeding polyps and reproductive polyps. The reproductive polyps produce motile bell-shaped medusae that make eggs and sperm. Each zygote becomes a ciliated planula larva that develops into the polyp stage.

Consult Figure 22.1 and note again that cnidarians are radially symmetrical and have two germ layers. How can radial symmetry benefit an animal?_____

_____

Figure 22.5 shows the anatomy of a hydra which will be studied as a typical cnidarian. Hydras exist only as sessile polyps; there is no alternate stage. Note the **tentacles** that surround the **mouth,** the large **gastrovascular cavity,** and the basal disk. A gastrovascular cavity has a single opening that is used both as an entrance for food and an exit for wastes.

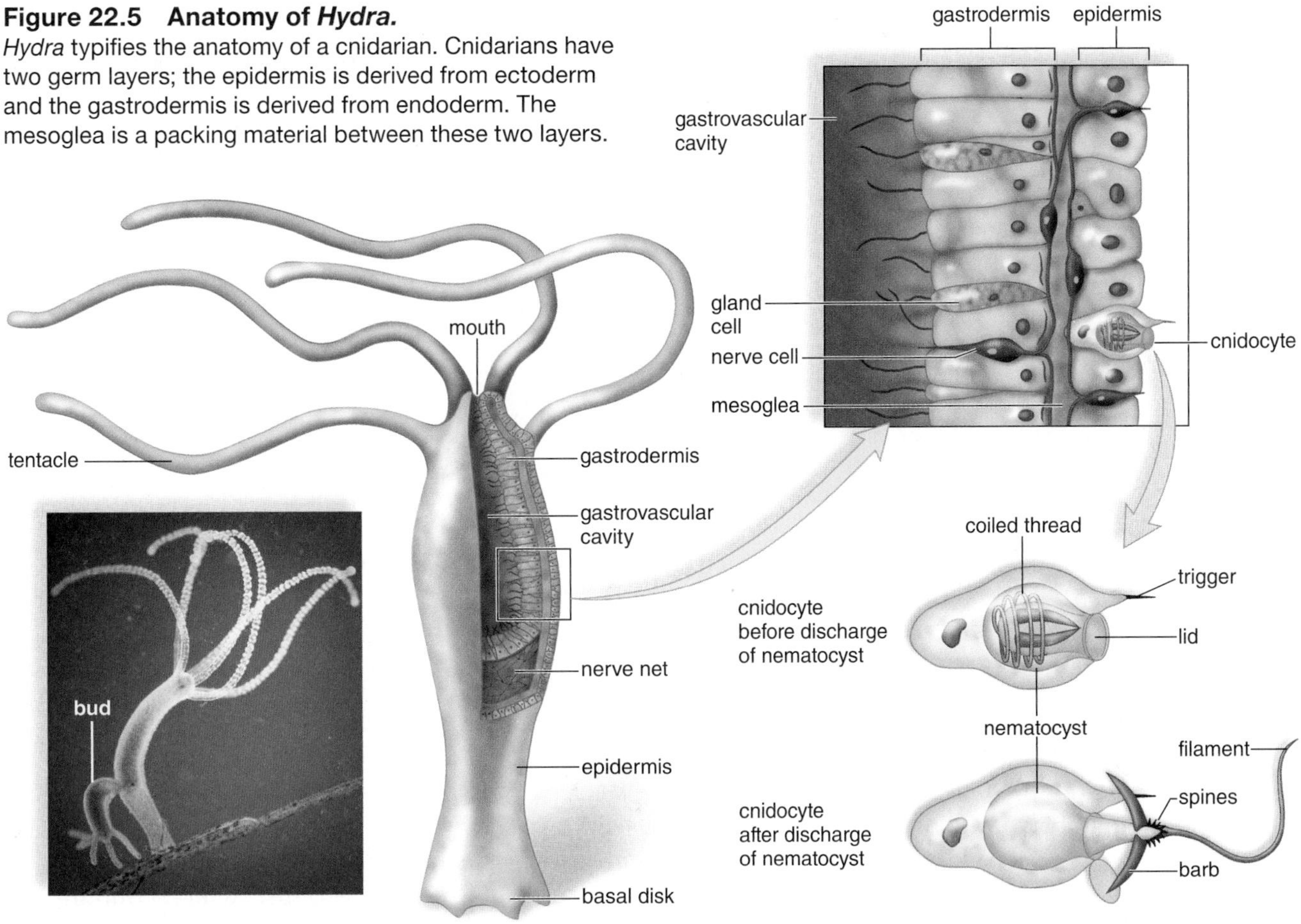

Figure 22.5 Anatomy of *Hydra*.

Hydra typifies the anatomy of a cnidarian. Cnidarians have two germ layers; the epidermis is derived from ectoderm and the gastrodermis is derived from endoderm. The mesoglea is a packing material between these two layers.

Observation: Cnidarians

Preserved Hydra

With the aid of a hand lens, examine preserved specimens of *Hydra*. Hydras typically reproduce asexually by budding (Fig. 22.5, *far left*). Do you see any evidence of buds that are developing directly into small hydras? _____ The body wall can also produce ovaries and testes that produce eggs and sperm. The testes are generally located near the attachment of the tentacles; the ovaries appear farther down on the trunk, toward the basal disk.

Prepared Slide of Hydra

Examine prepared slides of cross and longitudinal sections of *Hydra*. With the help of Figure 22.5, note the two tissue layers. A gelatinous material called mesoglea separates the epidermis from the gastrodermis, which lines the gastrovascular cavity. Switch to high power. Do you find any cells? _____ Describe them. _____

_____

Living Hydra

1. Observe a living *Hydra* in a small petri dish for a few minutes. What is the current behavior of your hydra? _____

_____

Most often a hydra is attached to a hard surface by its basal disk. A hydra can move, however, by turning somersaults:

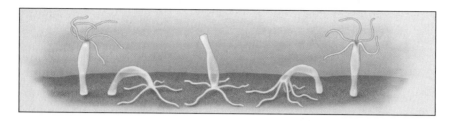

2. After a few minutes, tap the edge of the petri dish. What is the reaction of your hydra?

_____

_____

3. The tentacles of a hydra capture food, which is stuffed into the gastrovascular cavity (Fig. 22.6) where it is digested both externally in the cavity and internally in the cells that line the cavity. Place a small protozoan or crustacean near the tentacles of the hydra and observe it feeding.

4. Mount a living *Hydra* on a depression glass slide with a coverslip and examine a tentacle. Unique to cnidarians are specialized stinging cells, called **cnidocytes**, which give the phylum its name. Each cnidocyte has a fluid-filled capsule called a **nematocyst** (see Fig. 22.5, far right), which contains a long, spirally coiled hollow thread. The threads trap and/or sting prey. Note the cnidocytes as swellings on the tentacles. Add a drop of vinegar (5% acetic acid) and note what happens

to the cnidocytes. Did your hydra discard any nematocysts? _____

Describe. _____

_____

_____

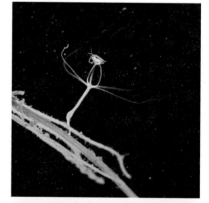

Of what benefit is it to *Hydra* to have cnidocysts? _____

_____

_____

Conclusions: Anatomy of Cnidarians

• The anatomy and behavior of a hydra aid its survival and its ability to reproduce. How does a hydra:

a. Acquire and digest food? _____

_____

b. Protect itself from predators? _____

_____

c. Reproduce asexually and sexually?

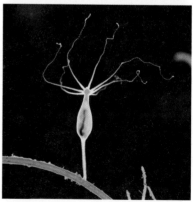

Figure 22.6   Hydra feeding.

_____

Figure 22.7 Cnidarian diversity.

Diversity of Cnidarians

Cnidarians consist of a large number of mainly marine animals (Fig. 22.7). Sea anemones, sometimes called the flowers of the sea are solitary polyps often found in coral reefs, areas of biological abundance in shallow tropical seas. Stony corals have a calcium carbonate skeleton that contributes greatly to the building of coral reefs. Portuguese man-of-war is a colony of modified polyps and medusae. Jellyfishes are a part of the zooplankton, suspended animals that serve as food for larger animals in the ocean.

Sea anemone, *Corynactis*

Cup coral, *Tubastrea*

Portuguese man-of-war, *Physalia*

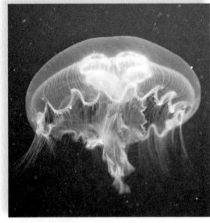
Jellyfish, *Aurelia*

22.4 Flatworms (Phylum Platyhelminthes)

Flatworms are bilaterally symmetrical animals that can be either free-living or parasitic. Free-living flatworms, called planarians, are more complex than the cnidarians. In addition to the germ layers ectoderm and endoderm, mesoderm is also present. Flatworms are usually **hermaphroditic**, which means they possess both male and female sex organs (Fig. 22.8c). Why is it advantageous for an animal to be hermaphroditic? _____

_____

Planarians practice cross-fertilization when the penis of one is inserted into the genital pore of the other. The fertilized eggs are enclosed in a cocoon and hatch as tiny worms in two or three weeks.

Planarians

Planarians such as *Dugesia* live in lakes, ponds, and streams, where they feed on small, living or dead organisms. In planarians, the three germ layers give rise to various organs aside from the reproductive organs (Fig. 22.8). The three-part **gastrovascular cavity** ramifies throughout the body; the excretory organs called **flame cells** (because their cilia reminded early investigators of a flickering flame of a candle) collect fluids from inside the body and send via a tube to an excretory pore; and the nervous system contains a brain and lateral nerve cords connected by transverse nerves. Therefore it is called **ladder-like**. Complete the labels in both 22.8b and d. Why would you expect an animal that lives in

fresh water to have a well-developed excretory system? _____

_____

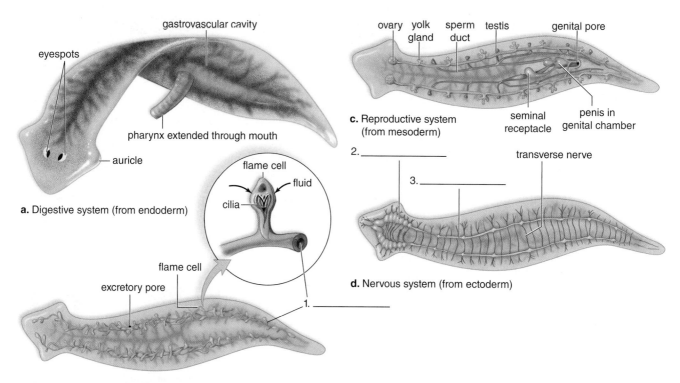

a. Digestive system (from endoderm)

gastrovascular cavity

eyespots

pharynx extended through mouth

auricle

flame cell

fluid

cilia

flame cell

excretory pore

1. _____

b. Excretory system (from mesoderm)

ovary   yolk   sperm   testis   genital pore
        gland   duct

c. Reproductive system
   (from mesoderm)

seminal   penis in
receptacle   genital chamber

2. _____   transverse nerve

3. _____

d. Nervous system (from ectoderm)

Figure 22.8   Planarian anatomy.
a. When a planarian extends the pharynx, food is sucked up into a gastrovascular cavity that branches throughout the body. **b.** The excretory system has flame cells. **c.** The reproductive system has both male (blue) and female (pink) organs. **d.** The nervous system looks like a ladder.

Observation: Planarians

Preserved Specimen

Examine a whole mount of a planarian that shows the branching gastrovascular cavity (Fig. 22.8*a*). What is the advantage of a gastrovascular cavity that ramifies through the body? _____

Prepared Slide

Examine a cross section of a planarian under the microscope. Can you locate the structures shown in Figure 22.9? Does a planarian have a body cavity?

_____ Explain. _____

_____

_____

_____

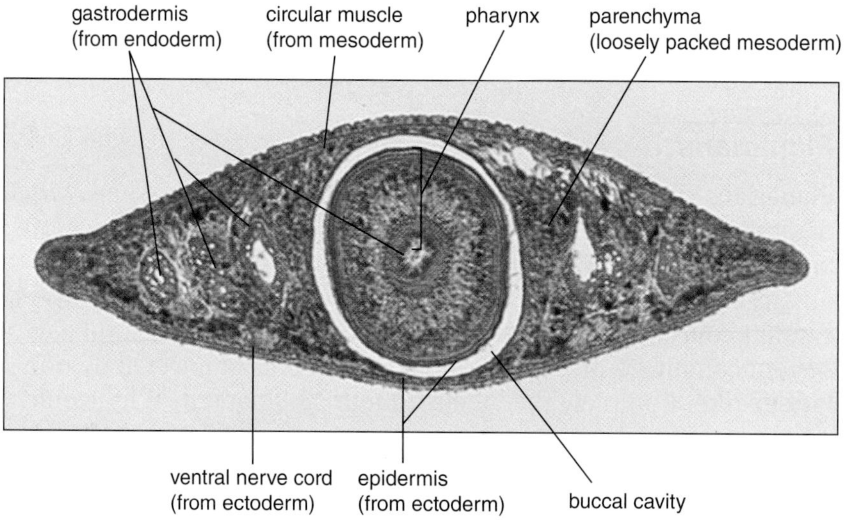

gastrodermis
(from endoderm)

circular muscle
(from mesoderm)

pharynx

parenchyma
(loosely packed mesoderm)

ventral nerve cord
(from ectoderm)

epidermis
(from ectoderm)

buccal cavity

Figure 22.9   Planarian micrograph.
Cross section of a planarian at the pharynx.

Living Specimen

1. Examine the behavior of a living planarian (Fig. 22.10) in a petri dish. Describe the behavior of the animal. _____

   _____

2. Does the animal move in a definite direction?_____ Why is it advantageous for a predator such as a planarian to have bilateral symmetry and a definite head region? _____

   _____

3. Gently touch the animal with a probe. What three types of cells must be present for flatworms to be able to respond to stimuli and move about? _____

   _____

   The auricles on the side of the head are sense organs. Flatworms have well-developed muscles and a nervous system consisting of a brain and nerves.

4. If a strong light is available, shine it on the animal. What part of the animal would be able to detect light? _____ How does the animal respond to the light? _____

5. Offer the worm some food, such as a small piece of liver, and describe its manner of eating.

   _____

   _____

   Roll the animal away from its food and note the pharynx extending from the body (Fig. 22.10).

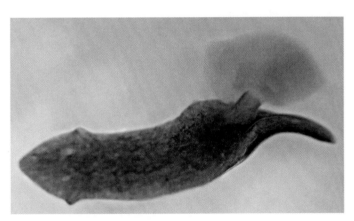

a. Planarian feeding

Figure 22.10   Micrograph of planarian feeding and the gastrovascular cavity.

b. Gastrovascular cavity

6. Transfer your worm to a concave depression slide and cover with a coverslip. Examine with a microscope and note the cilia on the ventral surface. Numerous gland cells secrete a mucous material that assists movement. Describe the mode of locomotion. _____

_____

_____

Conclusions: Planarians

- Planarians, with three germ layers, are more complex than cnidarians. Contrast a hydra with a planarian by stating in Table 22.1 significant organ differences between them.

- Planarians have no respiratory or circulatory system. As with cnidarians, each individual _____ takes care of its own needs for these two life functions.

Table 22.1	Contrasts Between a Hydra and a Planarian		
	Digestive System	Excretory System	Nervous Organization
Hydra			
Planarian			

Tapeworms

Tapeworms are parasitic flatworms known as cestodes. They live in the intestines of vertebrate animals, including humans (Fig. 22.11). The worms consist of a **scolex** (head), usually with suckers and hooks, and **proglottids** (segments of the body). Ripe proglottids detach and pass out with the host's feces, scattering fertilized eggs on the ground. If pigs or cattle happen to ingest these, larvae called bladder worms develop and eventually become encysted in muscle, which humans may then eat in poorly cooked or raw meat. A bladder worm that escapes from a cyst develops into a mature tapeworm attached to the intestinal wall.

1. How do humans get infected with the pig tapeworm? _____
_____

2. What is the function of a tapeworm's hooks and suckers? _____
_____

3. Proglottids mature into "bags of eggs." Given the life cycle of the tapeworm, why might a
tapeworm produce so many eggs? _____
_____

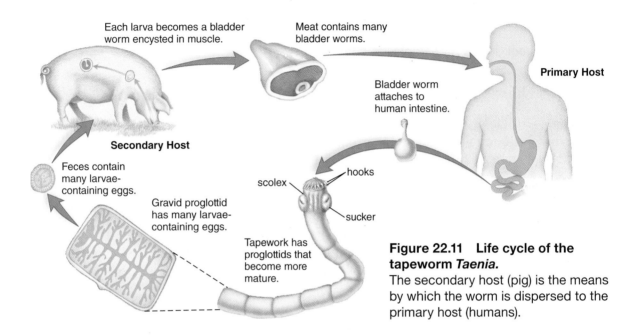

Each larva becomes a bladder worm encysted in muscle.

Meat contains many bladder worms.

Primary Host

Bladder worm attaches to human intestine.

Secondary Host

Feces contain many larvae-containing eggs.

Gravid proglottid has many larvae-containing eggs.

hooks

scolex

sucker

Tapework has proglottids that become more mature.

Figure 22.11   Life cycle of the tapeworm *Taenia*.
The secondary host (pig) is the means by which the worm is dispersed to the primary host (humans).

Observation: Tapeworm Anatomy

1. Examine a preserved specimen and/or slide of *Taenia pisiformis*, a tapeworm.
2. With the help of Figure 22.12, identify the scolex, with hooks and suckers, and the proglottids.

**Figure 22.12
Anatomy of *Taenia*.**
The adult worm is modified for its parasitic way of life. It consists of **a.** scolex and **b.** many proglottids, which become bags of eggs.

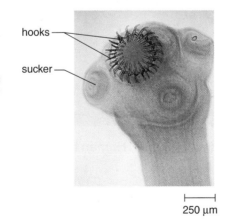

hooks

sucker

⊢————⊣
250 μm

a. Scolex functions to attach worm to host.

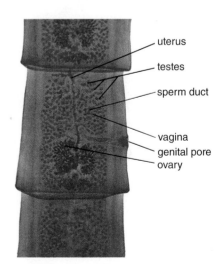

uterus

testes

sperm duct

vagina
genital pore
ovary

b. Proglottids function in reproduction.

Fluke Diversity and Anatomy

Flukes are parasitic flatworms known as trematodes. In general, the fluke body tends to be oval to elongate with no definite head except for an oral sucker at the anterior end. Usually there is also a ventral sucker for attachment to the host.

The many different types of flukes are usually designated by the type of organ they inhabit; for example, there are blood, liver, and lung flukes. Flukes have a life cycle that involves a secondary host (usually a water snail) and humans get infected with lung and liver flukes when they eat the snail or any other infected fish or shellfish without cooking thoroughly. The larvae of blood flukes bore through the skin of swimmers in tropical waters. Lung flukes cause chest pain and fever before coughing brings up fluke eggs, blood, and diseased tissue. Liver flukes result in an inflamed and diseased liver that is sensitive to touch. Blood flukes live in the larger blood vessels. The eggs of blood flukes have spines that allow them to enter the intestine or the bladder where they are released. Experts may be able to detect flukes in the stool or various samples of body fluids.

Observation: Fluke Anatomy

1. Examine a slide of a fluke such as the Chinese liver fluke (*Clonorchis sinensis*) using a dissecting microscope. Locate the oral and the ventral suckers. Can you also see the reproductive organs noted in Figure 22.13? _____

2. The human blood fluke, *Schistosoma mansoni*, is **dioecious**. Male and female organs are in separate individuals. Female worms are round, long, and slender (1.2–2.6 cm in length), while males are shorter and flattened. The males look cylindrical during copulation, however, because their body wall curves inward to form a canal in which the female resides.

   Examine a prepared slide of this blood fluke that shows a male and female fluke during copulation. Label the male and the female in Figure 22.14.

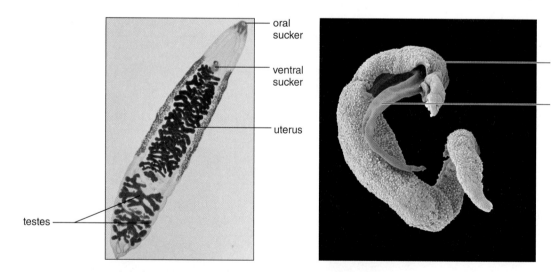

Figure 22.13 Chinese liver fluke.

Figure 22.14 Human blood fluke.

Planarians Versus Tapeworms and Flukes

Table 22.2 compares the adaptations of planarians, which are free-living, with tapeworms and flukes, which are parasitic. The parasitic flatworms have a way to attach themselves to their host and acquire nutrients. Therefore, the digestive system is reduced. On the other hand, when dispersal of offspring is a problem, the reproductive system expands.

Why would endoparasites such as flukes and tapeworms have a dispersal problem? _____

Table 22.2 Planarians Versus Parasitic Flatworms			
	Free-living	**Parasitic**	
	Planarians	*Flukes*	*Tapeworms*
Body wall	Ciliated epidermis	Specialized for absorption	Specialized for absorption
Cephalization	Head with eyespots and auricles	Oral suckers instead of head	Scolex with hooks and suckers
Nervous system	Nerves and brain	Reduced	Reduced
Digestive organ	Branched	Reduced	Absent
Reproductive organs	Hermaphroditic	Increased in volume	Extensively developed

22.5 Roundworms (Phylum Nematoda)

Like planarians, **roundworms** have three germ layers, bilateral symmetry, and various organs, including a well-developed nervous system. Both planarians and roundworms are nonsegmented; the body has no repeating units. In addition, roundworms have the following features:

1. **Complete digestive tract:** The digestive tract has both a mouth and an anus. What is the advantage of this? _____

2. **Pseudocoelom:** A body cavity, which allows space for the organs, is incompletely lined with mesoderm.

How would the presence of these two features lead to complexity? For example, how would a complete digestive system lead to a greater number of specialized organs such as both a small intestine that assists digestion and a large intestine that assists elimination?_____

How would a spacious body cavity (instead of packed mesoderm) promote a greater number of diverse internal organs such as a pancreas and a liver?_____

Roundworms are found in all aquatic habitats and in damp soil. Some even survive in hot springs, deserts, and cider vinegar. They parasitize (take nourishment from) both plants and animals. They are significant crop pests and also cause disease in humans. Both pinworms and hookworms are roundworms that cause intestinal difficulties; trichinosis and elephantiasis are also caused by roundworms. *Ascaris,* a large, primarily tropical intestinal parasite, is often studied as an example of this phylum.

Ascaris: *Dissection of Preserved Specimen*

1. Examine preserved specimens of *Ascaris,* both male and female (Fig. 22.15). In roundworms, the sexes are separate. The male is smaller and has a curved posterior end. Be sure to examine specimens of each sex.
2. Place a specimen of *Ascaris* in a dissecting pan, pinning the anterior and posterior ends to the wax so that the dorsal surface is exposed.
3. Carefully slit open the worm longitudinally, cutting along the middorsal line. Note the outer smooth, tough cuticle that covers the worm. Just like arthropods, roundworms molt, i.e., shed their outer covering.

Figure 22.15 Roundworm anatomy.
a. Photograph of male *Ascaris* **b.** Male reproductive system. **c.** Photograph of female *Ascaris*. **d.** Female reproductive system.

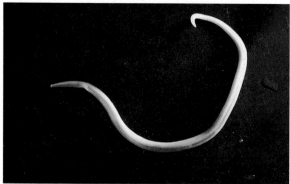

a. Male *Ascaris*

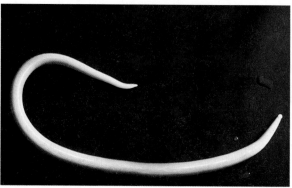

c. Female *Ascaris*

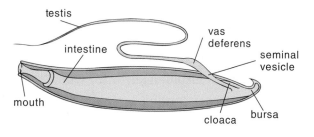

b. Male reproductive system

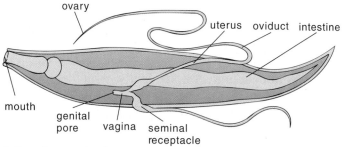

d. Female reproductive system

4. Pin the body open to expose the internal organs.
5. Add a small amount of water to the pan to prevent drying out. Note the body cavity, or pseudocoelom, now evident. Note also the inner tube (consisting of the digestive tract) and the outer tube (made up of the muscular body wall). Does this explain the phrase *tube-within-a-tube body plan?* _____
6. Examine the flattened digestive tract, which extends within the body cavity from the mouth to the anus. The anterior end forms a muscular pharynx, used to suck food materials into the mouth. The rest of the tract is a tube that passes to the anus.
7. Note the reproductive system. In the male, the reproductive structures are the **testis,** the **vas deferens,** and the **seminal vesicle** (Fig. 22.15*b*). These structures increase in size, from testes to vasa deferentia to seminal vesicles. In the female, the reproductive structures are the **ovary,** the **oviduct,** and the **uterus.** These structures increase in size, from ovaries to oviducts to uterus (Fig. 22.15*d*).

Trichinella

Trichinella is a parasitic roundworm that causes the disease **trichinosis**. When humans eat raw or undercooked pork infected with *Trichinella* cysts, juvenile worms are released in the digestive tract where they penetrate the wall of the small intestine and mature sexually. After male and female worms mate, females produce juvenile worms that migrate and form cysts in various human muscles (Fig. 22.16). A human with trichinosis has muscular aches and pains that can lead to death if the respiratory muscles fail.

Observation: Trichinella

1. Examine preserved, infected muscle or a slide of infected muscle, and locate the *Trichinella* cysts, which contain the juvenile worms.

2. How can trichinosis be prevented in humans?

   _____

3. How can pig farmers help to stamp out trichinosis so that humans are not threatened by the disease?_____

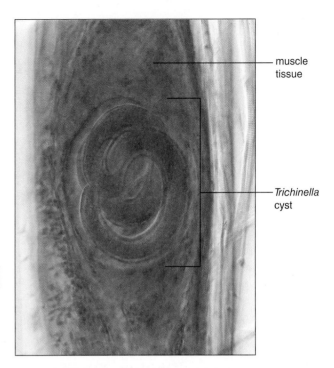

muscle tissue

Trichinella cyst

Figure 22.16   Larva of the roundworm *Trichinella* embedded in a muscle.
A larva coils in a spiral and is surrounded by a sheath derived from a muscle fiber.

Filarial Worm

A roundworm called a **filarial worm** infects lymphatic vessels and blocks the flow of lymph. The condition is called **elephantiasis** because when a leg is affected, it becomes massively swollen (Fig. 22.17).

Figure 22.17   Elephantiasis.

Vinegar Eels

Vinegar eels are tiny, free-living nematodes that can live in unpasteurized vinegar.

1. Examine live vinegar eels, and observe their active, whiplike swimming movements. This thrashing motion may be a result of nematodes having longitudinal muscles only; they lack circular muscles.
2. Select a few larger vinegar eels for further study, and place them in a small drop of vinegar on a clean microscope slide. If the eels are too active for study, you can slow them by briefly warming them or by adding methyl cellulose.
3. Try to observe the tubular digestive tract, which begins with the mouth and ends with the anus. Also, you may be able to see some of the reproductive organs, particularly in a large female vinegar eel.

Conclusion: Anatomy of Roundworms

• Nematodes are extremely plentiful, both in terms of their variety and overall number. From your knowledge of adaptive radiation, explain why there might be so many different types of nematodes.

_____

_____

22.6   Rotifers (Phylum Rotifera)

Rotifers are common and abundant freshwater animals. They are important constituents of the plankton of lakes, ponds, and streams, and are a significant food source for many species of fish and other animals.

Like roundworms, rotifers have a pseudocoelom and a complete digestive tract with a mouth and anus. The corona is a crown of cilia around the mouth. To some, the movement of the cilia resembles that of a rapid wheel; and in Latin, _rotifer_ means "wheel-bearer." The cilia draws water into the mouth and from there food is ground up by trophi (jaws) before entering the stomach from which nondigested remains pass through the cloaca and anus. Using Figure 22.18 as a guide, label Figure 22.19.

Figure 22.18   Live _Philodina_, a common rotifer.
The stomach contents are colored.

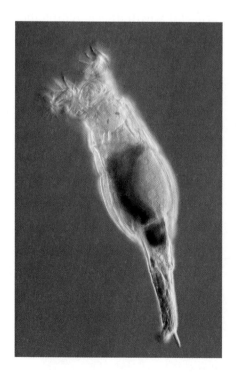

Living Specimen

1. Use a pipette to obtain a living rotifer specimen near a clump of vegetation. Place the liquid and rotifer on a concave depression slide. Do not add a cover slip. Study the animal's behavior and appearance.

2. Describe the rotifer's behavior. _____

    _____

3. Observe the elongated, cylindrical body that can be divided into three general regions: the head, the trunk, and a posterior foot. The rotifers have no true segmentation, but the cuticle covering the body can be divided into a number of superficial segments.

4. Observe the telescoping of the segments when the animal retracts its head.

5. Note also the large, ciliated **corona** at the anterior end. Most rotifers have a conspicuous corona that serves both for locomotion and for feeding. It creates a current that brings smaller microorganisms (e.g., algae, protozoans, bacteria) close enough to be swallowed.

6. Flame cells (there are several) are attached to tubules that take water and possibly metabolic wastes from the pseudocoelom to the cloaca. What other protostome studied today has both

    eyespots and flame cells? _____

7. With the help of Figure 22.19, try to identify some of the rotifer's internal organs. List those you

    observe here.

    _____   _____

    _____   _____

    _____   _____

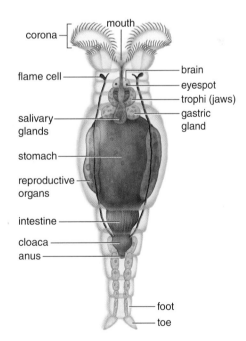

Figure 22.19   Rotifer anatomy.
Rotifers are only a few mm in length.

1. Distinguish invertebrate from vertebrate. _____

2. A hydra has radial symmetry but a planarian has bilateral symmetry. Explain. _____
_____

3. The animals studied today have some sort of defense. Contrast the defense of a sponge with that of a
hydra. _____
_____

4. If a cnidarian has both a polyp stage and a medusa stage, what is the life-cycle function of the medusa
stage? _____

5. Why would you expect an animal with bilateral symmetry to be more active than one with radial
symmetry? _____
_____

6. Name two contrasting anatomical features (other than symmetry) that distinguish a planarian from a
hydra.

Planarian	Hydra

7. Name a group of animals that is usually hermaphroditic and a group that is usually dioecious. _____
_____

8. How does the process of acquiring food in planarians differ from the process in tapeworms? _____
_____

9. Why would you expect a free-living roundworm to have a nervous system? _____
_____

10. Name two types of parasitic roundworms and one type of roundworm that is free-living. _____
_____

Biology Website

Enhance your study of the text and laboratory manual with study tools, practice tests, and virtual labs. Also ask your instructor about the resources available through ConnectPlus, including the media-rich eBook, interactive learning tools, and animations.

www.mhhe.com/maderbiology11

McGraw-Hill Access Science Website

An Encyclopedia of Science and Technology Online which provides more information including videos that can enhance the laboratory experience.

www.accessscience.com

23

Invertebrate Coelomates

Learning Outcomes

Introduction

The animals studied in this laboratory are **coelomates** because they have a body cavity completely lined with mesoderm, the last of the germ layers to appear during the evolution of animals. A coelom offers many advantages such as the digestive system and body wall can move independently; internal organs can become more complex; coelomic fluid can assist respiration, circulation, and excretion; and, in some animals, the coelom also serves as a hydrostatic skeleton because muscles can work against a fluid-filled cavity.

Among the coelomates, the molluscs, annelids, and arthropods are **protostomes**, animals in which the first *(protos)* embryonic opening becomes the mouth *(stoma),* while the echinoderms and the vertebrates, studied in Laboratory 24, are **deuterostomes**. In the deuterostomes, the first opening becomes the anus, and the second *(deutero)* opening becomes the mouth. The evolutionary relationship between protostomes and deuterostomes is shown in Figure 21.1.

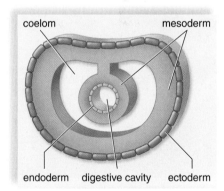

Coelomate (molluscs, annelids, arthropods, echinoderms, chordates)

23.1 Molluscs (Phylum Mollusca)

Most **molluscs** are marine, but there are also some freshwater and terrestrial molluscs (Fig. 23.1). Among molluscs, the grazing marine herbivores, known as **chitons**, have a body flattened dorsoventrally covered by a shell consisting of eight plates (Fig. 23.1*a*). The **bivalves** contain marine and freshwater sessile filter feeders, such as clams and scallops, with a body enclosed by a shell consisting of two valves (Fig. 23.1*b*). The **gastropods** contain marine, freshwater, and terrestrial species. In snails, the shell, if present, is coiled (Fig. 23.1*c*). The **cephalopods** contain marine active predators, such as squids and nautiluses. Tentacles are about the head (Fig. 23.1*d*).

All molluscs have a three-part body consisting of (1) a muscular **foot** specialized for various means of locomotion; (2) **visceral mass** that includes the internal organs; and (3) a **mantle,** a thin tissue that encloses the visceral mass and may secrete a shell. **Cephalization** is the development of a head region. On the lines provided in Figure 23.1, write *cephalization* or *no cephalization* as appropriate for this mollusc.

Figure 23.1   Molluscan diversity.
a. You can see the exoskeleton of this chiton but not its dorsally flattened foot. **b.** A scallop doesn't have a foot but it does have strong adductor muscles to close the shell. In this specimen, the edge of the mantel bears tentacles and many blue eyes. **c.** A gastropod, such as a snail, is named for the location of its large foot beneath the visceral mass. **d.** In a cephalopod, such as this nautilus, a funnel (its foot) opens in the area of the tentacles and allows it to move by jet propulsion.

a. Chitons, *Tonicella*

c. Snail, *Helix* is a gastropod.

b. Scallop, *Aequipecten* is a bivalve.

d. Nautilus, *Nautilus* is a cephalopod.

Anatomy of a Clam

Clams are bivalved because they have right and left shells secreted by the mantle. Clams have no head, and they burrow in sand by extending a **muscular foot** between the valves. Clams are **filter feeders** and feed on debris that enters the mantle cavity. In the visceral mass, the blood leaves the heart and enters sinuses (cavities) by way of anterior and posterior aortas. There are many different types of clams. The one examined here is the freshwater clam *Venus*.

External Anatomy

1. Examine the external shell (Fig. 23.2) of a preserved clam *(Venus)*. The shell is an **exoskeleton.**
2. Find the posterior and anterior ends. The more pointed end of the **valves** (the halves of the shell) is the posterior end.
3. Determine the clam's dorsal and ventral regions. The valves are hinged together dorsally.
4. What is the function of a heavy shell? _____

Figure 23.2 External view of the clam shell.

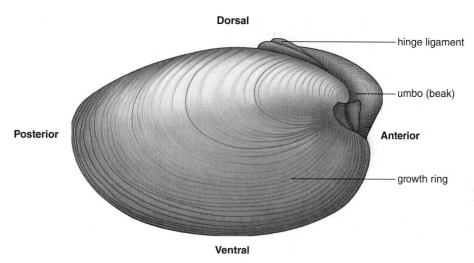

Dorsal

hinge ligament

umbo (beak)

Posterior

Anterior

growth ring

Ventral

Internal Anatomy

1. Place the clam in the dissecting pan, with the **hinge ligament** and **umbo** (blunt dorsal protrusion) down. Carefully separate the **mantle** from the right valve by inserting a scalpel into the slight

   opening of the valves. What is a mantle? _____

   _____

2. Insert the scalpel between the mantle and the valve you just loosened.
3. The **adductor muscles** hold the valves together. Cut the adductor muscles at the anterior and posterior ends by pressing the scalpel toward the dissecting pan. After these muscles are cut, the

   valve can be carefully lifted away. What is the advantage of powerful adductor muscles? _____

   _____

4. Examine the inside of the valve you removed. Note the concentric lines of growth on the outside, the hinge teeth that interlock with the other valve, the adductor muscle scars, and the mantle line. The inner layer of the shell is mother-of-pearl.
5. Examine the rest of the clam (Fig. 23.3) attached to the other valve. Notice the adductor muscles and the mantle, which lies over the visceral mass and foot.
6. Bring the two halves of the mantle together. Explain the term *mantle cavity.* _____

   _____

7. Identify the **incurrent** (more ventral) and **excurrent siphons** at the posterior end (Fig. 23.3).

   Explain how water enters and exits the mantle cavity. _____

   _____

Figure 23.3 Anatomy of a bivalve.

The mantle has been removed to reveal the internal organs. **a.** Drawing. **b.** Dissected specimen.

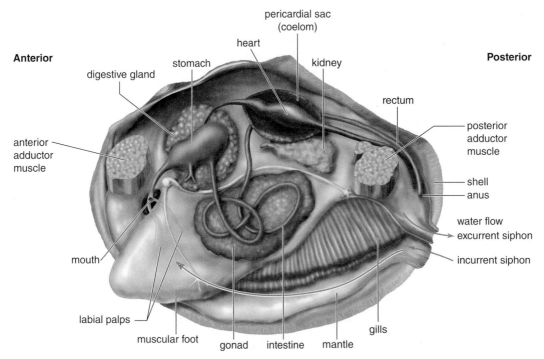

a.

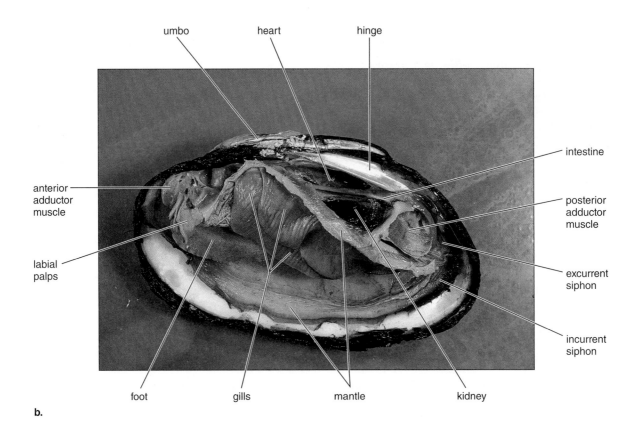

b.

8. Cut away the free-hanging portion of the mantle to expose the **gills.** Does the clam have a respiratory organ? _____

If so, what type of respiratory organ? _____

9. A mucous layer on the gills entraps food particles brought into the mantle cavity, and the cilia on the gills convey these food particles to the mouth. Why is the clam called a filter feeder?

_____

_____

10. The nervous system is composed of three pairs of ganglia (located anteriorly, posteriorly, and in the foot), all connected by nerves. The clam does not have a brain. A ganglion contains a limited number of neurons, whereas a brain is a large collection of neurons in a definite head region.

11. Identify the **foot,** a tough, muscular organ for locomotion, and the **visceral mass,** which lies above the foot and is soft and plump. The visceral mass contains the digestive and reproductive organs.

12. Identify the **labial palps** that channel food into the open mouth.

13. Identify the **anus,** which discharges into the excurrent siphon.

14. Find the **intestine** by its dark contents. Trace the intestine forward until it passes into a sac, the clam's only evidence of a coelom.

15. Locate the **pericardial sac (pericardium)** that contains the heart. The intestine passes through the heart. The heart pumps blood into the aortas, which deliver it to blood sinuses (open spaces) in the tissues.

A clam has an **open circulatory system.** Explain. _____

_____

16. Cut the visceral mass and the foot into exact left and right halves, and examine the cut surfaces. Identify the digestive glands, greenish-brown; the stomach, embedded in the digestive glands; and the intestine, which winds about in the visceral mass. Reproductive organs (gonads) are also present. Typically, clams have separate sexes.

Anatomy of a Squid

Squids are cephalopods because they have a well-defined head; the foot became the funnel surrounded by two and the many tentacles about the head. The head contains a brain and bears sense organs. The squid moves quickly by jet propulsion of water, which enters the mantle cavity by way of a space that encircles the head. When the cavity is closed off, water exits by means of the funnel. Then the squid moves rapidly in the opposite direction.

The squid seizes fish with its tentacles; the mouth has a pair of powerful, beaklike jaws and a **radula,** a beltlike organ containing rows of teeth. The squid has a **closed circulatory system** composed only of blood vessels and three hearts, one of which pumps blood to all the internal organs, while the other two pump blood to the gills located in the mantle cavity.

Observation: Anatomy of a Squid

1. Examine a preserved squid.

2. Refer to Figure 23.4 for help in locating the mouth (defined by beaklike jaws and containing a radula) and the tentacles and arms, which encircle the mouth.

3. Locate the head with its sense organs, notably the large, well-developed eye.

4. Find the funnel, where water exits from the mantle cavity, causing the squid to move backward.

5. If the squid has been dissected, note the gills, heart, and blood vessels. Confirm that the squid has a closed circulatory system.

Figure 23.4 Anatomy of a squid.

The squid is an active predator and lacks the external shell of a clam. It captures fish with its tentacles and bites off pieces with its jaws. A strong contraction of the mantle forces water out the funnel, resulting in "jet propulsion." **a.** Drawing of internal anatomy. **b.** Dissected specimen.

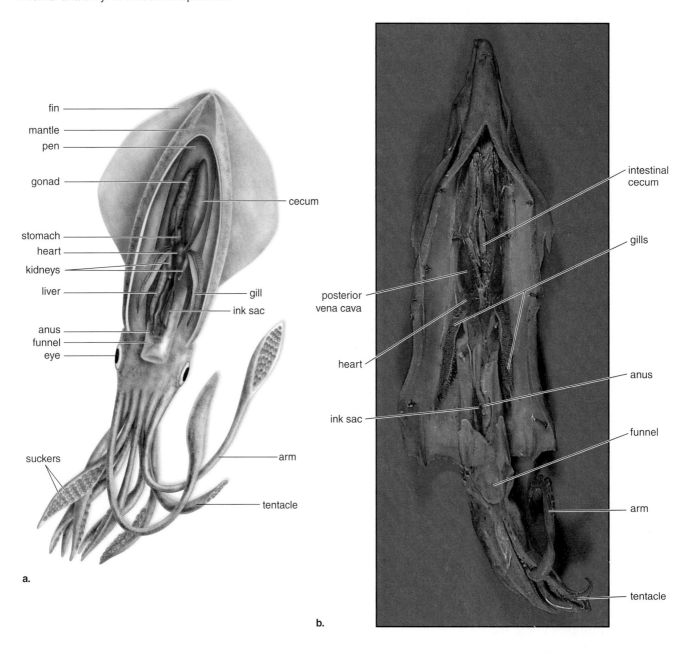

a.

b.

Conclusion: Comparison of Clam to Squid

- Compare clam anatomy to squid anatomy by completing Table 23.1.
- Use the information you entered into Table 23.1 to demonstrate how both clams and squids are adapted to their way of life.

_____

_____

_____

Table 23.1 Comparison of Clam to Squid

	Clam	Squid
Feeding mode		
Skeleton		
Circulation		
Cephalization		
Locomotion		

23.2 Annelids (Phylum Annelida)

Annelids are the **segmented** worms, so called because the body is divided into a number of segments and has a ringed appearance. The circular and longitudinal muscles work against the fluid-filled coelom to produce changes in width and length (Fig. 23.5). Therefore, annelids are said to have a **hydrostatic skeleton.**

Figure 23.5 Locomotion in the earthworm.
Contraction of first circular muscles and then longitudinal muscles allow the earthworm to move forward.

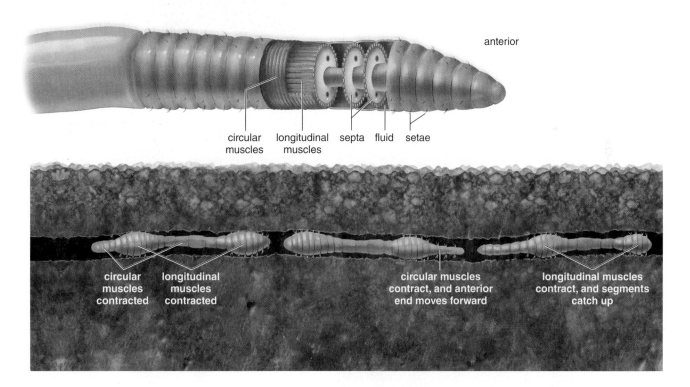

anterior

circular muscles • longitudinal muscles • septa • fluid • setae

circular muscles contracted • longitudinal muscles contracted

circular muscles contract, and anterior end moves forward

longitudinal muscles contract, and segments catch up

Among annelids, the **polychaetes** have many slender bristles called **setae;** the **earthworms,** which are oligochaetes, have fewer setae; and the **leeches** usually have no setae (Fig. 23.6). Polychaetes, almost all marine, are plentiful from the intertidal zone to the ocean depths. They are quite diverse, ranging from jawed forms that are carnivorous to fanworms that live in tubes and extend feathery filaments when filter feeding. Earthworms have a worldwide distribution in almost any soil, are often found in large numbers, and reach a length of as much as 3 meters. Leeches are known as bloodsuckers; the medicinal leech has been used in the practice of bloodletting.

Show that the annelids are the segmented worms by labeling a segment in Figure 23.6a, b, and d. In which group would you expect the animals to be predators based on the type of head region?

Figure 23.6 Annelid diversity.
Aside from **a.** earthworms, which are oligochaetes, there are **b.** clam worms and **c.** fanworms, both of which are polychaetes, and **d.** leeches. Note the obvious segmentation.

head region

head
(sense organs
and jaws)

a. Earthworms, *Lumbricus*, mating

b. Clam worm, *Nereis*

head region
(feathery arms)

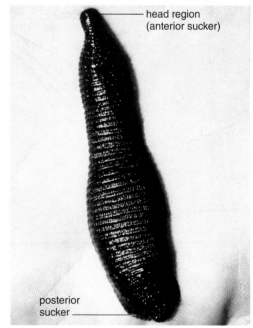

head region
(anterior sucker)

posterior
sucker

c. Giant fanworm, *Eudistylia*

d. Leech, *Hirudo*

Anatomy of the Earthworm

Earthworms are **segmented** in that the body has a series of ringlike segments. Earthworms have no head and burrow in the soil by alternately expanding and contracting muscles along the length of the body (see Fig. 23.5).

Earthworms are scavengers that feed on decaying organic matter in the soil. They have a well-developed coelomic cavity, providing room for a well-developed digestive tract and both sets of reproductive organs. Earthworms are **hermaphroditic**.

Observation: Anatomy of the Earthworm

External Anatomy

1. Examine a live or preserved specimen of an earthworm. Locate the small projection that sticks out over the mouth. Has cephalization occurred? _____ Explain. _____

_____

2. Count the total number of segments, beginning at the anterior end. The sperm duct openings are on segment 15 (Fig. 23.7). The enlarged section around a short length of the body is the **clitellum.** The clitellum secretes mucus that holds the worms together during mating. It also functions as a cocoon, in which fertilized eggs hatch and young worms develop. The anus is located on the worm's terminal segment.

3. Lightly pass your fingers over the earthworm's ventral and lateral sides. Do you feel the setae? _____

_____

Earthworms insert these slender bristles into the soil. Setae, along with circular and longitudinal muscles, enable the worm to locomote. Explain the action. _____

_____

Figure 23.7 External anatomy of an earthworm.
In this drawing, the segments are numbered.

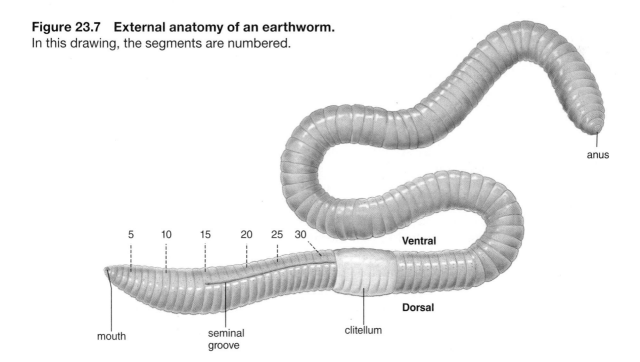

Internal Anatomy

1. Place a preserved earthworm on its ventral side in the dissecting pan. With a scalpel or razor blade, make a shallow incision slightly to the side of the blackish median dorsal blood vessel (Fig. 23.8a). Start your incision about ten segments after the clitellum, and proceed anteriorly to the mouth. If you see black ooze, you have accidentally cut the intestine.
2. Identify the thin partitions, the **septa,** between segments.
3. Lay out the **body wall,** and pin every tenth segment to the wax in your pan. Add water to prevent drying out. While alive, the earthworm body wall is always moist and this facilitates gas exchange across the body wall. Notice the earthworm has no respiratory organ.
4. An earthworm is a scavenger and feeds on **detritus** (organic matter) in the soil. Identify the digestive tract, which begins at the mouth and extends through body segments 1, 2, and 3 (Fig. 23.8a and b). It opens into the swollen, muscular, thick-walled **pharynx,** which extends from segment 3 to segment 6. The **esophagus** extends from the pharynx through segment 14 to the **crop.** Next is the **gizzard,** which lies in segments 17 through 28. The intestine extends from the gizzard to the anus. Does the digestive system show specialization of parts? _____

   Explain. _____

   _____

5. Identify the earthworm's circulatory system. The blood is always contained within vessels and never runs free. The **dorsal blood vessel** is readily seen along the dorsal side of the digestive tract. A series of aortic arches or "hearts" encircles the esophagus between segments 7 and 11, connecting the dorsal blood vessel with the ventral blood vessel. Does the earthworm have an open or closed circulatory system? _____ Explain. _____

   _____

6. Locate the earthworm's nervous system. The two-lobed brain is located dorsal to the pharynx in segment 3. Two nerves, one on each side of the pharynx, connect the brain to a ganglion that lies below the pharynx in segment 4. The **ventral nerve cord** then extends along the floor of the body cavity to the last segment.
7. Find the earthworm's excretory system, which consists of a pair of minute, coiled, white tubules, the **nephridia,** located in every segment except the first three and the last. Each nephridium opens to the outside by means of an excretory pore. Does the excretory system show that the earthworm is segmented? _____ Explain. _____

   _____

   _____

8. Identify the earthworm's reproductive system, including **seminal vesicles,** light-colored bodies in segments 9 through 12, which house maturing sperm that have been formed in two pairs of testes within them; **sperm ducts** that pass to openings in segment 15; and **seminal receptacles** (four small, white, special bodies that lie in segments 9 and 10), which store sperm received from another worm. **Ovaries** are located in segment 13 but are too small to be seen.
9. During mating, earthworms are arranged so that the sperm duct openings of one worm are just about, but not quite, opposite the seminal receptacle openings of the other worm. After being released, the sperm pass down a pair of seminal grooves on the ventral surface (see Fig. 23.7) and then cross over at the level of the seminal receptacles of the opposite worm. Once the worms separate, eggs and sperm are released into a cocoon secreted by the clitellum. Is the earthworm hermaphroditic? _____ Explain. _____

   _____

10. Does the earthworm have a respiratory system (i.e., gills or lungs)? _____ Review 3.

How does the worm exchange gases? _____

Why then would you expect an earthworm to lack an exoskeleton? _____

_____

Figure 23.8 Internal anatomy of an earthworm, dorsal view.
a. Drawing shows internal organs and a cross section. The segments are numbered.
b. Dissected specimen.

Virtual Lab Earthworm Dissection
A virtual lab called Earthworm Dissection is available on the *Biology* website **www.mhhe.com/ maderbiology11**. Follow the directions given to click and drag labels to external and internal illustrations of earthworm anatomy.

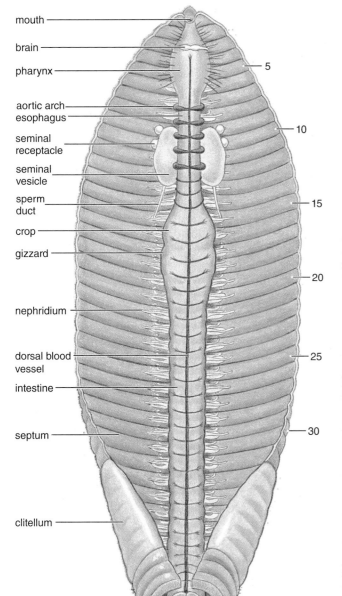

a.

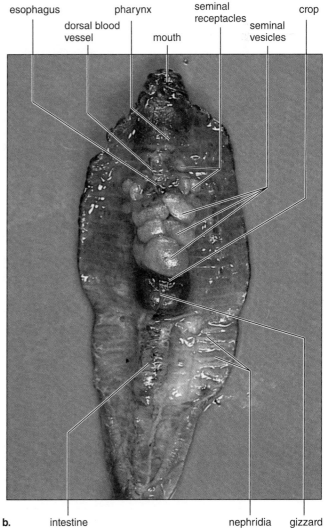

b.

1. Obtain a prepared slide of a cross section of an earthworm (Fig. 23.9). Examine the slide under the dissecting microscope and under the light microscope.
2. Identify the following structures in particular:
   a. **Body wall:** A thick outer circle of tissue, consisting of the **cuticle** and the **epidermis.**
   b. **Coelom:** A relatively clear space with scattered fragments of tissue.
   c. **Intestine:** An inner circle with a suspended fold.
   d. **Typhlosole:** A fold that increases the intestine's surface area.
   e. **Ventral nerve cord:** A white, threadlike structure.
3. Does the typhlosole help in nutrient absorption? _____ Explain. _____

Figure 23.9   Cross section of an earthworm.
Cross-section slide as it would appear under the microscope.

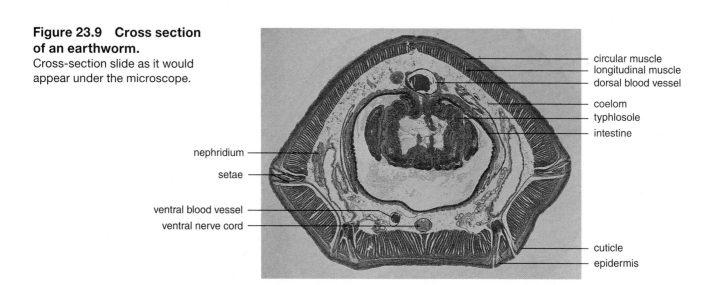

labels: nephridium, setae, ventral blood vessel, ventral nerve cord, circular muscle, longitudinal muscle, dorsal blood vessel, coelom, typhlosole, intestine, cuticle, epidermis

Conclusion: Comparison of Clam to Earthworm

- Complete Table 23.2 to compare the life style and anatomy of a clam to that of an earthworm.

Table 23.2 Comparison of Clam to Earthworm		
	Clam	**Earthworm**
Habitat (where they live)		
Feeding mode		
Skeleton		
Segmentation		
Circulation		
Respiratory organs		
Locomotion		
Reproductive organs		

23.3 Arthropods (Phylum Arthropoda)

Arthropods have paired, **jointed appendages** and a hard exoskeleton that contains **chitin**. The chitinous exoskeleton consists of hardened plates separated by thin, membranous areas that allow movement of the

Virtual Lab Arthropod Classification
A virtual lab called Arthropod Classification is available on the *Biology* website **www.mhhe.com/ maderbiology11**. Using the information provided in a reference guide students are asked to match five specimens to five classes of arthropods.

body segments and appendages. The hard exoskeleton of arthropods causes them to **molt**: as soon as they have grown larger, they shed their old exoskeleton and grow a new larger one.

Figure 23.10*a* features insects and relatives. Insects with three pairs of legs, with or without wings, and three distinct body regions comprise 95% of all arthropods. Millipedes have two pairs of legs per segment, while centipedes have one pair of legs per segment. Figure 23.10*b* features spiders and relatives. Spiders and scorpions have four pairs of legs, no antennae, and a cephalothorax (head and thorax are fused). The horseshoe crab is a living fossil. It has remained unchanged for thousands of years. The crustaceans (Fig. 23.10*c*), which include crabs, shrimp, and lobsters, have three to five pairs of legs, and two pairs of antennae. Barnacles are unusual, in that their legs are used to gather food.

For each animal in Figure 23.10, circle all obvious appendages.

Figure 23.10 Arthropod diversity.
a. Among arthropods insects, millipedes, and centipedes are possibly related. **b.** Spiders, scorpions, and horseshoe crabs are related. **c.** Crabs, shrimp, and barnacles, among others, are crustaceans.

a. Insects and relatives

Honeybee, *Apis mellifera*

Millipede, *Ophyiulus pilosus*

Centipede, *Scolopendra* sp.

b. Spider and relatives

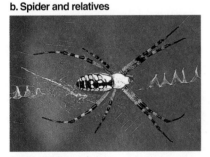

Spider, *Argiope rafaria*

Scorpion, *Hadrurus hirsutus*

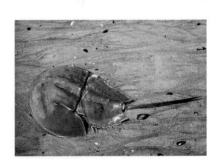

Horseshoe crab, *Limulus polyphemus*

c. Crustaceans

Crab, *Cancer productus*

Shrimp, *Stenopus* sp.

Barnacles, *Lepas anatifera*

Anatomy of a Crayfish

Crayfish also belong to the group of arthropods called **crustaceans**. Crayfish are adapted to an aquatic existence. They are known to be scavengers, but they also prey on other invertebrates. The mouth is surrounded by appendages modified for feeding, and there is a well-developed digestive tract. Dorsal, anterior, and posterior arteries carry **hemolymph** (blood plus lymph) to tissue spaces (hemocoel) and sinuses. Therefore, a crayfish has an open circulatory system.

Observation: Anatomy of a Crayfish

External Anatomy

1. In a preserved crayfish, identify the chitinous **exoskeleton.** With the help of Figure 23.11, identify the head, thorax, and abdomen. Arthropods are segmented but the segments have become specialized. Together, the head and thorax are called the **cephalothorax;** the cephalothorax is

Figure 23.11
Anatomy of a crayfish.
a. External anatomy.
b. Internal anatomy.

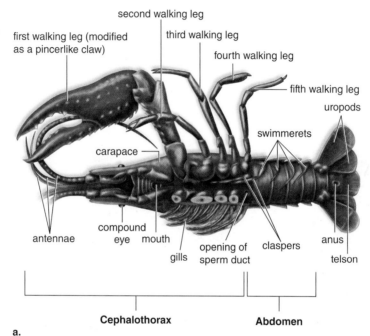

covered by the **carapace.** Has specialization of segments occurred? _____

Explain. _____

2. Find the **antennae,** which project from the head. At the base of each antenna, locate a small, raised nipple containing an opening for the **green glands,** the organs of excretion. Crayfish excrete a liquid nitrogenous waste.

3. Locate the **compound eyes,** composed of many individual units for sight. Do crayfish demonstrate cephalization? _____ Explain. _____

4. Identify the six pairs of appendages around the mouth for handling food.

5. Find the five pairs of walking legs attached to the cephalothorax. The most anterior pair is modified as pincerlike claws. In the female, identify the **seminal receptacles,** a swelling located between the bases of the third and fourth pairs of walking legs. Sperm from the male are deposited in the seminal receptacles. In the male, identify the opening of the sperm duct located at the base of the fifth walking leg.

6. Locate the five pairs of **swimmerets** on the abdomen. Crayfish use their swimmerets to swim and move water over their gills. In males, the anterior two pairs are stiffened and folded forward. They are **claspers** that aid in the transfer of sperm during mating. In females, clusters of eggs and larvae attach to the swimmerets. Find the last abdominal segment, which bears a pair of broad, fan-shaped **uropods** that, together with a terminal extension of the body, form a tail.

7. Has specialization of appendages occurred? _____ Explain. _____

Internal Anatomy

1. Cut away the lateral surface of the carapace with scissors to expose the **gills** (Fig. 23.11*b*). Observe that the gills occur in distinct, longitudinal rows. How many rows of gills are there in your specimen? _____ The gills are attached to the base of certain appendages. Which ones?

_____

2. Remove a gill with your scissors by cutting it free near its point of attachment, and place it in a watch glass filled with water. Observe the numerous gill filaments arranged along a central axis.

3. Carefully cut away the dorsal surface of the carapace with scissors and a scalpel. The epidermis that adheres to the exoskeleton secretes the exoskeleton. Remove any epidermis adhering to the internal organs.

4. Identify the diamond-shaped heart lying in the middorsal region. A crayfish has an open circulatory system. Carefully remove the heart.

5. Locate the **gonads** anterior to the heart in both the male and female. The gonads are tubular structures bilaterally arranged in front of the heart and continuing behind it as a single mass. In the male, the testes are highly coiled, white tubes.

6. Find the **mouth;** the short, tubular **esophagus;** and the two-part **stomach,** with the attached **digestive gland,** that precedes the intestine.

7. Identify the **green glands,** two excretory structures just anterior to the stomach, on the ventral segment wall.

8. Remove the thoracic contents previously identified.

9. Identify the **brain** in front of the esophagus. The brain is connected to the ventral nerve cord by a pair of nerves that pass around the esophagus.

10. Remove the animal's entire digestive tract, and float it in water. Observe the various parts, especially the connections of the digestive gland to the stomach.

11. Cut through the stomach, and notice in the anterior region of the stomach wall the heavy, toothlike projections, called the **gastric mill,** which grind up food. Do you see any grinding stones ingested by the crayfish? _____

If possible, identify what your specimen had been eating. _____

Anatomy of a Grasshopper

The grasshopper is an **insect.** All insects have a head, a thorax, and an abdomen. Their appendages always include (1) three pairs of jointed legs and usually (2) two antennae as sensory organs. Grasshoppers are adapted to live on land. Wings and jumping legs are suitable for locomotion on land; **Malpighian tubules** save water by secreting a solid nitrogenous waste; the **tracheae** are tiny tubules that deliver air directly to the muscles; and the male has a penis with attached claspers to deliver sperm to the seminal receptacles of a female so they do not dry out.

Observation: Anatomy of a Grasshopper

External Anatomy

1. Obtain a preserved grasshopper *(Romalea),* and study its external anatomy with the help of Figure 23.12*a.* Identify the head, thorax, and abdomen.

2. Use a hand lens or dissecting microscope to examine the grasshopper's special sense organs of the **head.** Identify the **antennae** (a pair of long, jointed feelers), the **compound eyes,** and the three, dotlike **simple eyes.** The labial palps, labeled in Figure 23.12*a,* have sense organs for tasting food.

3. Note the sturdy **mouthparts,** which are used for chewing plant material. A grasshopper's mouth parts are quite different from those of a piercing and sucking insect.

4. Locate the leathery **forewings** and the inner, membraneous **hindwings** attached to the **thorax.**

   Which pair of legs is used for jumping? _____ How many segments does each leg

   have? _____

5. Is locomotion in the grasshopper adapted to land? _____

   Explain. _____

   _____

6. In the **abdomen,** identify the **tympana** (sing., **tympanum**), one on each side of the first abdominal segment (Fig. 23.12*a*). The grasshopper detects sound vibrations with these membranes.

Figure 23.12   Female grasshopper.
a. External anatomy. **b.** Internal anatomy.

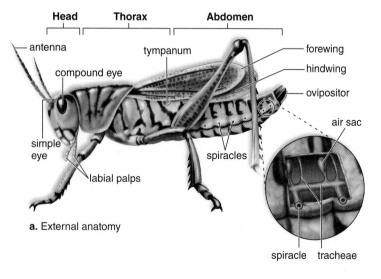

a. External anatomy

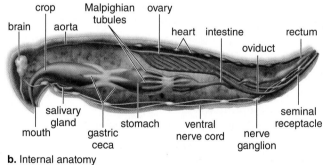

b. Internal anatomy

Figure 23.13 Grasshopper genitalia.
a. Females have an ovipositor, and **b.** males have claspers at the distal end of the penis.

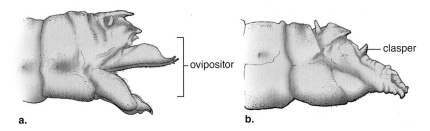

a. b.

7. Locate the **spiracles,** along the sides of the abdominal segments. These openings allow air to enter the tracheae, which constitute the respiratory system.
8. Find the **ovipositors** (Figs. 23.12*a* and 23.13*a*), four curved and pointed processes projecting from the abdomen of the female. These are used to dig a hole in which eggs are laid. The male has a **penis** with **claspers** used during copulation (Fig. 23.13*b*).

Internal Anatomy

Observe a longitudinal section of a grasshopper if available on demonstration. Try to locate the structures shown in Figure 23.12*b*. What is the function of Malpighian tubules?_____

Conclusion: Comparison of Crayfish to Grasshopper

Compare the adaptations of a crayfish to those of a grasshopper by completing Table 23.3. Put a star beside each item that indicates an adaptation to life in the water (crayfish) and to life on land (grasshopper). Check with your instructor to see if you identified the maximum number of adaptations.

Table 23.3 Comparison of Crayfish to Grasshopper		
	Crayfish	**Grasshopper**
Locomotion		
Respiration		
Sense organs		
Nervous system		

External reproductive features
    Male
    Female		

Grasshopper Metamorphosis

Metamorphosis means a change, usually a drastic one, in form and shape. Grasshoppers undergo *incomplete metamorphosis,* a gradual change in form rather than a drastic change. The immature stages of the grasshopper are called **nymphs,** and they are recognizable as grasshoppers even though they differ somewhat in shape and form (Fig. 23.14*a*). Some insects undergo what is called *complete metamorphosis,* in which case they have three stages of development: **larvae, pupa,** and **adult** (Fig. 23.14*b*). Metamorphosis occurs during the pupa stage when the animal is enclosed within a hard covering. The animals best known for complete metamorphosis are the butterfly and the moth, whose larval stage is called a caterpillar and whose pupa stage is the cocoon; the adult is the butterfly or moth.

Figure 23.14 Grasshopper metamorphosis.
During **a.** incomplete metamorphosis of a grasshopper, a series of **nymphs** leads to a full-grown grasshopper.
b. complete metamorphosis of a moth, a series of larvae lead to pupation. The **adult** hatches out of the pupa.

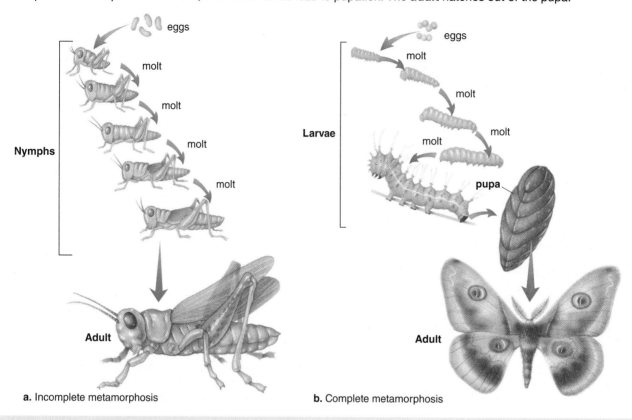

a. Incomplete metamorphosis **b.** Complete metamorphosis

Observation: Grasshopper Metamorphosis

1. Use Figure 23.14 to add grasshopper and moth to Table 23.4.
2. Examine any specific life cycle displays or plastomounts that illustrate complete and incomplete metamorphosis and add these examples to Table 23.4.

Table 23.4 Insect Metamorphosis	
Specimen	**Complete or Incomplete Metamorphosis**

Conclusions: Insect Metamorphosis

- In insects with incomplete metamorphosis, do the nymphs or the adult have better developed wings? _____ What is the benefit of wings to an insect?_____

  _____

- What stage is missing when an insect has incomplete metamorphosis? _____ What happens during this stage? _____

- What form, the larvae or the adult, disperses new individuals in flying insects that exhibit complete metamorphosis? _____

  How is this a benefit? _____

  _____

- In insects that undergo complete metamorphosis, the larvae and the adults utilize different food sources and habitats. Why might this be a benefit? _____

  _____

23.4 Echinoderms (Phylum Echinodermata)

The echinoderms are the only invertebrate group that shares deuterostome development with the vertebrates (see Fig. 22.15). Unlike the vertebrates, all **echinoderms** are marine, and they dwell on the seabed, either attached to it, like sea lilies, or creeping slowly over it. The name *echinoderm* means "spiny-skinned," and most members of the group have defensive spines on the outside of their bodies. The spines arise from an **endoskeleton** composed of calcium carbonate plates. The endoskeleton supports the body wall and is covered by living tissue that may be soft (as in sea cucumbers) or hard (as in sea urchins).

Especially note that (1) adult echinoderms are radially symmetrical, with generally five points of symmetry arranged around the axis of the mouth but the larvae are bilaterally symmetrical and (2) the echinoderms' most unique feature is their **water vascular system.** In those echinoderms in which the arms make contact with the substratum, the **tube feet** associated with the water vascular system are used for locomotion. In other echinoderms, the tube feet are used for gas exchange and food gathering.

Echinoderms belong to one of five groups: sea lilies and feather stars; sea stars; brittle stars; sea urchins and sand dollars; and sea cucumbers. When appropriate in Figure 23.15, write "obvious radial symmetry" or "radial symmetry not obvious" beside each photo.

Figure 23.15 Echinoderm diversity.

a. Sea lily, *Comonthino* sp. _____

b. Sea star, *Pentaceraster cumingi* _____

c. Brittle stars, *Ophiopholis aculeata* _____

d. Sea urchin, *Stronglocentrotus pranciscanus* _____

e. Sand dollar, *Dendraster excentricus* _____

f. Sea cucumber, *Pseudocolochirus* sp. _____

Anatomy of a Sea Star

Sea stars (starfish) usually have five arms that radiate from a central disk. The mouth is normally oriented downward, and when sea stars feed on clams, they use the suction of their tube feet to force the shells open a crack. Then they evert the cardiac portion of the stomach, which releases digestive juices into the mantle cavity. Partially digested tissues are taken up by the pyloric portion of the stomach; digestion continues in this portion of the stomach and in the digestive glands found in the arms.

Observation: Anatomy of a Sea Star

External Anatomy

1. Place a preserved sea star in a dissecting pan so that the aboral side is uppermost.
2. With the help of Figure 23.16, identify the **central disk** and five arms. What type of symmetry does an adult sea star have? _____

Figure 23.16   Anatomy of a sea star.
a. Diagram and **b.** image of dissected sea star. Both show the aboral side. **c.** Image of dissected cut arm. **d.** Canals and tube feet of water vascular system. Seen from aboral side.

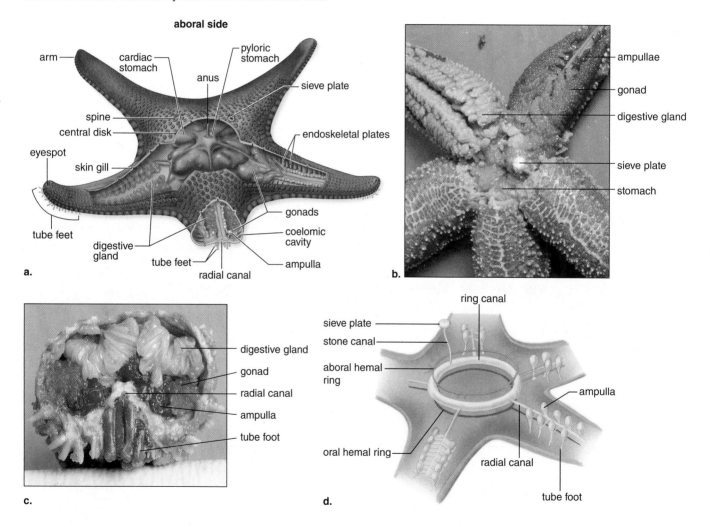

3. With the oral side uppermost, find the mouth, located in the center and protected by spines. Why is this side of the sea star called the oral side? _____

4. Locate the groove that runs along the middle of each arm and the **tube feet** (suctionlike disks) in rows on either side of the groove. Pluck away the tube feet from one area.

   How many rows of feet are there? _____ In the valley of each ambulacral groove, identify the **radial canal** that extends to the tip of each arm.

5. Turn the sea star over to its aboral side.

6. Locate the anal opening (Fig. 23.16). Why is this side of the sea star called the aboral side? _____

   _____

7. Lightly run your fingers over the surface spines. These extend from the calcium carbonate plates that lie buried in the body wall beneath the surface. The plates form an endoskeleton of the animal.

8. Identify the **sieve plate (madreporite),** a brownish, circular spot between two arms where water enters the water vascular system.

Internal Anatomy

1. Place the sea star so that the aboral side is uppermost. Refer to Figure 23.16a and b as you dissect the sea star following these instructions:

2. Cut the tip of one of the arms and, with scissors, carefully cut through the body wall along each side of this arm.

3. Carefully lift up the upper body wall. Separate any internal organs that may be adhering so that all internal organs are left intact.

4. Cut off the body wall near the central disk, but leave the sieve plate (madreporite) in place.

5. Remove the body wall of the central disk, being careful not to injure the internal organs.

6. Identify the digestive system. The mouth leads into a short **esophagus,** which is connected to the saclike **cardiac stomach.** When a sea star eats, the cardiac stomach sticks out through the sea star's mouth and starts digesting the contents of a clam or oyster. Above the cardiac stomach is the **pyloric stomach,** which leads to a short intestine. Each arm contains one pair of **digestive glands.**

   To which stomach do the digestive glands attach? _____

7. Cut off a portion of an arm, and examine the cut edge (Fig. 23.16c). Identify the digestive glands, ambulacral groove, radial canal, ampullae, and tube feet.

8. Remove the digestive glands of one arm.

9. Identify the **gonads** extending into the arm. What is the function of gonads? _____

   _____

   It is not possible to distinguish male sea stars from females by this observation.

10. Remove both stomachs.

11. In the **water vascular system** (Fig. 23.16d), you have already located the sieve plate and tube feet. Now try to identify the following components:
    a. **Stone canal:** Takes water from the sieve plate to the ring canal.
    b. **Ring canal:** Surrounds the mouth and takes water to the radial canals.
    c. **Radial canals:** Send water into the ampulla. When the ampullae contract, water enters the tube feet. Each tube foot has an inner muscular sac called an ampulla. The ampulla contracts and forces water into the tube foot.

    What is the function of the water vascular system? _____

    _____

1. List two advantages of having a coelom. _____

_____

2. What is a protostome, deuterostome? (Explain what these terms mean.) _____

_____

3. What are the three general characteristics of molluscs? _____

_____

4. Compare the locomotion of a clam to the locomotion of a squid. How is each suitable to the way the animal acquires food? _____

_____

5. Give evidence that earthworms are segmented by stating an organ that occurs in most every segment.

_____

6. Externally, arthropods have specialized segments compared to annelids. Explain. _____

_____

_____

7. Compare respiratory organs in the crayfish and the grasshopper. How are these suitable to the habitat of each? _____

_____

8. All animals studied in this laboratory have a coelom and a complete digestive tract with a mouth and anus. Why does this allow these animals to have specialized parts of the digestive tract? _____

_____

9. For each of the following characteristics, name an animal with the characteristic, and state the characteristic's advantages:

   a. Closed circulatory system _____ , _____

   _____

   b. Respiratory organ _____ , _____

   _____

   c. Jointed appendages _____ , _____

   _____

   d. Exoskeleton _____ , _____

10. In general, describe the water vascular system of echinoderms. _____

_____

Biology Website

Enhance your study of the text and laboratory manual with study tools, practice tests, and virtual labs. Also ask your instructor about the resources available through ConnectPlus, including the media-rich eBook, interactive learning tools, and animations.

www.mhhe.com/maderbiology11

McGraw-Hill Access Science Website

An Encyclopedia of Science and Technology Online which provides more information including videos that can enhance the laboratory experience.

www.accessscience.com

24

The Vertebrates

Introduction

Vertebrates are **chordates.** All chordates have (1) a dorsal tubular nerve cord; (2) a dorsal supporting rod, called a **notochord**, at some time in their life history; (3) a postanal tail (e.g., tailbone or coccyx); and, in chordates that breathe by means of gills, (4) pharyngeal pouches that become gill slits. In terrestrial chordates, these pouches are modified for other purposes.

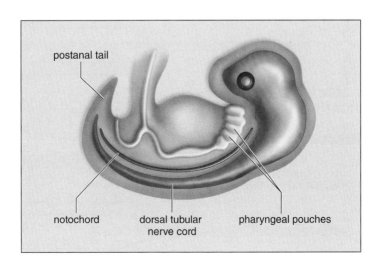

24.1 Evolution of Chordates

The evolutionary tree of the chordates (Fig. 24.1) shows that some chordates, notably the **tunicates** and **lancelets**, are not vertebrates. These chordates retain the notochord and are called the **invertebrate chordates**. Explain the term invertebrate chordates: _____

The other animal groups in Figure 24.1 are **vertebrates** in which the notochord has been replaced by the vertebral column. Fishes include three groups: the **jawless fishes** were the first to evolve, followed by the **cartilaginous fishes** and then the **bony fishes**. The bony fishes include the **ray-finned** fishes (the largest group of vertebrates) and the **lobe-finned fishes**. The first lobe-finned fishes had a bony skeleton, fleshy appendages, and a lung. These lobe-finned fishes lived in shallow pools and gave rise to the amphibians.

What three features called out in Figure 24.1 evolved among fishes? _____

_____

The terrestrial vertebrates are all **tetrapods** because they have four limbs. The limbs of tetrapods are _____ appendages just like those of arthropods. Amphibians still return to the water to reproduce, but **reptiles** are fully adapted to life on land because among other features they produce an **amniotic egg**. The amniotic egg is so named because the embryo is surrounded by an amniotic membrane that encloses amniotic fluid. Therefore, amniotes develop in an aquatic environment of their own making. Do all animals develop in a water environment? _____ Explain. _____

_____

In placental mammals, such as humans, the fertilized egg develops inside the female, where the unborn receives nutrients via the placenta. Reptiles (including birds) and mammals have many other adaptations that are suitable to living on land as we will stress in later sections.

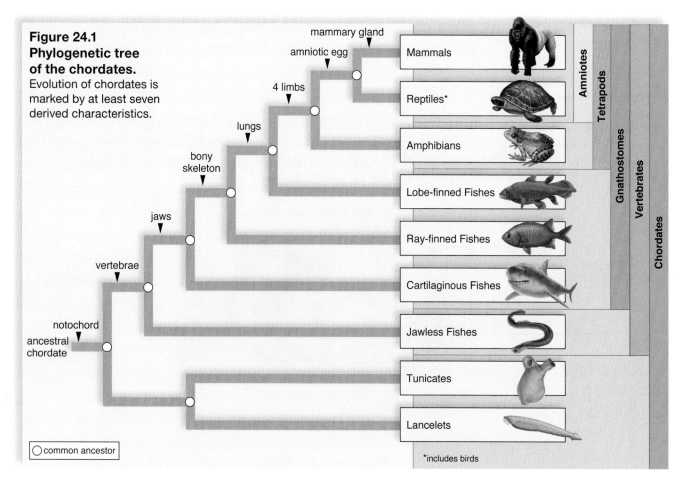

**Figure 24.1
Phylogenetic tree
of the chordates.**
Evolution of chordates is
marked by at least seven
derived characteristics.

mammary gland

amniotic egg

4 limbs

lungs

bony
skeleton

jaws

vertebrae

notochord

ancestral
chordate

Mammals

Reptiles*

Amphibians

Lobe-finned Fishes

Ray-finned Fishes

Cartilaginous Fishes

Jawless Fishes

Tunicates

Lancelets

Amniotes

Tetrapods

Gnathostomes

Vertebrates

Chordates

○ common ancestor

*includes birds

24.2   Invertebrate Chordates (Phylum Chordata)

Among the chordates, two major groups contain invertebrates, and the others contain the vertebrates.

Invertebrate Chordates

The two types of invertebrate chordates are urochordates and cephalochordates.

1. **Urochordates.** The tunicates, or sea squirts (Fig. 24.2), come in varying sizes and shapes, but all have incurrent and excurrent siphons. **Gill slits** are the only remaining chordate characteristic in adult tunicates. Examine any examples of tunicates on display.

Figure 24.2   Urochordates.
The gill slits of a tunicate are the only chordate characteristic remaining in the adult.

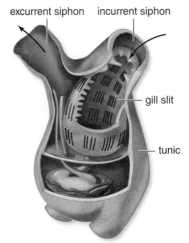

excurrent siphon   incurrent siphon

gill slit

tunic

2. **Cephalochordates.** Lancelets, also known as amphioxus *(Branchiostoma)*, are small, fishlike animals that occur in shallow marine waters in most parts of the world. They spend most of their time buried in the sandy bottom, with only the anterior end projecting.

Observation: Lancelet Anatomy

Preserved Specimen

1. Examine a preserved lancelet (Fig. 24.3).
2. Identify the **caudal fin** (enlarged tail) used in locomotion, the **dorsal fin,** and the short **ventral fin.**
3. Examine the lancelet's V-shaped muscles.

Figure 24.3   Anatomy of the lancelet, *Branchiostoma.*
Lancelets feed on microscopic particles filtered out of the constant stream of water that enters the mouth and exits through the gill slits into a protective atrium formed by body folds. The water exits at the atriopore.

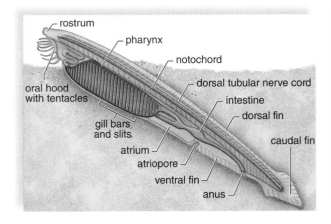

rostrum
pharynx
notochord
oral hood with tentacles
dorsal tubular nerve cord
intestine
gill bars and slits
dorsal fin
atrium
caudal fin
atriopore
ventral fin
anus

4. Find the tentacled **oral hood,** located anterior to the mouth and covering a vestibule. Water entering the mouth is channeled into the **pharynx,** where food particles are trapped before the water exits at the **atriopore.** Lancelets are filter feeders. Has cephalization occurred? _____ Explain. _____

_____

Prepared Slide

Examine a prepared cross section of a lancelet (Fig. 24.4), and note three chordate characteristics: notochord, dorsal tubular nerve cord, and gill slits.

Figure 24.4   Lancelet cross section.
Cross-section slide as it would appear under a microscope.

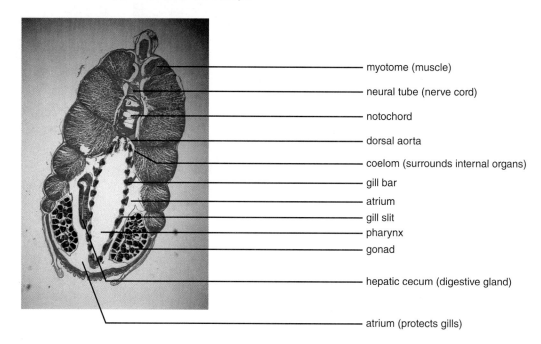

- myotome (muscle)
- neural tube (nerve cord)
- notochord
- dorsal aorta
- coelom (surrounds internal organs)
- gill bar
- atrium
- gill slit
- pharynx
- gonad
- hepatic cecum (digestive gland)
- atrium (protects gills)

24.3   Vertebrates (Subphylum Vertebrata)

In vertebrates (Fig. 24.5), the embryonic notochord is replaced by a vertebral column composed of individual vertebrae that protect the nerve cord. The internal jointed skeleton consists not only of a vertebral column, but also of a skull that encloses and protects the well-developed brain. There is an extreme degree of cephalization with complex sense organs. The eyes develop as outgrowths of the brain. The ears are primarily equilibrium devices in aquatic vertebrates; and in land vertebrates, they function as sound wave receivers.

> **Virtual Lab Mammals**   A virtual lab called Mammals is available on the *Biology* website **www.mhhe.com/maderbiology11**. After opening the virtual lab, click on any one of ten mammalian orders to bring up specific information about that order. Now you are ready to click on one of five skulls to see an enlarged version and a pop-up menu that allows you to choose an order for that skull.

The vertebrates are extremely motile and have well-developed muscles and usually paired appendages. They have bilateral symmetry and are segmented, as witnessed by the vertebral column. There is a large body cavity, a complete gut with both a mouth and anus (or instead a cloacal opening), and the circulatory system consists of a well developed heart and many blood vessels. They have an efficient means of extracting oxygen from water (gills) or air (lungs) as appropriate. The kidneys are important excretory and water-regulating organs that conserve or rid the body's water as necessary. The sexes are generally separate, and reproduction is usually sexual.

Figure 24.5 Vertebrate groups.

Cartilaginous fishes
Lack operculum and swim bladder; tail fin usually asymmetrical: (sharks, skates, and rays)

Blue shark

Bony fishes
Operculum; swim bladder or lungs; tail fin usually symmetrical: lung-fishes, lobe-finned fishes, and ray-finned fishes (herring, salmon, sturgeon, eels, and sea horse)

Blueback butterflyfish

Amphibians
Tetrapods with nonamniotic egg; nonscaly skin; some show metamorphosis; three-chambered heart: (salamanders, frogs, and toads)

Northern leopard frog

Reptiles
Tetrapods with amniotic egg; scaly skin: (snakes, lizards, turtles, and tortoises)

Pearl River redbelly turtle

Birds
Now grouped with reptiles; tetrapods with feathers; bipedal with wings; double circulation: (sparrows, penguins, and ostriches)

Scissor-tailed flycatcher

Mammals
Tetrapods with hair, mammary glands; double circulation; teeth differentiated: monotremes (spiny anteater and duckbill platypus), marsupials (opossum and kangaroo), and placental mammals (whales, rodents, dogs, cats, elephants, horses, bats, and humans)

Grey fox

Anatomy of the Frog

In this laboratory the anatomy of the frog will be considered typical of vertebrates in general. Frogs are amphibians, a group of animals in which metamorphosis occurs. Metamorphosis includes a change in structure, as when an aquatic tadpole becomes a frog with lungs and limbs (Fig. 24.6). Amphibians were the first vertebrates to be adapted to living on land; however, they typically return to the water to reproduce. Underline every structure mentioned in the following Observation that represents an adaptation to a land environment.

Figure 24.6 External frog anatomy.

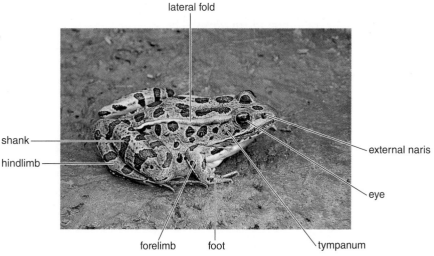

lateral fold

shank

hindlimb

external naris

eye

forelimb   foot   tympanum

Observation: External Anatomy of the Frog

1. Place a preserved frog *(Rana pipiens)* in a dissecting tray.
2. Identify the bulging eyes, which have a nonmovable upper and lower lid but can be covered by a **nictitating membrane** that serves to moisten the eye.
3. Locate the **tympanum** behind each eye (Fig. 24.6). What is the function of a tympanum? _____

   _____

4. Examine the external **nares** (sing., **naris,** or **nostril**). Insert a probe into an external naris, and observe that it protrudes from one of the paired small openings, the internal nares (Fig. 24.7),

   inside the mouth cavity. What is the function of the nares? _____

5. Identify the paired limbs. The bones of the fore- and hindlimbs are the same as in all tetrapods, in that the first bone articulates with a girdle and the limb ends in phalanges. The hind feet have five

   phalanges, and the forefeet have only four phalanges. Which pair of limbs is longest? _____

   How does a frog locomote on land? _____

   _____

   What is a frog's means of locomotion in the water? _____

   _____

Observation: Internal Anatomy of the Frog

> **Virtual Lab Frog Dissection** A virtual lab called Frog Dissection is available on the *Biology* website **www.mhhe.com/maderbiology11**. Allow ample time to complete this virtual lab with audio that allows a complete dissection of the frog systems. If highlighted buttons/arrows do not lead you to the next section, click on the menu provided.

Mouth

1. Open your frog's mouth very wide (Fig. 24.7), cutting the angles of the jaws if necessary.
2. Identify the tongue attached to the lower jaw's anterior end.
3. Find the **auditory (eustachian) tube** opening in the angle of the jaws. These tubes lead to the ears. Auditory tubes equalize air pressure in the ears.
4. Examine the **maxillary teeth** located along the rim of the upper jaw. Another set of teeth—**vomerine teeth**—is present just behind the midportion of the upper jaw.

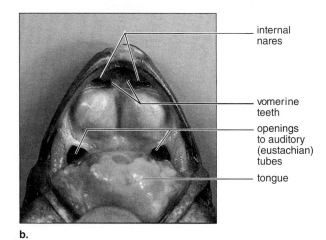

Figure 24.7 Mouth cavity of a frog.
a. Drawing. **b.** Dissected specimen.

5. Locate the **glottis,** a slit through which air passes into and out of the **trachea,** the short tube from glottis to lungs. What is the function of a glottis? _____

_____

6. Identify the **esophagus,** which lies dorsal and posterior to the glottis and leads to the stomach.

Opening the Frog

1. Place the frog ventral side up in the dissecting pan. Lift the skin with forceps, and use scissors to make a large, circular cut to remove the skin from the abdominal region as close to the limbs as possible. Cut only skin, not muscle.
2. Now, remove the muscles by cutting through them in the same circular fashion. At the same time, cut through any bones you encounter. A vein, called the abdominal vein, will be slightly attached to the internal side of the muscles.
3. Identify the **coelom,** or body cavity. Recall from laboratory 23 that vertebrates are deuterostomes in which the first embryonic opening becomes the anus and the second opening becomes the mouth.
4. If your frog is female, the abdominal cavity is likely to be filled by a pair of large, transparent **ovaries,** each containing hundreds of black and white eggs. Gently lift the left ovary with forceps, and find its place of attachment. Cut through the attachment, and remove the ovary in one piece.

Respiratory System and Liver

1. Insert a probe into the glottis, and observe its passage into the trachea. Enlarge the glottis by making short cuts above and below it. When the glottis is spread open, you will see a fold on either side; these are the vocal cords used in croaking.
2. Identify the **lungs,** two small sacs on either side of the midline and partially hidden under the liver (Fig. 24.8). Sequence the organs in the respiratory tract to trace the path of air from the external

   nares to the lungs. _____

_____

3. Locate the **liver,** the large, prominent, dark-brown organ in the midventral portion of the trunk (Fig. 24.8). Between the right half and left half of the liver, find the **gallbladder.**

Circulatory System

1. Lift the liver gently. Identify the **heart,** covered by a membranous covering (the **pericardium**). With forceps, lift the covering, and gently slit it open. The heart consists of a single, thick-walled **ventricle** and two (right and left) anterior, thin-walled **atria.**
2. Locate the three large veins that join together beneath the heart to form the **sinus venosus.** (To lift the heart, you may have to snip the slender strand of tissue that connects the atria to the pericardium.) Blood from the sinus venosus enters the right atrium. The left atrium receives blood from the lungs.
3. Find the **conus arteriosus,** a single, wide arterial vessel leaving the ventricle and passing ventrally over the right atrium. Follow the conus arteriosus forward to where it divides into three branches on each side. The middle artery on each side is the **systemic artery,** which fuses behind the heart to become the **dorsal aorta.** The dorsal aorta transports blood through the body cavity and gives off many branches. The **posterior vena cava** begins between the two kidneys and returns blood to the sinus venosus. Which vessel lies above (ventral to) the other? _____

_____

Digestive Tract

1. Identify the **esophagus,** a very short connection between the mouth and the stomach. Lift the left liver lobe, and identify the stomach, whitish and J-shaped. The **stomach** connects with the esophagus anteriorly and with the small intestine posteriorly.

2. Find the **small intestine** and the **large intestine,** which enters the **cloaca.** The cloaca lies beneath the pubic bone and is a general receptacle for the intestine, the reproductive system, and the urinary system. It opens to the outside by way of the anus. Sequence the organs in the digestive tract to trace the path of food from the mouth to the cloaca. _____

Accessory Glands

1. You identified the liver and gallbladder previously. Now try to find the **pancreas,** a yellowish tissue near the stomach and intestine.
2. Lift the stomach to see the **spleen,** a small, pea-shaped body.

Figure 24.8   Internal organs of a female frog, ventral view.

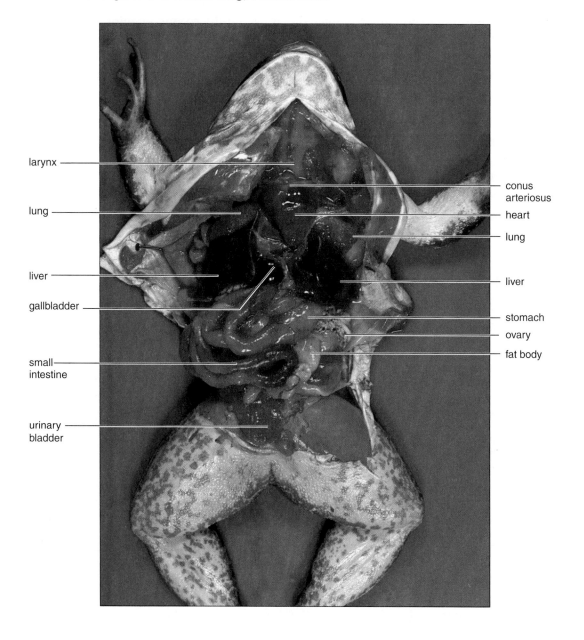

Urogenital System

1. Identify the **kidneys,** long, narrow organs lying against the dorsal body wall (Fig. 24.9).
2. Locate the **testes** in a male frog (Fig. 24.9). Testes are yellow, oval organs attached to the anterior portions of the kidneys. Several small ducts, the **vasa efferentia,** carry sperm into kidney ducts that also carry urine from the kidneys to the cloaca. **Fat bodies,** which store fat, are attached to the testes.
3. Locate the ovaries in a female frog. The ovaries are attached to the dorsal body wall (Fig. 24.10). Fat bodies are also attached to the ovaries. Highly coiled **oviducts** lead to the cloaca. The ostium (opening) of the oviduct is dorsal to the liver.
4. Find the **mesonephric ducts**—thin, white tubes that carry urine from the kidney to the cloaca. In female frogs, you will have to remove the left ovary to see the mesonephric ducts.
5. Locate the **cloaca.** You will need to split through the bones of the pelvic girdle in the midventral line and carefully separate the bones and muscles to find the cloaca.
6. Identify the urinary bladder attached to the ventral wall of the cloaca. In frogs, urine backs up into the bladder from the cloaca.
7. Explain the term *urogenital system.* _____
   _____
8. The cloaca receives material from (1) _____ ,
   (2) _____ , and (3) _____ .
9. Compare the frog's urogenital system to the human urinary system, which in females has no connection to the genital system. Beside each organ listed on the right tell how the comparable frog organ differs from that of a human.

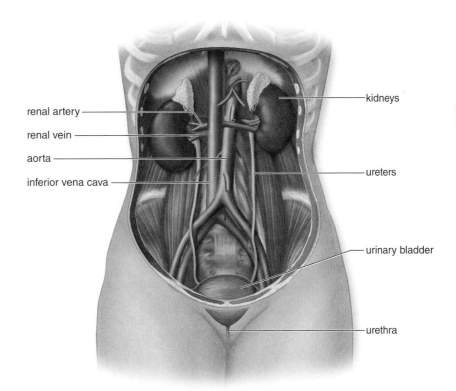

renal artery

renal vein

aorta

inferior vena cava

kidneys

ureters

urinary bladder

urethra

Figure 24.9   Urogenital system of a male frog.
a. Drawing. **b.** Dissected specimen.

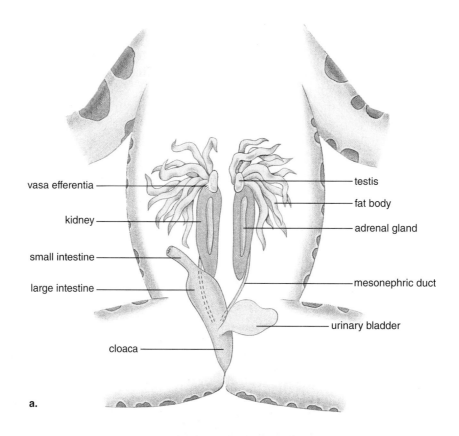

vasa efferentia

kidney

small intestine

large intestine

cloaca

testis

fat body

adrenal gland

mesonephric duct

urinary bladder

a.

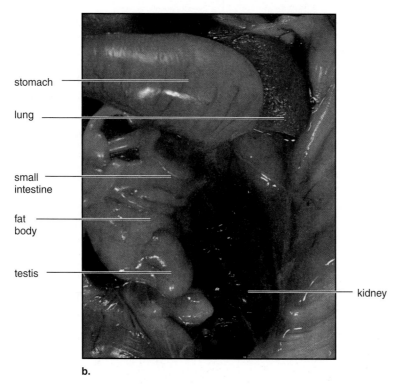

stomach

lung

small intestine

fat body

testis

kidney

b.

Figure 24.10 Urogenital system of a female frog.
a. Drawing. b. Dissected specimen.

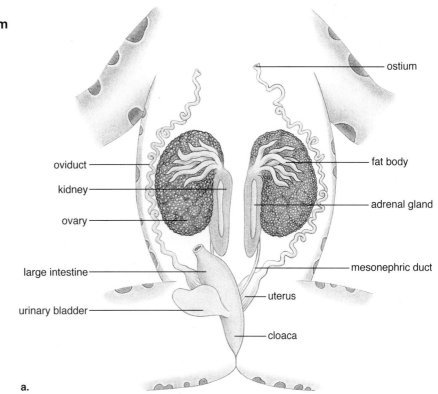

ostium

oviduct

kidney

ovary

fat body

adrenal gland

large intestine

urinary bladder

mesonephric duct

uterus

cloaca

a.

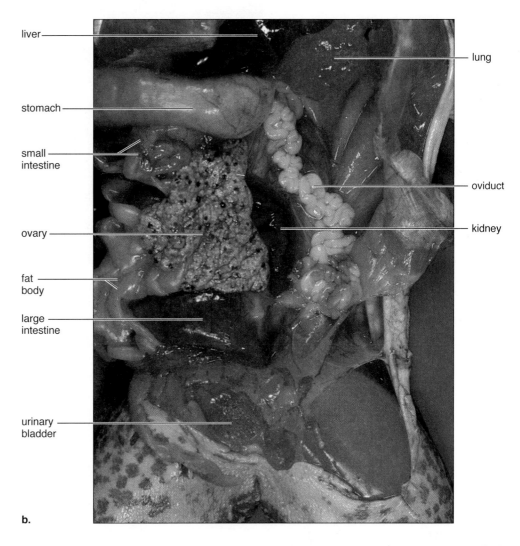

liver

stomach

small intestine

ovary

fat body

large intestine

urinary bladder

lung

oviduct

kidney

b.

Nervous System

In the frog demonstration dissection, identify the **brain,** lying exposed within the skull. With the help of Figure 24.11, find the major parts of the brain.

Figure 24.11   Frog brain, dorsal view.
a. Drawing. **b.** Dissected specimen.

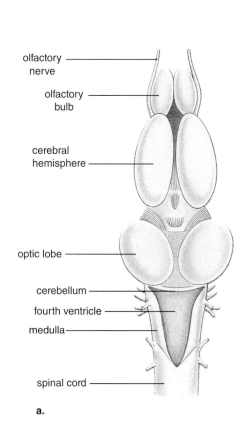

a.

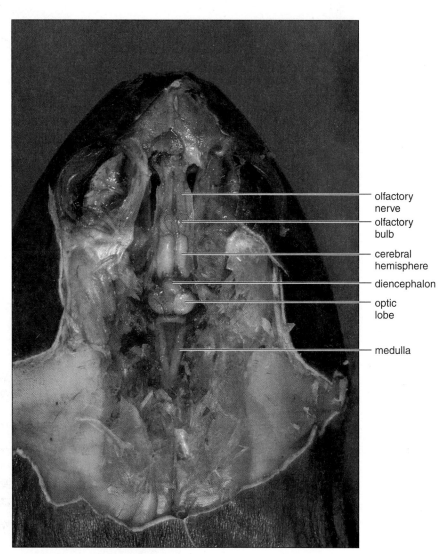

b.

24.4 Comparative Vertebrate Anatomy

In addition to the frog (amphibian), examine the perch (fish), pigeon (reptile), and rat (mammal) on display.

Observation: External Anatomy of Vertebrates

1. Compare the external features of the perch, frog, pigeon, and rat by examining specimens in the laboratory. Answer the following questions and record your observations in Table 24.1.
   - **a.** Is the skin smooth, scaly, hairy, or feathery?
   - **b.** Is there any external evidence of segmentation?
   - **c.** Are all forms bilaterally symmetrical?
   - **d.** Is the body differentiated into regions?
   - **e.** Is there a well-defined neck?
   - **f.** Is there a postanal tail?
   - **g.** Are there nares (nostrils)?
   - **h.** Is there a cloaca opening, or are there urogenital and anal openings?
   - **i.** Are eyelids present? How many?
   - **j.** How many appendages are there? (Fins are considered appendages.) (Fig. 24.12)

Table 24.1   Comparison of External Features

	Perch	Frog	Pigeon	Rat
a. Skin				
b. Segmentation				
c. Symmetry				
d. Regions				
e. Neck				
f. Postanal tail				
g. Nares				
h. Cloaca				
i. Eyelids				
j. Appendages				

2. Evidence that birds are reptiles: Birds
   a. have feathers, which are modified scales.
   b. have scales on their feet.
   c. and reptiles both lay eggs.
   d. and reptiles have similar internal organs.
   e. and reptiles also show some skeletal (skull) similarities.

Which of these can you substantiate by external examination? _____

_____

3. The perch, which lives in fresh water, and the pigeon and rat, which live on land, have a nearly impenetrable covering. Why is this an advantage in each case? _____

_____

4. A frog uses its skin for breathing. Is the skin of a frog thick and dry or thin and moist? Explain.

_____

Figure 24.12   Perch anatomy.
All fins are shown except the pectoral fin.

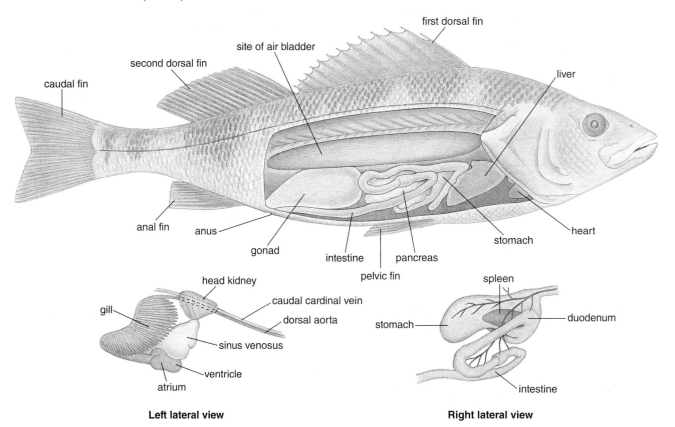

Left lateral view   Right lateral view

Figure 24.13 Cardiovascular systems in vertebrates.

a. In a fish, the blood moves in a single loop. The heart has a single atrium and ventricle, which pumps the blood into the gill region, where gas exchange takes place. **b.** Amphibians have a double-loop system in which the heart pumps blood to both the gills and the body. **c.** In birds and mammals, the right side pumps blood to the lungs, and the left side pumps blood to the body.

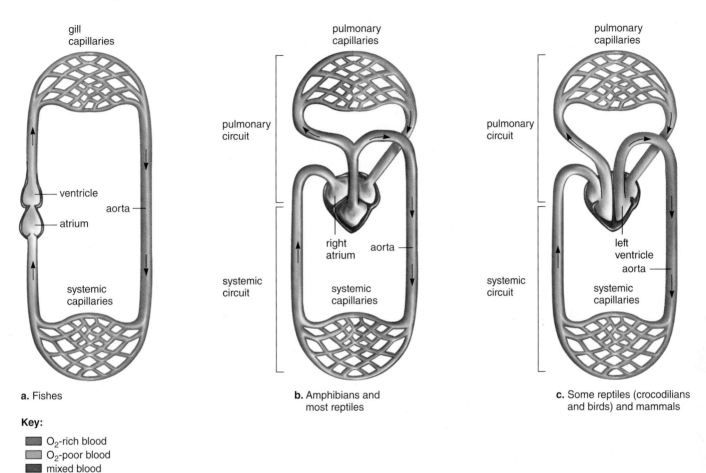

a. Fishes

b. Amphibians and most reptiles

c. Some reptiles (crocodilians and birds) and mammals

Key:
- ▮ O₂-rich blood
- ▯ O₂-poor blood
- ▮ mixed blood

Observation: Internal Anatomy of Vertebrates

1. Examine the internal organs of the perch, frog, pigeon (see Fig. 24.14) and rat (see Fig. 24.15).
2. If necessary, make a median longitudinal incision in the ventral body wall, from the jaws to the anus. The body cavity is called a coelom because it is completely lined by mesoderm (see page 311).
3. Which of these animals has a **diaphragm** dividing the body cavity into **thorax** and **abdomen?**

Circulatory Systems

Study heart models for a fish, amphibian, bird, and mammal. Trace the path of the vessel that leaves the ventricle(s), and determine whether the animals have a **pulmonary system** (Fig. 24.13). The word *pulmonary* comes from the Latin *pulmonarius,* meaning "lungs."

Figure 24.14   Pigeon anatomy.

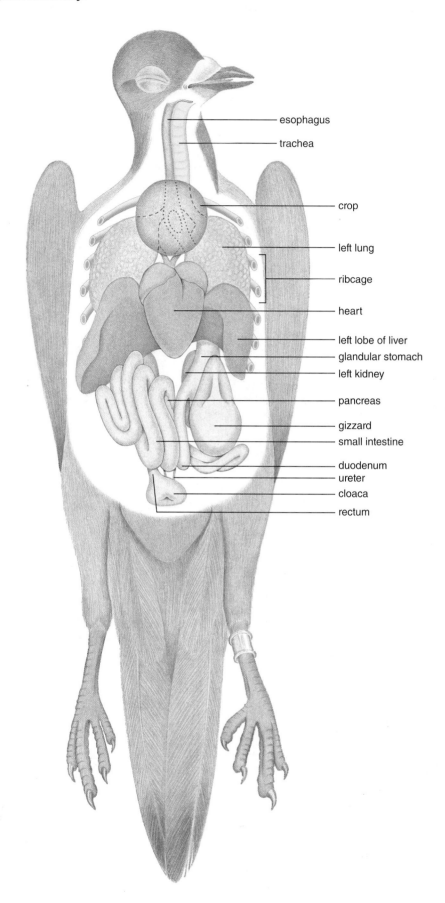

esophagus

trachea

crop

left lung

ribcage

heart

left lobe of liver

glandular stomach

left kidney

pancreas

gizzard

small intestine

duodenum

ureter

cloaca

rectum

Figure 24.15 Rat anatomy.

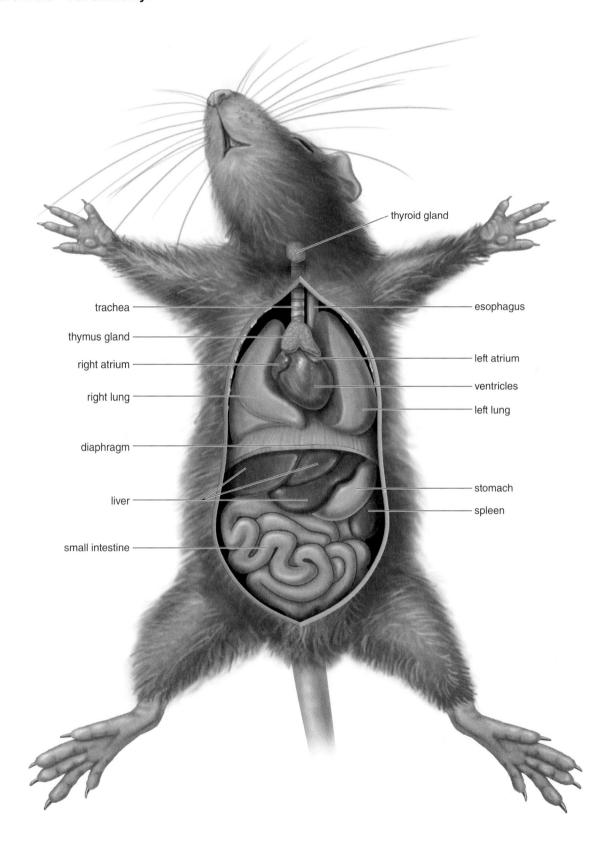

thyroid gland

trachea

thymus gland

right atrium

right lung

diaphragm

liver

small intestine

esophagus

left atrium

ventricles

left lung

stomach

spleen

Complete Table 24.2.

Table 24.2 Comparative Circulatory Systems		
Animal	**Number of Heart Chambers**	**Pulmonary Circuit (Yes or No)**
Perch		
Frog		
Pigeon		
Rat		

1. Do fish have a blood vessel that returns blood from the gills to the heart? _____ Would you expect blood pressure to be high or low after blood has moved through the gills? _____

2. What animals studied have pulmonary vessels that take blood from the heart to the respiratory organ and back to the heart? _____ What is the advantage of a pulmonary circuit? _____

3. Which of these animals has a four-chambered heart? _____
What is the advantage of having separate ventricles? _____
_____

4. The circulatory system distributes the heat of muscle contraction in birds and mammals. Is the anatomy of birds and mammals conducive to maintaining a warm internal temperature? _____ Explain. _____
_____

Respiratory Systems

Compare the respiratory systems of the perch, frog, pigeon, and rat, and complete Table 24.3 by checking the anatomical features that appear in each animal.

Table 24.3 Respiratory Systems						
	Gills	**Trachea**	**Lungs**	**Rib Cage***	**Diaphragm**	**Air Sacs**
Perch						
Frog						
Pigeon						
Rat						

*A rib cage consists of ribs plus a sternum. Some ribs are connected to the sternum, which lies at the midline in the anterior portion of the rib cage.

1. Among the animals studied, only a perch breathes by _____ . Can the particular respiratory organ be related to the environment of the animals? _____ Explain. _____
_____

2. Knowing that gills are attached to the pharynx (throat in humans), explain why fish have no trachea. _____

_____

3. A rib cage is present in the rat and pigeon but missing in the frog. Can this difference be related to the fact that frogs breathe by positive pressure, while birds and mammals breathe by negative pressure? _____ (A frog swallows air and then pushes the air into its lungs; in birds and mammals, the thorax expands first, and then the air is drawn in.) Explain. _____

_____

4. A diaphragm is present only in mammals (e.g., rat). Of what benefit is this feature to the expansion of lungs in mammals? _____

_____

_____

5. Air sacs (not shown in Fig. 24.14) are present only in birds. This feature allows air to pass one way through the lungs of a bird and greatly increases the bird's ability to extract oxygen from the air.

Inhalation: Air bypasses lungs and enters posterior air sacs.

Exhalation continues: Air passes through lungs and enters anterior air sacs.

Digestive Systems and Urogenital Systems

1. All the vertebrates have a stomach, small intestine, and large intestine where food is processed. They also have a liver and pancreas. Which is the larger, more prominent organ (liver or pancreas) in the pigeon and rat? _____

2. All vertebrates have an anus. As noted in Table 24.1, which vertebrates studied have a cloaca (receptacle for the urogenital and digestive systems? _____

In the other vertebrates studied the urogenital and digestive systems are separate.

3. Urogenital systems. All the animals have gonads and kidneys. Which of these organs is involved urine production? _____

As we shall learn later, the kidneys help maintain the proper balance of fluid and salts in the blood.

Which of these organs is involved in reproduction. _____
The sexes are separate in vertebrates: females have ovaries and males have testes. In reptiles and mammals, males usually have a penis to pass sperm to the female. What is the chief biological benefit of the penis in terrestrial animals? _____

1. What are the four characteristics of all chordates?

   a. _____

   b. _____

   c. _____

   d. _____

2. Give an example of a

   Fish _____

   Amphibian _____

   Reptile (i.e., bird) _____

   Mammal _____

3. A lancelet isn't a vertebrate. Explain. _____
   _____

4. Which groups of vertebrates are fully adapted to life on land?
   _____

5. Complete each of the following sentences: In a frog, . . .

   a. the glottis allows air to enter the _____ .

   b. the esophagus allows food to enter the _____ .

   c. the cloaca receives material from the _____ , _____ , and _____ .

   d. in males sperm reach the cloaca by way of the _____ .

6. What is the major difference between the heart of a frog and that of a rat? _____
   _____

7. A pulmonary circuit is seen in vertebrate animals adapted to life on land. Explain. _____
   _____

8. What is the major difference between the respiratory system of a perch and that of a frog, a pigeon,
   and a rat? _____

9. Name a major difference between bird and mammalian lungs. _____
   _____

10. When frogs mate, the male presses his cloaca up against the female's cloaca. Why does this pass sperm
    to the female? _____

25

Animal Organization

Learning Outcomes

25.1 Tissue Level of Organization
- Identify slides and models of various types of epithelium.   356–58
- Tell where particular types of epithelium are located in the body, and state a function.   358
- Identify slides and models of various types of connective tissue.   359–62
- Tell where particular connective tissues are located in the body and state a function.   362
- Identify slides and models of three types of muscular tissue.   362–64
- Tell where each type of muscular tissue is located in the body, and state a function.   364
- Identify a slide and model of a neuron.   365
- Tell where nervous tissue is located in the body, and state a function.   365

25.2 Organ Level of Organization
- Identify a slide of the intestinal wall and the layers in the wall. State a function for each tissue.   366
- Identify a slide of skin and the two regions of skin. State a function for each region of skin.   367

Introduction

Humans, as well as all living things, are made up of **cells.** Groups of cells that have the same structural characteristics and perform the same functions are called **tissues.** Figure 25.1 shows the four categories of tissues in the human body. An **organ** is composed of different types of tissues, and various organs form **organ systems.** Humans thus have the following levels of biological organization: cells → tissues → organs → organ systems.

The photomicrographs of tissues in this laboratory were obtained by viewing prepared slides with a light microscope. Preparation required the following sequential steps:

1. **Fixation:** The tissue is immersed in a preservative solution to maintain the tissue's existing structure.
2. **Embedding:** Water is removed with alcohol, and the tissue is impregnated with paraffin wax.
3. **Sectioning:** The tissue is cut into extremely thin slices by an instrument called a microtome. When the section runs the length of the tissue, it is called a longitudinal section (l.s.); when the section runs across the tissue, it is called a cross section (c.s.).
4. **Staining:** The tissue is immersed in dyes that stain different structures. The most common dyes are hematoxylin and eosin stains (H & E). They give a differential blue and red color to the basic and acidic structures within the tissue. Other dyes are available for staining specific structures.

Figure 25.1 The major tissues in the human body.

The many kinds of tissues in the human body are grouped into four types: epithelial tissue, muscular tissue, nervous tissue, and connective tissue.

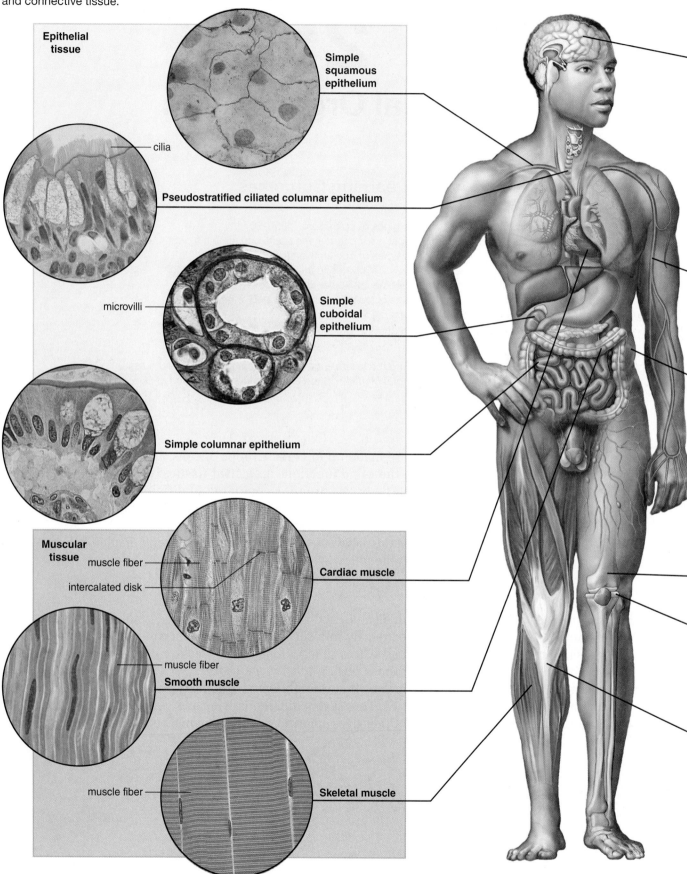

Epithelial tissue

Simple squamous epithelium

cilia

Pseudostratified ciliated columnar epithelium

microvilli

Simple cuboidal epithelium

Simple columnar epithelium

Muscular tissue

muscle fiber

intercalated disk

Cardiac muscle

muscle fiber

Smooth muscle

muscle fiber

Skeletal muscle

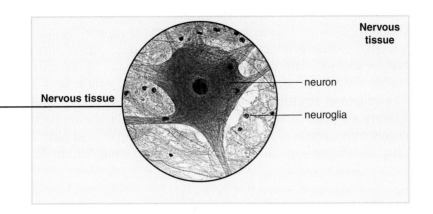

Nervous tissue

Nervous
tissue

neuron

neuroglia

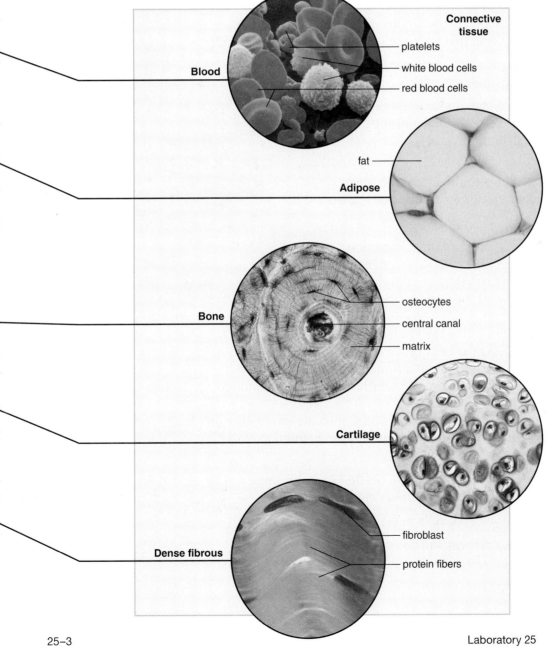

Connective
tissue

Blood

platelets

white blood cells

red blood cells

fat

Adipose

Bone

osteocytes

central canal

matrix

Cartilage

fibroblast

Dense fibrous

protein fibers

25.1 Tissue Level of Organization

Epithelial tissue (epithelium) forms a continuous layer, or sheet, over the entire body surface and most of the body's inner cavities. Externally, it forms a covering that protects the animal from infection, injury, and drying out. Some epithelial tissues produce and release secretions. Others absorb nutrients.

The name of an epithelial tissue includes two descriptive terms: the shape of the cells and the number of layers. The three possible shapes are *squamous, cuboidal,* and *columnar.* With regard to layers, an epithelial tissue may be simple or stratified. **Simple** means that there is only one layer of cells; **stratified** means that cell layers are placed on top of each other. Some epithelial tissues are **pseudostratified,** meaning that they only appear to be layered. Epithelium may also have cellular extensions called **microvilli** or hairlike extensions called **cilia.** In the latter case, "ciliated" may be part of the tissue's name.

Observation: Simple and Stratified Squamous Epithelium

Simple Squamous Epithelium

Simple squamous epithelium is a single layer of thin, flat, many-sided cells, each with a central nucleus. It lines internal cavities, the heart, and all the blood vessels. It also lines parts of the urinary, respiratory, and male reproductive tracts.

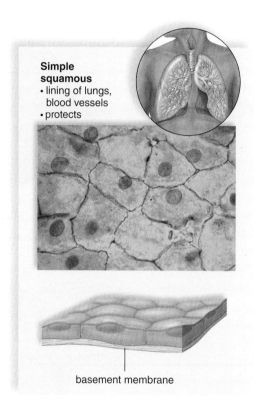

Simple squamous
• lining of lungs, blood vessels
• protects

basement membrane

1. Study a model or diagram of simple squamous epithelium. What does squamous mean? _____

2. Examine a prepared slide of squamous epithelium. Under low power, note the close packing of the flat cells. What shapes are the cells?

   _____

3. Under high power, examine an individual cell, and identify the plasma membrane, cytoplasm, and nucleus.

Stratified Squamous Epithelium

As would be expected from its name, stratified squamous epithelium consists of many layers of cells. The innermost layer produces cells that are first cuboidal or columnar in shape, but as the cells push toward the surface, they become flattened.

The outer region of the skin, called the epidermis, is stratified squamous epithelium. As the cells move toward the surface, they flatten, begin to accumulate a protein called **keratin,** and eventually die. Keratin makes the outer layer of epidermis tough, protective, and able to repel water.

The linings of the mouth, throat, anal canal, and vagina are stratified epithelium. The outermost layer of cells surrounding the cavity is simple squamous epithelium. In these organs, this layer of cells remains soft, moist, and alive.

1. Either now or when you are studying skin in Section 25.2, examine a slide of skin and find the portion of the slide that is stratified squamous epithelium.

2. Approximately how many layers of cells make up this portion of skin? _____

3. Which layers of cells best represent squamous epithelium? _____

Observation: Simple Cuboidal Epithelium

Simple cuboidal epithelium is a single layer of cube-shaped cells, each with a central nucleus. It is found in tubules of the kidney and in the ducts of many glands, where it has a protective function. It also occurs in the secretory portions of some glands—that is, where the tissue produces and releases secretions.

1. Study a model or diagram of simple cuboidal epithelium.
2. Examine a prepared slide of simple cuboidal epithelium. Move the slide until you locate cube-shaped cells that line a lumen (cavity). Are these cells ciliated? _____

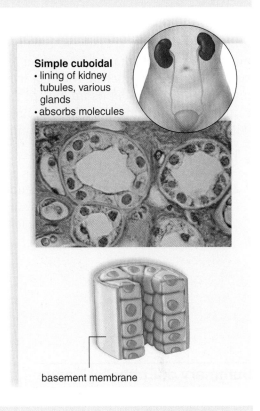

Simple cuboidal
• lining of kidney tubules, various glands
• absorbs molecules

basement membrane

Observation: Simple Columnar Epithelium

Simple columnar epithelium is a single layer of tall, cylindrical cells, each with a nucleus near the base. This tissue, which lines the digestive tract from the stomach to the anus, protects, secretes, and allows absorption of nutrients.

1. Study a model or diagram of simple columnar epithelium.
2. Examine a prepared slide of simple columnar epithelium. Find tall and narrow cells that line a lumen. Under high power, focus on an individual cell. Identify the plasma membrane, the cytoplasm, and the nucleus. Epithelial tissues are attached to underlying tissues by a basement membrane composed of extracellular material containing protein fibers.
3. The tissue you are observing contains mucus-secreting cells. Search among the columnar cells until you find a **goblet cell,** so named because of its goblet-shaped, clear interior. This region contains mucus, which may be stained a light blue. In the living animal, the mucus is discharged into the gut cavity and protects the lining from digestive enzymes.

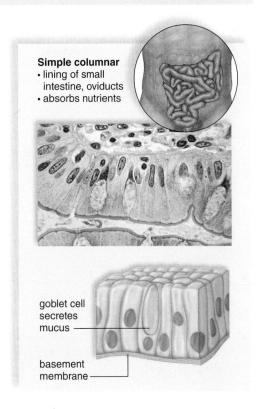

Simple columnar
• lining of small intestine, oviducts
• absorbs nutrients

goblet cell secretes mucus

basement membrane

Pseudostratified ciliated columnar epithelium

Pseudostratified ciliated columnar epithelium appears to be layered, while actually all cells touch the basement membrane. Many cilia are located on the free end of each cell. In the human trachea, the cilia wave back and forth, moving mucus and debris up toward the throat so that it cannot enter the lungs. Smoking destroys these cilia, but they will grow back if smoking is discontinued.

1. Study a model or diagram of pseudostratified ciliated columnar epithelium.
2. Examine a prepared slide of pseudostratified ciliated columnar epithelium. Concentrate on the part of the slide that resembles the model. Identify the cilia.

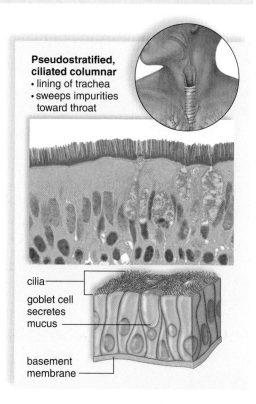

Pseudostratified, ciliated columnar
- lining of trachea
- sweeps impurities toward throat

cilia

goblet cell secretes mucus

basement membrane

Summary of Epithelial Tissue

Complete Table 25.1 to summarize your study of epithelial tissue.

Table 25.1 Epithelial Tissue			
Type	**Structure**	**Function**	**Location**
Simple squamous			Walls of capillaries, lining of blood vessels, air sacs of lungs, lining of internal cavities
Stratified squamous	Innermost layers are cuboidal or columnar; outermost layers are flattened	Protection, repel water	
Simple cuboidal		Secretion, absorption	
Simple columnar	Columnlike—tall, cylindrical nucleus at base		Lining of uterus, tubes of digestive tract
Pseudostratified ciliated columnar		Protection, secretion, movement of mucus and sex cells	

Connective Tissue

Connective tissue joins different parts of the body together. There are four general classes of connective tissue: connective tissue proper, bone, cartilage, and blood. All types of connective tissue consist of cells surrounded by a matrix that usually contains fibers. Elastic fibers are composed of a protein called elastin. Collagenous fibers contain the protein collagen.

Observation: Connective Tissue

There are several different types of connective tissue. The accompanying illustrations will help you understand the name of each tissue, where it occurs, and its functions. We will study loose fibrous connective tissue, dense fibrous connective tissue, adipose tissue, bone, cartilage, and blood. **Loose fibrous connective tissue**, so named because it has space between components, supports epithelium and also many internal organs, such as muscles, blood vessels, and nerves. Its presence allows organs to expand. **Dense fibrous connective tissue** contains many collagenous fibers packed closely together, as in tendons, which connect muscles to bones, and in ligaments, which connect bones to other bones at joints.

1. Examine a slide of loose fibrous connective tissue, and compare it to the figure below *(left)*.

   What is the function of loose fibrous connective tissue? _____

2. Examine a slide of dense fibrous connective tissue, and compare it to the figure below *(right)*.

   What two kinds of structures in the body contain dense fibrous connective tissue? _____

   _____

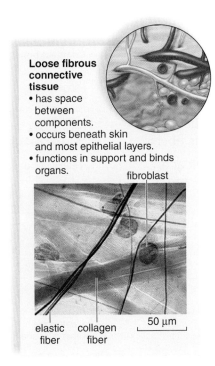

Loose fibrous connective tissue
• has space between components.
• occurs beneath skin and most epithelial layers.
• functions in support and binds organs.

fibroblast

elastic fiber   collagen fiber   50 μm

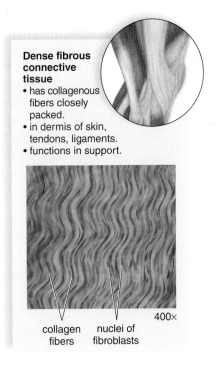

Dense fibrous connective tissue
• has collagenous fibers closely packed.
• in dermis of skin, tendons, ligaments.
• functions in support.

collagen fibers   nuclei of fibroblasts   400×

Observation: Adipose Tissue

In **adipose tissue,** the cells have a large, central, fat-filled vacuole that causes the nucleus and cytoplasm to be at the perimeter of the cell. Adipose tissue occurs beneath the skin, where it insulates the body, and around internal organs, such as the kidneys and heart. It cushions and helps protect these organs.

1. Examine a prepared slide of adipose tissue. Why is the nucleus pushed to one side? _____

_____

2. State a location for adipose tissue in the body.

_____

_____

What are two functions of adipose tissue at this location?

_____

_____

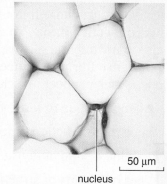

Adipose tissue
- cells are filled with fat.
- occurs beneath skin, around heart and other organs.
- functions in insulation, stores fat.

50 μm

nucleus

Observation: Compact Bone

Compact bone is found in the bones that make up the skeleton. It consists of **osteons** (Haversian system) with a **central canal,** and concentric rings of spaces called **lacunae,** connected by tiny crevices called **canaliculi.** The central canal contains a nerve and blood vessels, which service bone. The lacunae contain bone cells called **osteocytes,** whose processes extend into the canaliculi. Separating the lacunae is a matrix that is hard because it contains minerals, notably calcium salts. The matrix also contains collagenous fibers.

1. Study a model or diagram of compact bone. Then look at a prepared slide and identify the central canal, lacunae, and canaliculi.

2. What is the function of the central canal and canaliculi?

_____

_____

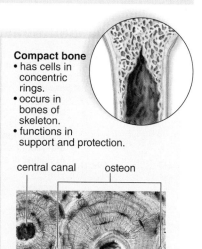

Compact bone
- has cells in concentric rings.
- occurs in bones of skeleton.
- functions in support and protection.

central canal   osteon

320×

osteocyte   canaliculi
within a lacuna

Observation: Hyaline Cartilage

In **hyaline cartilage,** cells called **chondrocytes** are found in twos or threes in lacunae. The lacunae are separated by a flexible matrix containing weak collagenous fibers.

1. Study the diagram and photomicrograph of hyaline cartilage in the figure at right. Then study a prepared slide of hyaline cartilage, and identify the matrix, lacunae, and chondrocytes.

2. Compare compact bone and hyaline cartilage. Which of these

   types of connective tissue is more organized? _____

   Why? _____

3. Which of these two types of connective tissue lends more

   support to body parts? _____

   _____

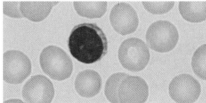

Hyaline cartilage
• has cells in lacunae.
• occurs in nose and walls of respiratory passages; at ends of bones, including ribs.
• functions in support and protection.

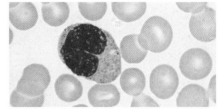

chondrocyte within lacunae   matrix   50 µm

Observation: Blood

Blood is a connective tissue in which the matrix is an intercellular fluid called **plasma. Red blood cells** (erythrocytes) have a biconcave appearance and lack a nucleus. These cells carry oxygen combined with the respiratory pigment hemoglobin. **White blood cells** (leukocytes) have a nucleus and are typically larger than the more numerous red blood cells. These cells fight infection.

1. Study a prepared slide of human blood. With the help of Figure 25.2, identify the red blood cells and the white blood cells, which appear faint because of the stain.

2. Try to identify a neutrophil, which has a multilobed nucleus, and a lymphocyte, which is the smallest of the white blood cells, with a spherical or slightly indented nucleus.

Figure 25.2 Blood cells.
Red blood cells are more numerous than white blood cells. White blood cells can be separated into five distinct types. If you have blood work done that includes a complete blood count (CBC), the doctor is getting a count of each of these types of WBCs. (*a–e:* Magnification 1,050×)

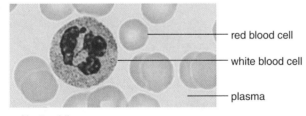

— red blood cell

— white blood cell

— plasma

a. Neutrophil

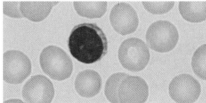

b. Lymphocyte

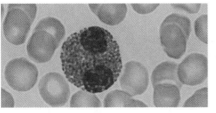

c. Eosinophil

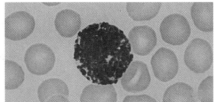

d. Basophil

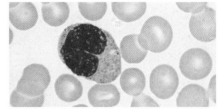

e. Monocyte

Summary of Connective Tissue

1. Complete Table 25.2 to summarize your study of connective tissue.

Table 25.2	Connective Tissue		
Type	**Structure**	**Function**	**Location**
Loose fibrous connective tissue			Between the muscles; beneath the skin; beneath most epithelial layers
Dense fibrous connective tissue		Binds organs together, binds muscle to bones, binds bone to bone	
Adipose			Beneath the skin; around the kidney and heart; in the breast
Compact bone		Support, protection	
Hyaline cartilage	Cells in lacunae		Nose; ends of bones; rings in walls of respiratory passages; between ribs and sternum
Blood	Red and white cells floating in plasma		Blood vessels

2. Working with others in a group, decide how the structure of each connective tissue suits its function.

Loose fibrous connective tissue _____

Dense fibrous connective tissue _____

Adipose tissue _____

Compact bone _____

Hyaline cartilage _____

Blood _____

Muscular Tissue

Muscular (contractile) tissue is composed of cells called muscle fibers. Muscular tissue has the ability to contract, and contraction usually results in movement. The body contains skeletal, cardiac, and smooth muscle.

Skeletal muscle occurs in the muscles attached to the bones of the skeleton. The contraction of skeletal muscle is said to be **voluntary** because it is under conscious control. Skeletal muscle is striated; it contains light and dark bands. The striations are caused by the arrangement of contractile filaments (actin and myosin filaments) in muscle fibers. Each fiber contains many nuclei, all peripherally located.

1. Study a model or diagram of skeletal muscle, and note that striations are present. You should see several muscle fibers, each marked with striations.
2. Examine a prepared slide of skeletal muscle. The striations may be difficult to make out, but bringing the slide in and out of focus may help.

Skeletal muscle
• has striated cells with multiple nuclei.
• occurs in muscles attached to skeleton.
• functions in voluntary movement of body.

striation   nucleus

Cardiac muscle is found only in the heart. It is called **involuntary** because its contraction does not require conscious effort. Cardiac muscle is striated in the same way as skeletal muscle. However, the fibers are branched and bound together at **intercalated disks,** where their folded plasma membranes touch. This arrangement aids communication between fibers.

1. Study a model or diagram of cardiac muscle, and note that striations are present.
2. Examine a prepared slide of cardiac muscle. Find an intercalated disk. What is the function of cardiac muscle?

_____

_____

Cardiac muscle
• has branching, striated cells, each with a single nucleus.
• occurs in the wall of the heart.
• functions in the pumping of blood.
• is involuntary.

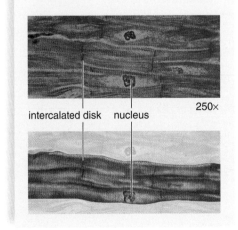

intercalated disk   nucleus   250×

Smooth muscle is sometimes called **visceral muscle** because it makes up the walls of the internal organs, such as the intestines and the blood vessels. Smooth muscle is involuntary because its contraction does not require conscious effort.

1. Study a model or diagram of smooth muscle, and note the shape of the cells and the centrally placed nucleus. Smooth muscle has spindle-shaped cells. What does *spindle-shaped* mean? _____

   _____

2. Examine a prepared slide of smooth muscle. Distinguishing the boundaries between the different cells may require you to take the slide in and out of focus.

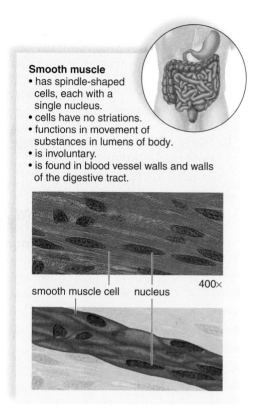

Smooth muscle
• has spindle-shaped cells, each with a single nucleus.
• cells have no striations.
• functions in movement of substances in lumens of body.
• is involuntary.
• is found in blood vessel walls and walls of the digestive tract.

smooth muscle cell   nucleus   400×

Summary of Muscular Tissue

1. Complete Table 25.3 to summarize your study of muscular tissue.

Table 25.3	Muscular Tissue		
Type	**Striations (Yes or No)**	**Branching (Yes or No)**	**Conscious Control (Yes or No)**
Skeletal			
Cardiac			
Smooth			

2. How does it benefit an animal that skeletal muscle is voluntary, while cardiac and smooth muscle are involuntary? _____

   _____

Nervous Tissue

Nervous tissue is found in the brain, spinal cord, and nerves. Nervous tissue receives and integrates incoming stimuli before conducting nerve impulses, which control the glands and muscles of the body. Nervous tissue is composed of two types of cells: **neurons** that transmit messages and **neuroglia** that support and nourish the neurons. Motor neurons, which take messages from the spinal cord to the muscles, are often used to exemplify typical neurons. Motor neurons have several **dendrites,** processes that take signals to a **cell body,** where the nucleus is located, and an **axon** that takes nerve impulses away from the cell body.

Observation: Nervous Tissue

1. Study a model or diagram of a neuron, and then examine a prepared slide. Most likely, you will not be able to see neuroglia because they are much smaller than neurons.
2. Identify the dendrites, cell body, and axon in Figure 25.3 and label the micrograph. Long axons are called nerve fibers.
3. Explain the appearance and function of the parts of a motor neuron:
   a. Dendrites _____
   _____
   b. Cell body _____
   _____
   c. Axon _____
   _____

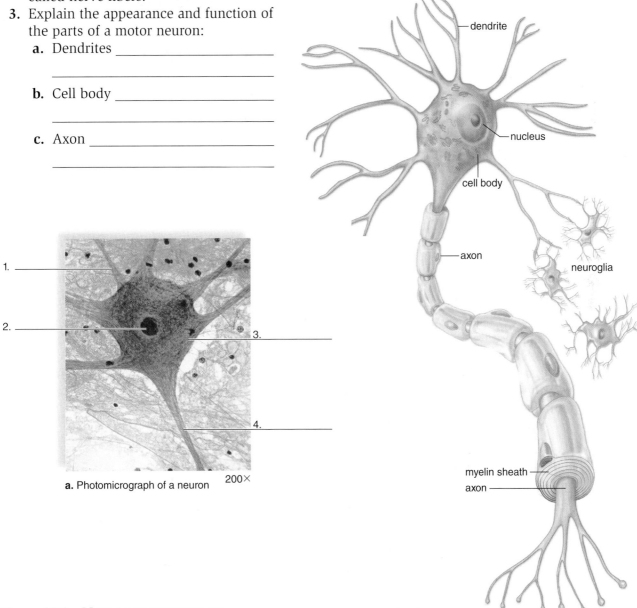

Figure 25.3 Motor neuron anatomy.

a. Photomicrograph of a neuron   200×

b. Drawing

25.2 Organ Level of Organization

Organs are structures composed of two or more types of tissue that work together to perform particular functions. You may tend to think that a particular organ contains only one type of tissue. For example, muscular tissue is usually associated with muscles and nervous tissue with the brain. However, muscles and the brain also contain other types of tissue—for example, loose connective tissue and blood. Here we will study the compositions of two organs—the intestine and the skin.

Intestine

The **intestine,** a part of the digestive system, processes food and absorbs nutrient molecules.

Observation: Intestinal Wall

Study a slide of a cross section of intestinal wall. With the help of Figure 25.4, identify the following layers:

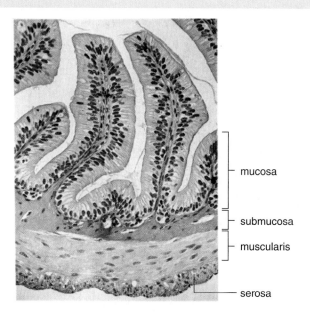

Figure 25.4   The intestinal wall.
A cross section reveals the various layers of the intestinal wall, noted to the right of this photomicrograph. (Magnification ×25)

1. **Mucosa** (mucous membrane layer): This layer, which lines the central lumen (cavity), is made up of columnar epithelium overlying the submucosa. This epithelium is glandular—that is, it secretes mucus from goblet cells and digestive enzymes from the rest of the epithelium. The membrane is arranged in deep folds (fingerlike projections) called **villi,** which increase the small intestine's absorptive surface.
2. **Submucosa** (submucosal layer): This loose fibrous connective tissue layer contains nerve fibers, blood vessels, and lymphatic vessels. The products of digestion are absorbed into these blood and lymphatic vessels.
3. **Muscularis** (smooth muscle layer): Circular muscular tissue and then longitudinal muscular tissue are found in this layer. Rhythmic contraction of these muscles causes **peristalsis,** a wavelike motion that moves food along the intestine.
4. **Serosa** (serous membrane layer): In this layer, a thin sheet of loose fibrous connective tissue underlies a thin, outermost sheet of squamous epithelium. This membrane is part of the **peritoneum,** which lines the entire abdominal cavity.

In Table 25.4, list the types of tissue found in the layers of the intestinal wall.

Table 25.4 Tissues of the Intestinal Wall			
	Mucosa	**Submucosa**	**Muscularis**	**Serosa**
Tissue(s)				

Skin

The skin covers the entire exterior of the human body. Skin functions include protection, water retention, sensory reception, body temperature regulation, and vitamin D synthesis.

Observation: Skin

Study a model or diagram and also a prepared slide of the skin. Identify the two skin regions and the subcutaneous layer.

1. **Epidermis:** This region is composed of stratified squamous epithelial cells. The outer cells of the epidermis are nonliving and create a waterproof covering that prevents excessive water loss. These cells are always being replaced because an inner layer of the epidermis is composed of living cells that constantly produce new cells.

2. **Dermis:** This region is a connective tissue containing blood vessels, nerves, sense organs, and the expanded portions of oil (sebaceous) and sweat glands and hair follicles.

   List the structures you can identify on your slide:

   _____

   _____

3. **Subcutaneous layer:** This is a layer of loose connective tissue and adipose tissue that lies beneath the skin proper and serves to insulate and protect inner body parts.

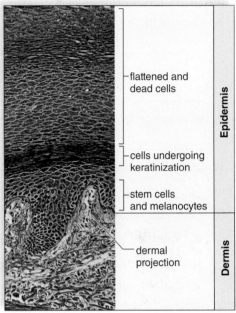

flattened and dead cells

cells undergoing keratinization

stem cells and melanocytes

dermal projection

Epidermis

Dermis

Photomicrograph of skin

Laboratory Review 25

1. Name four major types of tissues, and state a general function for each.

   _____

   _____

   _____

   _____

2. Describe the shapes of three types of epithelial tissue, and state a function for each.

   _____

   _____

   _____

3. Describe the appearance of three types of muscular tissue, and state a function for each.

_____

_____

_____

4. What is meant by the expression *involuntary muscle?* _____

_____

5. List the three parts of a neuron, and state a function for each part in a motor neuron.

_____

_____

_____

6. Why are certain types of connective tissue called support tissues?

_____

7. Describe how you would recognize a slide of compact bone.

_____

8. The outer region of skin consists of stratified squamous epithelium. Define these terms:

   a. Stratified _____

   b. Squamous _____

   c. Epithelium _____

9. Identify the tissues a.–d. below as epithelial tissue, muscular tissue, nervous tissue, or connective tissue.

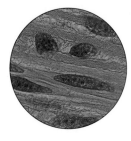  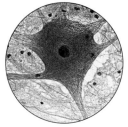 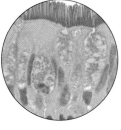

a. _____   b. _____   c. _____   d. _____

10. What is in the shape of the cell in 9a? _____

***Biology* Website**

Enhance your study of the text and laboratory manual with study tools, practice tests, and virtual labs. Also ask your instructor about the resources available through ConnectPlus, including the media-rich eBook, interactive learning tools, and animations.

   www.mhhe.com/maderbiology11

McGraw-Hill Access Science Website

An Encyclopedia of Science and Technology Online which provides more information including videos that can enhance the laboratory experience.

   www.accessscience.com

26

Basic Mammalian Anatomy I

Learning Outcomes

Introduction

In this laboratory, you will dissect a fetal pig. Alternately your instructor may choose to have you observe a pig that has already been dissected. Both pigs and humans are mammals; therefore, you will be studying mammalian anatomy. The period of pregnancy, or gestation, in pigs is approximately 17 weeks (compared with an average of 40 weeks in humans). The piglets used in class will usually be within 1 to 2 weeks of birth.

The pigs may have a slash in the right neck region, indicating the site of blood drainage. A red latex solution may have been injected into the **arterial system,** and a blue latex solution may have been injected into the **venous system** of the pigs. If so, when a vessel appears red, it is an artery, and when a vessel appears blue, it is a vein.

As a result of this laboraory, you should gain an appreciation of which organs work together. For example, the liver and the pancreas aid the digestion of fat in the small intestine.

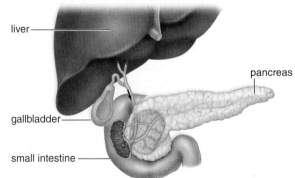

26.1 External Anatomy

Mammals are characterized by the presence of mammary glands and hair. Mammals also occur in two distinct sexes, males and females, often distinguishable by their external **genitals,** the reproductive organs.

Both pigs and humans are placental mammals, which means that development occurs within the uterus of the mother. An **umbilical cord** stretches externally between the fetal animal and the **placenta,** where carbon dioxide and organic wastes are exchanged for oxygen and organic nutrients.

Pigs and humans are tetrapods—that is, they have four limbs. Pigs walk on all four of their limbs; in fact, they walk on their toes, and their toenails have evolved into hooves. In contrast, humans walk only on the feet of their legs.

Observation: External Anatomy

Body Regions and Limbs

1. Place your animal in a dissecting pan, and observe the following body regions: the rather large head; the short, thick neck; the cylindrical trunk with two pairs of appendages (forelimbs and hindlimbs); and the short tail (Fig. 26.1*a*). The tail is an extension of the vertebral column.

> ⚠ **Latex gloves** Wear protective latex gloves when handling preserved animal organs.

2. Examine the four limbs, and feel for the joints of the digits, wrist, elbow, shoulder, hip, knee, and ankle.
3. Determine which parts of the forelimb correspond to your upper arm, elbow, lower arm, wrist, and hand.
4. Do the same for the hindlimb, comparing it with your leg.
5. The pig walks on its toenails, which would be like a ballet dancer on "tiptoe." Notice how your heel touches the ground when you walk. Where is the heel of the pig? _____

Umbilical Cord

1. Locate the umbilical cord arising from the ventral (toward the belly) portion of the abdomen.
2. Note the cut ends of the umbilical blood vessels. If they are not easily seen, cut the umbilical cord near the end and observe this new surface.
3. What is the function of the umbilical cord? _____

_____

Nipples and Hair

1. Locate the small **nipples,** the external openings of the **mammary glands.** The nipples are *not* an indication of sex, since both males and females possess them. How many nipples does your pig have? _____ When is it advantageous for a pig to have so many nipples?

_____

2. Can you find hair on your pig? _____ Where? _____

_____

Directional Terms for Dissecting Fetal Pig

Anterior: toward the head end	Ventral: toward the belly
Posterior: toward the hind end	Dorsal: toward the back

Figure 26.1 External anatomy of the fetal pig.

a. Body regions and limbs. b, c. The sexes can be distinguished by the external genitals.

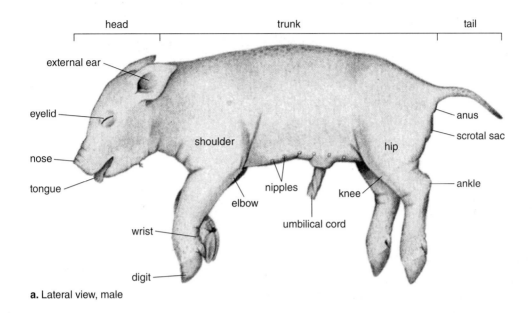

a. Lateral view, male

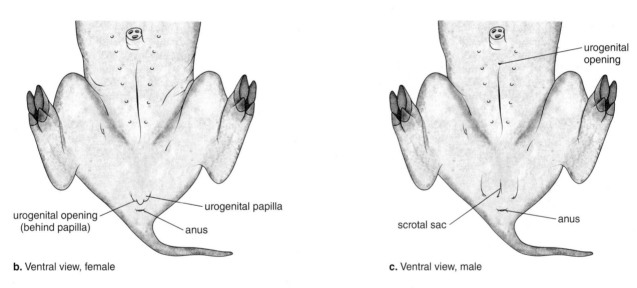

b. Ventral view, female

c. Ventral view, male

Anus and External Genitals

1. Locate the **anus** under the tail. The anus is an opening for what system in the body? _____

2. In females, locate the **urogenital opening,** just anterior to the anus, and a small, fleshy **urogenital papilla** projecting from the urogenital opening (Fig. 26.1*b*).

3. In males, locate the urogenital opening just posterior to the umbilical cord (Fig. 26.1*c*). The duct leading to it runs forward from between the legs in a long, thick tube, the **penis,** which can be felt under the skin. In males, the urinary system and the genital system are always joined.

4. You are responsible for identifying pigs of both sexes. What sex is your pig? _____
   Be sure to look at a pig of the opposite sex that another group of students is dissecting.

26.2 Oral Cavity and Pharynx

The **oral cavity** is the space in the mouth that contains the tongue and the teeth. The **pharynx** is dorsal to the oral cavity and has three openings: The **glottis** is an opening through which air passes on its way to the **trachea** (the windpipe) and lungs. The **esophagus** is a portion of the digestive tract that leads through the neck and thorax to the stomach. The **nasopharynx** leads to the nasal passages.

Observation: Oral Cavity and Pharynx

Oral Cavity

1. Insert a sturdy pair of scissors into one corner of the specimen's mouth, and cut posteriorly (toward the hind end) for approximately 4 cm. Repeat on the opposite side.
2. Place your thumb on the tongue at the front of the mouth, and gently push downward on the lower jaw. This will tear some of the tissue in the angles of the jaws so that the mouth will remain partly open (Fig. 26.2).
3. Note small, underdeveloped teeth in both the upper and lower jaws. Other embryonic, nonerupted teeth may also be found within the gums. The teeth are used to chew food.
4. Examine the tongue, which is partly attached to the lower jaw region but extends posteriorly and is attached to a bony structure at the back of the oral cavity (Fig. 26.2). The tongue manipulates food for swallowing.
5. Locate the hard and soft palates (Fig. 26.2). The **hard palate** is the ridged roof of the mouth that separates the oral cavity from the nasal passages. The **soft palate** is a smooth region posterior to the hard palate. An extension of the soft palate—the **uvula**—hangs down into the throat in humans. (A pig does not have a uvula.)

Figure 26.2 Oral cavity of the fetal pig.
The roof of the oral cavity contains the hard and soft palates, and the tongue lies above the floor of the oral cavity.

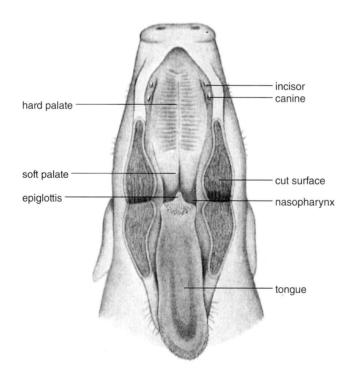

Pharynx

1. Push down on the tongue until you open the jaws far enough to see a slightly pointed flap of tissue pointing dorsally (toward the back) (Fig. 26.2). This flap is the **epiglottis,** which covers the glottis. The **glottis** leads to the trachea (Fig. 26.3*a*).
2. Posterior and dorsal to the glottis, find the opening into the **esophagus,** a tube that takes food to the stomach. Note the proximity of the glottis and the opening to the esophagus. Each time the pig—or a human—swallows, the epiglottis instantly closes to keep food and fluids from going into the lungs via the trachea.
3. Insert a blunt probe into the glottis, and note that it enters the trachea. Remove the probe, insert it into the esophagus, and note the position of the esophagus beneath the trachea.
4. Make a midline cut in the soft palate from the epiglottis to the hard palate. Then make two lateral cuts at the edge of the hard palate.
5. Posterior to the soft palate, locate the openings to the nasal passages.
6. Explain why it is correct to say that the air and food passages cross in the pharynx.

_____

_____

Figure 26.3   Air and food passages in the fetal pig.
The air and food passages cross in the pharynx. **a.** Drawing. **b.** Dissection of specimen.

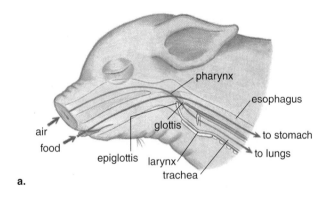

a.

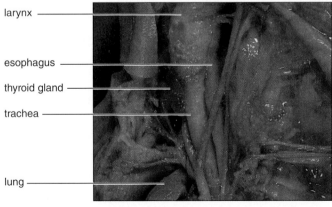

b.

26.3 Thoracic and Abdominal Incisions

First, prepare your pig according to the following directions, and then make thoracic and abdominal incisions so that you will be able to study the internal anatomy of your pig.

Preparation of Pig for Dissection

1. Place the fetal pig on its back in the dissecting pan.
2. Tie a cord around one forelimb, and then bring the cord around underneath the pan to fasten back the other forelimb.
3. Spread the hindlimbs in the same way.
4. With scissors always pointing up (never down), make the following incisions to expose the thoracic and abdominal cavities. The incisions are numbered on Figure 26.4 to correspond with the following steps.

Thoracic Incisions

1. Starting at the **diaphragm,** a structure that separates the thoracic cavity from the abdominal cavity, cut anteriorly until you reach the hairs in the throat region.
2. Make two lateral cuts, one on each side of the midline incision anterior to the forelimbs, taking extra care not to damage the blood vessels around the heart.
3. Make two lateral cuts, one on each side of the midline just posterior to the forelimbs and anterior to the diaphragm, following the ends of the ribs. Pull back the flaps created by these cuts (do not remove them) to expose the **thoracic cavity.** List the organs you find in the thoracic cavity.

_____

Abdominal Incisions

4. With scissors pointing up, cut posteriorly from the diaphragm to the umbilical cord.
5. Make a flap containing the umbilical cord by cutting a semicircle around the cord and by cutting posteriorly to the left and right of the cord.
6. Make two cuts, one on each side of the midline incision posterior to the diaphragm. Examine the diaphragm, attached to the chest wall by radially arranged muscles. The central region of the diaphragm, called the **central tendon,** is a membranous area.
7. Make two more cuts, one on each side of the flap containing the umbilical cord and just anterior to the hindlimbs. Pull back the side flaps created by these cuts to expose the **abdominal cavity.**
8. Lifting the flap with the umbilical cord requires cutting the **umbilical vein.** Before cutting the umbilical vein, tie a thread on each side of where you will cut to mark the vein for future reference.
9. Rinse out your pig as soon as you have opened the abdominal cavity. If you have a problem with excess fluid, obtain a disposable plastic pipette to suction off the liquid.
10. Anatomically, the diaphragm separates what two cavities? _____
11. List the organs you find in the abdominal cavity. _____

_____

Figure 26.4 Ventral view of the fetal pig indicating incisions.

These incisions are to be made preparatory to dissecting the internal organs. They are numbered here in the order they should be done.

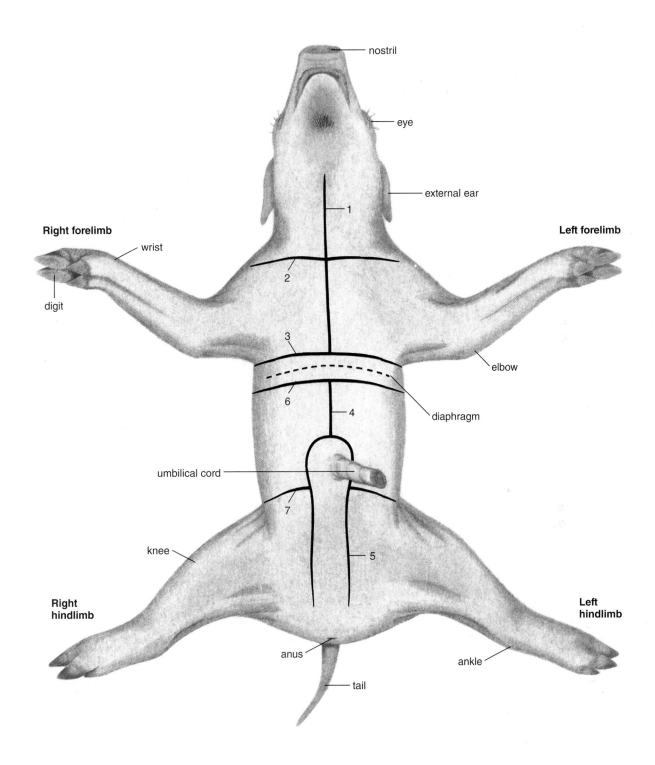

26.4 Neck Region

You will locate several organs in the neck region. Use Figures 26.3b and 26.5 as a guide, but *keep all the flaps on your pig* so you can close the thoracic and abdominal cavities at the end of the laboratory session.

The **thymus gland** is a part of the lymphatic system. Certain white blood cells called T (for thymus) lymphocytes mature in the thymus gland and help us fight disease. The **larynx,** or voice box, sits atop the **trachea,** or windpipe. The esophagus is a portion of the digestive tract that leads to the stomach. The **thyroid gland** secretes hormones that travel in the blood and act upon other body cells. These hormones (e.g., thyroxine) regulate the rate at which metabolism occurs in cells.

Observation: Neck Region

Thymus Gland

1. Move the skin apart in the neck region just below the hairs mentioned earlier. If necessary, cut the body wall laterally to make flaps. You will most likely be viewing exposed muscles.
2. *Cut through and clear away muscle* to expose the thymus gland, a diffuse gland that lies among the muscles. Later you will notice that the glandular thymus flanks the thyroid and overlies the heart (Fig. 26.5). The thymus is particularly large in fetal pigs, since their immune systems are still developing.

Larynx, Trachea, and Esophagus

1. Probe down into the deeper layers of the neck. Medially (toward the center), beneath several strips of muscle, you will find the hard-walled larynx and the trachea, parts of the respiratory passage to be examined later in Laboratory 27. Dorsal to the trachea, find the esophagus.
2. Open the mouth and insert a probe into the glottis and esophagus from the pharynx to better understand the orientation of these two organs.

Thyroid Gland

Locate the thyroid gland just posterior to the larynx, lying ventral to (on top of) the trachea.

26.5 Thoracic Cavity

As previously mentioned, the body cavity of mammals, including human beings, is divided by the diaphragm into the thoracic cavity and the abdominal cavity. The heart and lungs are in the thoracic cavity (Figs. 26.5 and 26.6). The **heart** is a pump for the cardiovascular system, and the **lungs** are organs of the respiratory system where gas exchange occurs.

Observation: Thoracic Cavity

Heart and Lungs

1. If you have not yet done so, fold back the chest wall flaps. To do this, you will need to tear the thin membranes that divide the thoracic cavity into three compartments: the **left pleural cavity** containing the left lung, the **right pleural cavity** containing the right lung, and the **pericardial cavity** containing the heart.
2. Examine the lungs. Locate the four lobes of the right lung and the three lobes of the left lung. The trachea, dorsal to the heart, divides into the **bronchi,** which enter the lungs. Later (Laboratory 27), when the heart is removed, you will be able to see the trachea and bronchi.
3. Sequence the organs of the respiratory tract to trace the path of air from the nasal passages to the lungs.

_____

_____

Figure 26.5 Internal anatomy of the fetal pig.

The major organs are featured in this drawing. In the fetal pig, a red color tells you a vessel is an artery, and a blue color tells you it is a vein. (It does not tell you whether this vessel carries O₂-rich or O₂-poor blood.) Contrary to this drawing, *keep all the flaps on your pig* so you can close the thoracic and abdominal cavities at the end of the laboratory session.

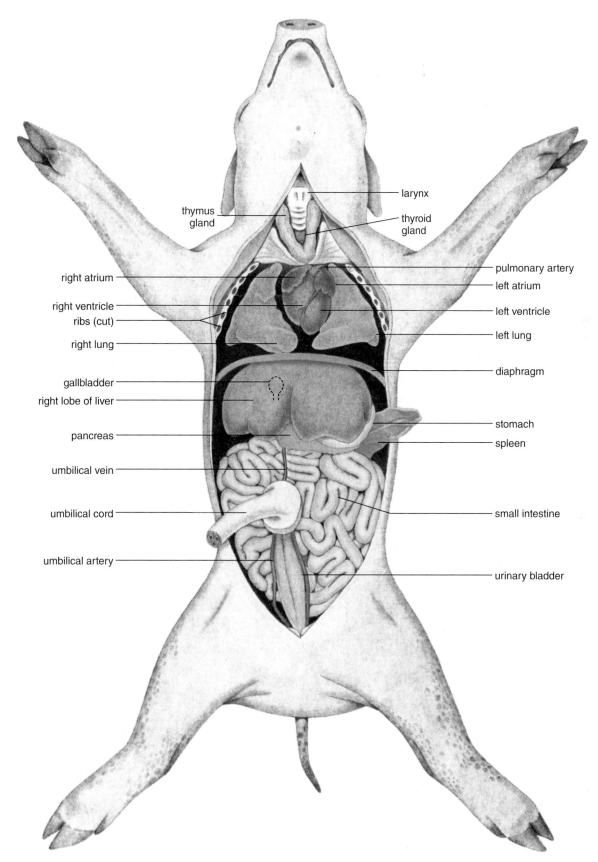

26.6 Abdominal Cavity

The abdominal wall and organs are lined by a membrane called **peritoneum,** consisting of epithelium supported by connective tissue. Double-layered sheets of peritoneum, called **mesenteries,** project from the body wall and support the organs.

Observation: Abdominal Cavity

If your particular pig is partially filled with dark, brownish material, take your animal to the sink and rinse it out. This material is clotted blood. Consult your instructor before removing any red or blue latex masses, since they may enclose organs you will need to study.

Liver

The **liver,** the largest organ in the abdomen (see Fig. 26.6), performs numerous vital functions, including (1) disposing of worn-out red blood cells, (2) producing bile, (3) storing glycogen, (4) maintaining the blood glucose level, and (5) producing blood proteins.

1. Locate the liver, a large, brown organ. Its anterior surface is smoothly convex and fits snugly into the concavity of the diaphragm.
2. Name several functions of the liver. _____

_____

Stomach and Spleen

The organs of the digestive tract include the stomach, small intestine, and large intestine. The **stomach** (see Fig. 26.5) stores food and has numerous gastric glands. These glands secrete a juice that digests protein. The **spleen** (see Fig. 26.5) is a lymphoid organ in the lymphatic system that contains both white and red blood cells. It purifies blood and disposes of worn-out red blood cells.

1. Push aside and identify the stomach, a large sac dorsal to the liver on the left side.
2. Locate the point near the midline of the body where the **esophagus** penetrates the diaphragm and joins the stomach.
3. Find the spleen, a long, flat, reddish organ attached to the stomach by mesentery.
4. The stomach is a part of what system? _____

What is its function? _____

5. The spleen is a part of what system? _____

What is its function? _____

Small Intestine

The **small intestine** is the part of the digestive tract that receives secretions from the pancreas and gallbladder. Besides being an area for the digestion of all components of food, carbohydrate, protein, and fat, the small intestine absorbs the products of digestion: glucose, amino acids, glycerol, and fatty acids.

1. Look posteriorly where the stomach makes a curve to the right and narrows to join the anterior end of the small intestine called the **duodenum.**
2. From the duodenum, the small intestine runs posteriorly for a short distance and is then thrown into an irregular mass of bends and coils held together by a common mesentery.
3. The small intestine is a part of what system? _____

What is its function? _____

Figure 26.6 Internal anatomy of the fetal pig.

Most of the major organs are shown in this photograph. The stomach has been removed. The spleen, gallbladder, and pancreas are not visible. *Do not* remove any organs or flaps from your pig.

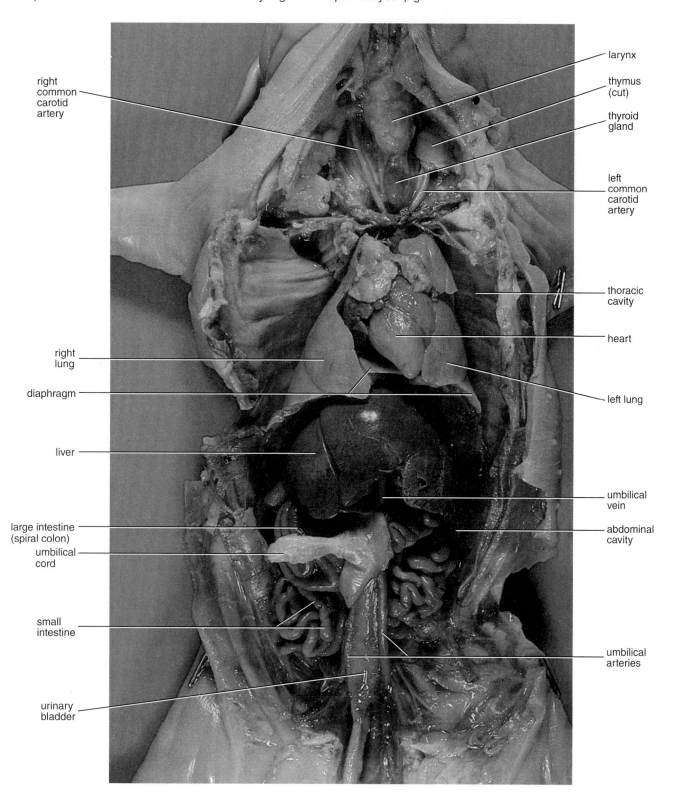

Gallbladder and Pancreas

The **gallbladder** stores and releases bile, which aids the digestion of fat. The **pancreas** (see Fig. 26.5) is both an exocrine and an endocrine gland. As an exocrine gland, it produces and secretes pancreatic juice, which digests all the components of food in the small intestine. Both bile and pancreatic juice enter the duodenum by way of ducts. As an endocrine gland, the pancreas secretes the hormones insulin and glucagon into the bloodstream. Insulin and glucagon regulate blood glucose levels.

1. Locate the **bile duct,** which runs in the mesentery stretching between the liver and the duodenum. Find the gallbladder, embedded in the liver on the underside of the right lobe. It is a small, greenish sac.
2. Lift the stomach and locate the pancreas, the light-colored, diffuse gland lying in the mesentery between the stomach and the small intestine. The pancreas has a duct that empties into the duodenum of the small intestine.
3. What is the function of the gallbladder? _____
4. What is the function of the pancreas? _____

_____

Large Intestine

The **large intestine** is the part of the digestive tract that absorbs water and prepares feces for defecation at the anus. The first part of the large intestine, called the **cecum**, has a projection called the vermiform (meaning wormlike) appendix.

1. Locate the distal (far) end of the small intestine, which joins the large intestine posteriorly, in the left side of the abdominal cavity (right side in humans). At this junction, note the cecum, a blind pouch.
2. Compare the large intestine of your pig to Figure 26.7. The organ does not have the same appearance in humans.
3. Follow the main portion of the large intestine, known as the **colon,** as it runs from the point of juncture with the small intestine into a tight coil (spiral colon), then out of the coil anteriorly, then posteriorly again along the midline of the dorsal wall of the abdominal cavity. In the pelvic region, the **rectum** is the last portion of the large intestine. The rectum leads to the **anus.**
4. The large intestine is a part of what system? _____
5. What is the function of the large intestine? _____
6. Sequence the organs of the digestive system to trace the path of food from the mouth to the anus. _____

_____

Storage of Pigs

1. Before leaving the laboratory, place your pig in the plastic bag provided.
2. Expel excess air from the bag, and tie it shut.
3. Write your *name* and *section* on the tag provided, and attach it to the bag. Your instructor will indicate where the bags are to be stored until the next laboratory period.
4. Clean the dissecting tray and tools, and return them to their proper location.
5. Wipe off your goggles.
6. Wash your hands.

26.7   Human Anatomy

Humans and pigs are both mammals, and their organs are similar. A human torso model shows the exact location of the organs in human beings (see Fig. 26.7). You should learn to associate each human organ with its particular system. Four systems are color-coded in Figure 26.7.

Figure 26.7   Human internal organs.
The dotted lines indicate the full shape of an organ that is partially covered by another organ.

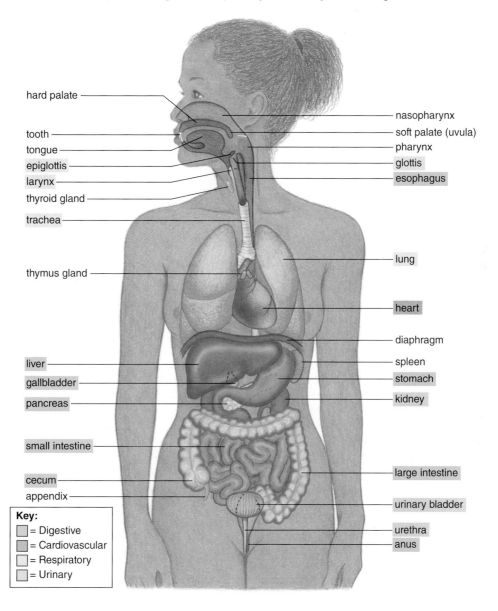

hard palate

tooth
tongue
epiglottis
larynx
thyroid gland
trachea

thymus gland

liver
gallbladder
pancreas

small intestine

cecum
appendix

nasopharynx
soft palate (uvula)
pharynx
glottis
esophagus

lung

heart

diaphragm
spleen
stomach
kidney

large intestine

urinary bladder

urethra
anus

Key:
☐ = Digestive
☐ = Cardiovascular
☐ = Respiratory
☐ = Urinary

Observation: Human Torso

1. Examine a human torso model, and using Figure 26.7 as a guide, locate the same organs you have just dissected in the fetal pig.

2. In your studies so far, have you seen any major differences between pig internal anatomy and human internal anatomy? _____

1. What two features indicate that a pig is a mammal? _____
   _____

2. Sequence the following organs to trace the path of air: lungs, nasal passages, nasopharynx, trachea, bronchi, glottis. _____
   _____

3. What difficulty would probably arise if a person were born without an epiglottis?
   _____
   _____

4. What two cavities studied in this laboratory hold the internal organs?
   _____

5. Name two principal organs in the thoracic cavity, and give a function for each.
   _____
   _____

6. What difficulty would arise if a person were born without a thymus gland?
   _____
   _____

7. Name the largest organ in the abdominal cavity and list several functions.
   _____
   _____

8. A large portion of the abdominal cavity is taken up with digestive organs. What are these organs?
   _____
   _____

9. Why is it proper to associate the gallbladder with the liver?
   _____
   _____

10. Where would you find the pancreas?
    _____
    _____

Biology **Website**

Enhance your study of the text and laboratory manual with study tools, practice tests, and virtual labs. Also ask your instructor about the resources available through ConnectPlus, including the media-rich eBook, interactive learning tools, and animations.

www.mhhe.com/maderbiology11

McGraw-Hill Access Science Website

An Encyclopedia of Science and Technology Online which provides more information including videos that can enhance the laboratory experience.

www.accessscience.com

27

Basic Mammalian Anatomy II

Learning Outcomes

Introduction

The heart and blood vessels form the cardiovascular system. The heart is a double pump that keeps the blood flowing in one direction—away from and then back to the heart. The blood vessels transport blood and its contents, serve the needs of the body's cells by carrying out exchanges with them, and direct blood flow to those systemic tissues that most require it at the moment. After removing the heart and examining the heart chambers you will have an opportunity to remove and study the lungs and the small intestine. The latter will expose the urinary and reproductive systems. These two systems are so closely associated in mammals that they are often considered together as the urogenital system. In this laboratory, we will focus first on dissecting the urinary and reproductive systems in the fetal pig. We will then compare the anatomy of the reproductive systems in pigs with those in humans.

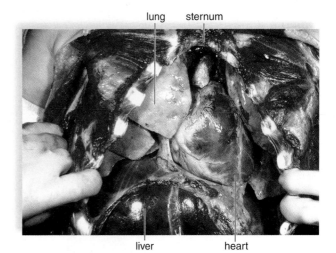

lung   sternum

liver   heart

27.1 Cardiovascular System

The **cardiovascular system** includes the heart and two major circular pathways: the **pulmonary circuit** to and from the lungs and the **systemic circuit** to and from the body's organs.

Heart

The mammalian heart has a **right** and **left atrium** and a **right** and **left ventricle.** *To tell the left from the right side, mentally position the heart so that it corresponds to your own body.* Contraction of the heart pumps the blood through the heart and out into the arteries. The right side of the heart sends blood through the smaller pulmonary circuit, and the left side of the heart sends blood through the much larger systemic circuit.

Observation: Heart

Heart Model

1. Study a heart model (Fig. 27.1) or a preserved heart, and identify the four chambers of the heart: right atrium, right ventricle, left atrium, left ventricle.

2. Which ventricle is more muscular? _____ Why is this appropriate? _____

Figure 27.1   External view of mammalian heart.
Externally, notice the coronary arteries and cardiac veins that serve the heart itself.

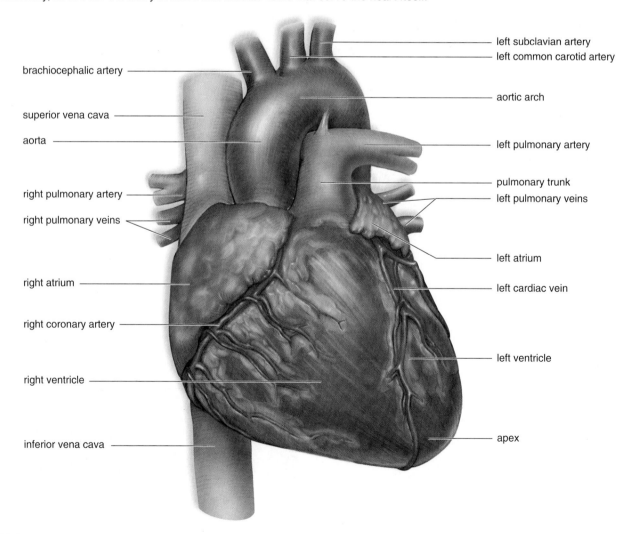

3. Identify the vessels connected to the heart. Locate the:
   a. **Pulmonary trunk,** which leaves the ventral side of the heart from the top of the right ventricle and then passes forward diagonally before branching into the right and left pulmonary arteries.
   b. **Aorta,** which arises from the anterior end of the left ventricle, just dorsal to the origin of the pulmonary trunk. The aorta soon bends to the animal's left as the **aortic arch.**
   c. **Venae cavae.** The anterior (superior) and posterior (inferior) venae cavae enter the right atrium. They bring blood from the head and body, respectively, to the heart.

   d. **Pulmonary veins,** which return blood from the lungs to the left atrium. Why are the pulmonary veins colored red in Figure 27.2? _____

   _____

4. Locate the valves of the heart (Fig. 27.2). Remove the ventral half of the heart model or the ventral half of the heart. Identify the:
   a. **Right atrioventricular (tricuspid) valve** and the **left atrioventricular (bicuspid, mitral) valve.**
   b. **Pulmonary semilunar valve** (in the base of the pulmonary trunk).
   c. **Aortic semilunar valve** (in the base of the aorta).
   d. **Chordae tendineae** that hold the atrioventricular valves in place while the heart contracts. These extend from the papillary muscles.

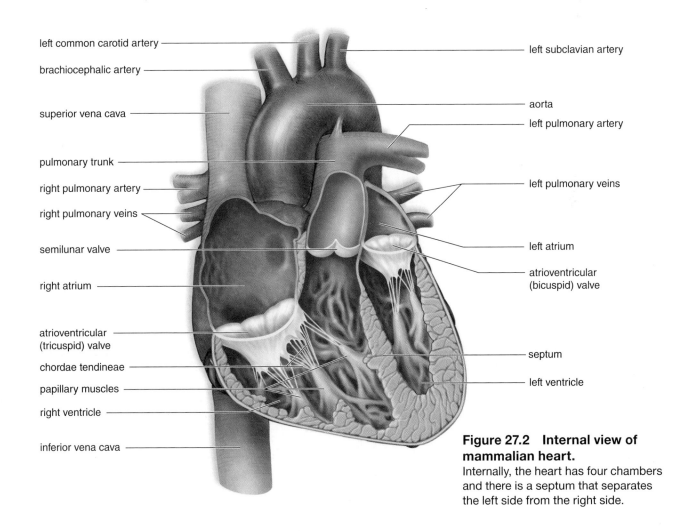

Figure 27.2   Internal view of mammalian heart.
Internally, the heart has four chambers and there is a septum that separates the left side from the right side.

Tracing the Path of Blood Through the Heart

To demonstrate that O_2-poor blood is kept separate from O_2-rich blood, trace the path of blood from the right side of the heart to the aorta by filling in the blanks with the names of blood vessel valves. In the adult mammal, note that blood passes through the lungs to go from the right side of the heart to the left side.

From Venae Cavae   **From Lungs**

_____   _____

_____ valve   _____

_____   _____ valve

_____ valve   _____

_____   _____ valve

_____

To Lungs   **To Aorta**

Blood Vessels

Blood that leaves the heart enters one of two sets of blood vessels: the **pulmonary circuit**, which takes blood from the heart to the lungs and from the lungs to the heart, and the **systemic circuit**, which takes blood from the heart to the body proper and from the body proper to the heart.

Pulmonary Circuit

1. Sequence the blood vessels in the pulmonary circuit to trace the path of blood from the right ventricle to the left atrium of the heart:

   Right ventricle of the heart

   _____

   _____

   Lungs

   _____

   Left atrium of the heart

2. Which of these blood vessels contains O_2-rich blood? _____

Observation: Systemic Circuit

1. The major blood vessels in the systemic system are the **aorta** (takes blood away from the heart) and the **venae cavae** (takes blood to the heart). Otherwise, with the help of Figure 27.3, complete Table 27.1.

Table 27.1 Other Blood Vessels in the Systemic Circuit		
Body Part	**Artery**	**Vein**
Head		
Front legs in pig (arms in humans)		
Kidney		
Hind legs		

Figure 27.3 Overview of the fetal pig arteries and veins.
Arteries and veins are found in all parts of the body. In this drawing, only veins are labeled on the left, and only arteries are labeled on the right.

⚠ **Latex gloves** Wear protective latex gloves when handling preserved animal organs.

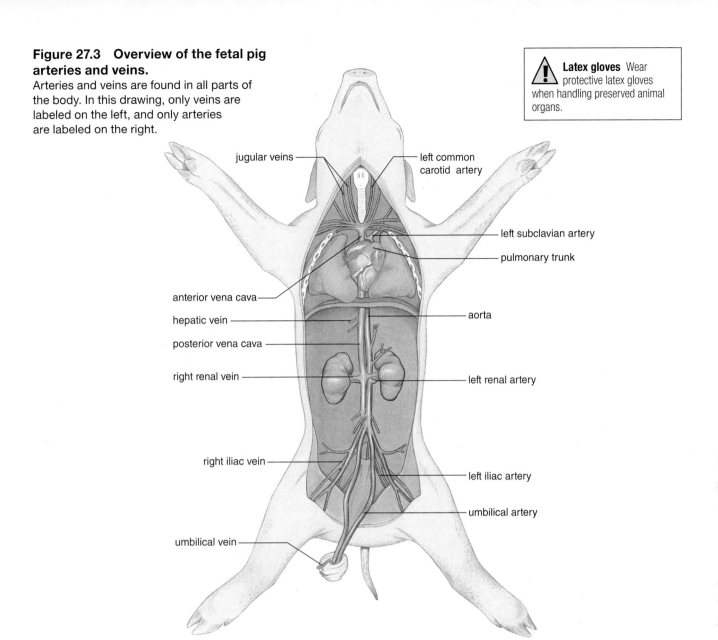

2. With the help of Figure 27.3 and Table 27.1, sequence the blood vessels in the systemic circuit to trace the path of blood from the heart to the kidneys and from the kidneys to the heart.

Left ventricle

_____

_____

Kidneys

_____

_____

Right atrium

27.2 Respiratory, Digestive, and Urinary Systems

You will now have the opportunity to examine more carefully the organs of the respiratory and digestive systems you merely observed in Laboratory 26. You will also study the urinary system.

Respiratory System

The respiratory system contains the lungs and those structures that conduct air to (and from) the lungs (Fig. 27.4). Name two locations from which air can enter the glottis (see Fig. 26.3*a*). _____

The lungs contain alveoli, air sacs surrounded by capillaries. Here, gas exchange by diffusion brings oxygen into the blood and takes carbon dioxide out of the blood.

Observation: Organs of the Respiratory System

1. Open a pig's mouth, insert your blunt probe into the **glottis,** and explore the pathway of air in normal breathing by carefully working the probe down through the **larynx** to the level of the bronchi.
2. If you wish, you can slit open the larynx along its midline and observe the small, paired, lateral flaps inside, known as **vocal cords.** These are not yet well-developed in this fetal animal.
3. Remove the heart of your pig by cutting the blood vessels that enter and exit heart. Name these blood vessels as you cut them. You will now see the trachea and how it divides into two bronchi. Carefully cut the trachea crosswise just below the larynx. Do not cut the esophagus, which lies just dorsal to the trachea.
4. Lift out the entire portion of the respiratory system you have just freed: trachea, bronchi, bronchioles, and lungs (Fig. 27.4). Place them in a small container of water. Holding the trachea with your forceps, gently but firmly stroke the lung repeatedly with the blunt wooden base of one of your probes. If you work carefully, the alveolar tissue will be fragmented and rubbed away, leaving the branching system of air tubes and blood vessels.

Figure 27.4   Dorsal view of lungs and associated air tubes.
This drawing also shows the heart and associated blood vessels.

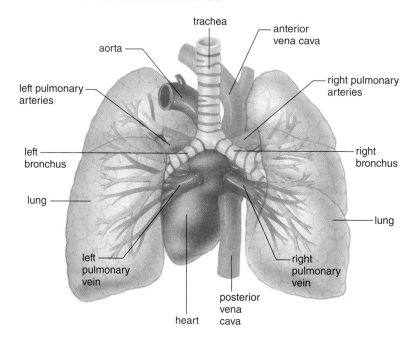

Digestive System

The digestive system contains organs in which food is ingested, digested, absorbed, and prepared for elimination. Sequence the organs *large intestine, mouth, small intestine, esophagus, stomach,* and *anus,* to trace the path of food from its entrance into the body to the elimination of the remains of digestion:

Observation: Organs of the Digestive System

1. With removal of the heart and lungs, you now have a better view of the esophagus. Open the mouth again, and insert a blunt probe into the **esophagus** (see Fig. 26.3). Then trace the esophagus to the stomach.
2. Open one side of the **stomach,** and examine its interior surface. Does it appear smooth or rough?

3. At the lower end of the stomach, find the **pyloric sphincter,** the muscle that surrounds the entrance to the upper region of the small intestine, the **duodenum.** The pyloric sphincter regulates the entrance of material into the duodenum from the stomach.
4. Find the gallbladder, which is a small, greenish sac dorsal to the liver you also observed in Laboratory 26, pages 379–80. Dissect more carefully now to find the **bile duct system** (see Fig. 26.6) that conducts the bile to the duodenum.
5. Locate the rest of the **small intestine** and the **large intestine** (colon and rectum) first observed in Laboratory 26, pages 378–80. To complete your study, sever the coiled small intestine just below the duodenum, and sever the colon at the point where it joins the **rectum**.
6. Carefully cut the mesenteries holding the coils of the small intestine so that it can be laid out in a straight line. As defined previously, mesenteries are double-layered sheets of membrane that project from the body wall and support the organs.
7. Measure and record in meters the length of the intestinal tract. _____ m
8. Considering that nutrient molecules are absorbed into the blood vessels of the small intestine, why would such a great length be beneficial to the body? _____
9. Slit open a short portion of the intestines, and note the corrugated texture of the interior lining.

Urinary System in Pigs

The urinary system consists of the **kidneys**, which produce urine; the **ureters**, which transport urine to the **urinary bladder**, where urine is stored; and the **urethra**, which transports urine to the outside. In males, the urethra also transports sperm during copulation.

Observation: Urinary System in Pigs

During this dissection, compare the urinary system structures of male and female fetal pigs. Later in this laboratory period, exchange specimens with a neighboring team for a more thorough inspection.

1. The large, paired kidneys (Fig. 27.5) are reddish organs covered by **peritoneum,** a membrane that anchors them to the dorsal wall of the abdominal cavity, sometimes called the **peritoneal cavity.** Clean the peritoneum away from one of the kidneys, and study it more closely.
2. Locate the **ureters,** which leave the kidneys and run posteriorly under the peritoneum (Fig. 27.5).
3. Clear the peritoneum away, and follow a ureter to the **urinary bladder,** which normally lies in the ventral portion of the abdominal cavity. The urinary bladder is on the inner surface of the flap of tissue to which the umbilical cord was attached.

Figure 27.5 Urinary system of the fetal pig.
In both sexes, urine is made by the kidneys, transported to the bladder by the ureters, stored in the bladder, and then excreted from the body through the urethra.

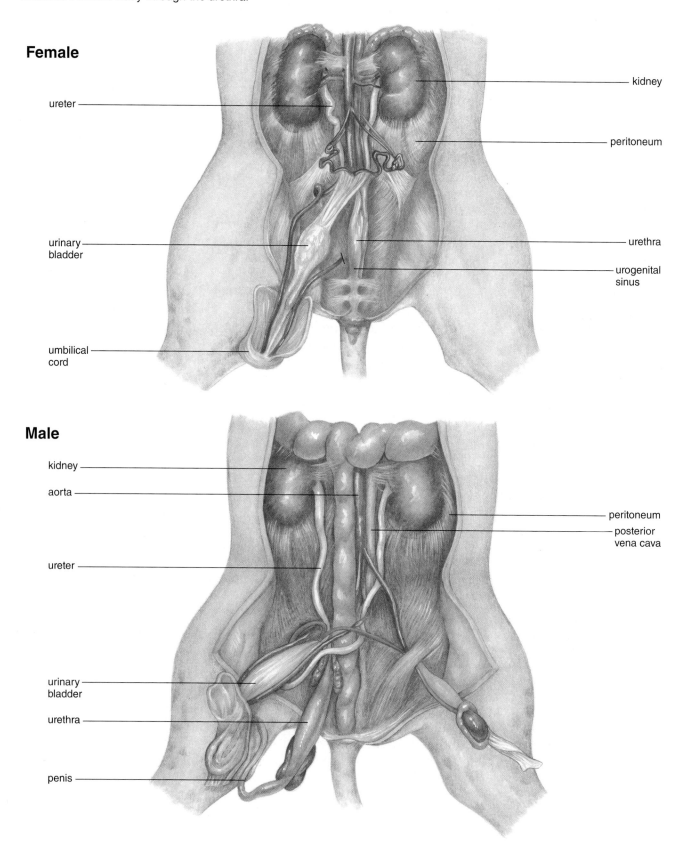

Female

ureter

kidney

peritoneum

urinary bladder

urethra

urogenital sinus

umbilical cord

Male

kidney

aorta

peritoneum

posterior vena cava

ureter

urinary bladder

urethra

penis

4. The **urethra** arises from the bladder posteriorly and joins the urogenital sinus. Follow the urethra until it passes from view into the ring formed by the pelvic girdle.

5. Sequence the organs in the urinary system to trace the path of urine from its production to its exit.

---

6. Using a scalpel, section one of the kidneys in place, cutting it lengthwise (Fig. 27.6). At the center of the medial portion of the kidney is an irregular, cavity-like reservoir, the **renal pelvis.** The outermost portion of the kidney (the **renal cortex**) shows many small striations perpendicular to the outer surface. The cortex and the more even-textured **renal medulla** contain **nephrons** (excretory tubules), microscopic organs that produce urine.

Figure 27.6   Anatomy of the kidney.
A kidney has a renal cortex, renal medulla, renal pelvis, and microscopic tubules called nephrons.

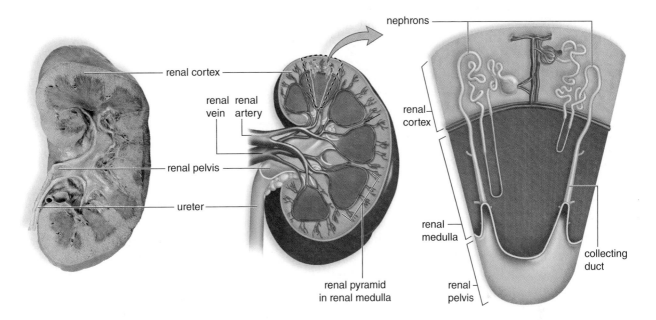

27.3   Male Reproductive System

The **male reproductive system** consists of the **testes** (sing., testis), which produce sperm, and the **epididymides** (sing., epididymis), which store sperm before they enter the **vasa deferentia** (sing., vas deferens). Just prior to ejaculation, sperm leave the vasa deferentia and enter the **urethra,** located in the penis. The **penis** is the male organ of sexual intercourse. **Seminal vesicles,** the **prostate gland,** and the **bulbourethral glands** (Cowper's glands) add fluid to semen (sperm plus fluids) after sperm reach the urethra. Table 27.2 summarizes the male reproductive organs.

The testes begin their development in the abdominal cavity, just anterior and dorsal to the kidneys. Before birth, however, they gradually descend into paired **scrotal sacs** within the scrotum, suspended anterior to the anus. Each scrotal sac is connected to the body cavity by an **inguinal canal,** the opening of which can be found in your pig. The passage of the testes from the body cavity into the scrotal sacs is called the descent of the testes and it occurs in human males. The testes in most of the male fetal pigs being dissected will probably be partially or fully descended.

While doing this dissection, consult Table 27.2 for the function of the male reproductive organs.

Table 27.2   Male Reproductive Organs and Functions

Organ	Function
Testis	Produces sperm and sex hormones
Epididymis	Stores sperm as they mature
Vas deferens	Stores sperm and conducts sperm to urethra
Urethra	Conducts sperm to urogenital opening
Seminal vesicle Prostate gland Bulbourethral glands	Contributes secretions to semen
Penis	Organ of copulation

Inguinal Canal, Testis, Epididymis, and Vas Deferens

1. Locate the opening of the left inguinal canal, which leads to the left scrotal sac (Fig. 27.7).
2. Expose the canal and sac by making an incision through the skin and muscle layers from a point over this opening back to the left scrotal sac.
3. Open the sac, and find the testis. Note the much-coiled tubule—the epididymis—that lies alongside a testis. An epididymis is continuous with a vas deferens, which runs toward the abdominal cavity.
4. Each vas deferens loops over an umbilical artery and ureter and unites with the urethra as it leaves the urinary bladder. We will dissect this juncture below.

Penis, Urethra, and Accessory Glands

1. Cut through the ventral skin surface just posterior to the umbilical cord. This will expose the rather undeveloped penis, which contains a long portion of the urethra.
2. Lay the penis to one side, and then cut down through the ventral midline, laying the legs wide apart in the process (Fig. 27.8). The cut will pass between muscles and through pelvic cartilage (bone has not developed yet). Do not cut any of the ducts or tracts in the region.
3. You will now see the urethra ventral to the rectum. It is somewhat heavier in the male due to certain accessory glands:
   a. Bulbourethral glands (Cowper's glands), about 1 cm in diameter, are further along the urethra and are more prominent than the other accessory glands.
   b. The prostate gland, about 4 mm across and 3 mm thick, is located on the dorsal surface of the urethra, just posterior to the juncture of the bladder with the urethra. It is often difficult to locate and is not shown in Figures 27.7 and 27.8.
   c. Small, paired seminal vesicles may be seen on either side of the prostate gland.
4. Trace the urethra as it leaves the bladder. When it nears the end of the abdominal cavity, it turns rather abruptly and runs anterorally just under the skin where you have just dissected it. This portion of the urethra is within the penis.
5. Now you should be able to see the vasa deferentia enter the urethra. If necessary, free these ducts from surrounding tissue to see them enter the urethra near the location of the prostate gland. In males, the urethra transports both sperm and urine to the urogenital opening (Fig. 26.1c).
6. Sequence the organs in the male reproductive system to trace the path of sperm from the organ of

   production to the penis. _____

_____

Figure 27.7 Male reproductive system of the fetal pig.

In males, the urinary system and the reproductive system are joined. The vasa deferentia (sing., vas deferens) enter the urethra, which also carries urine.

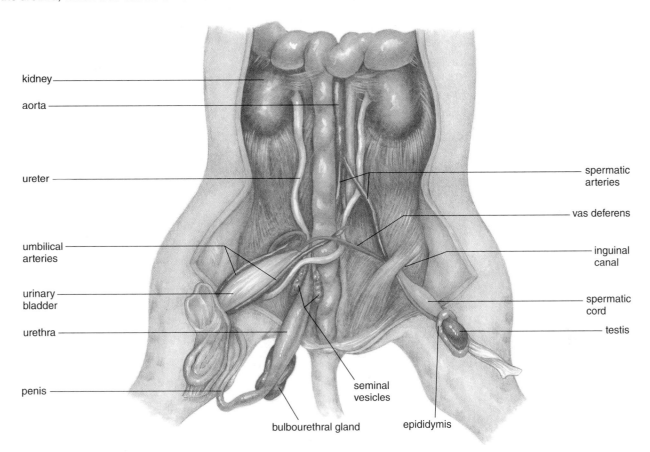

Figure 27.8 Photograph of the male reproductive system of the fetal pig.

Compare the diagram in Figure 27.7 to this photograph to help identify the structures of the male urogenital system.

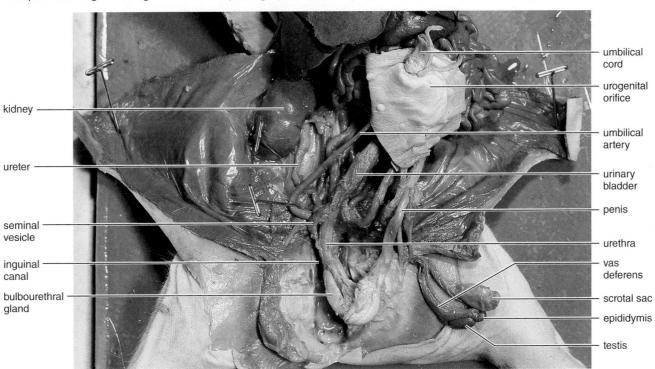

Comparison of Male Fetal Pig and Human Male

Use Figure 27.9 to help you compare the male pig reproductive system with the human male reproductive system. Complete Table 27.3, which compares the location of the penis in these two mammals.

Figure 27.9   Human male urogenital system.
In the fetal pig, but not in the human male, the penis lies beneath the skin and exits at a urogenital opening.

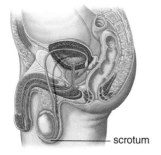

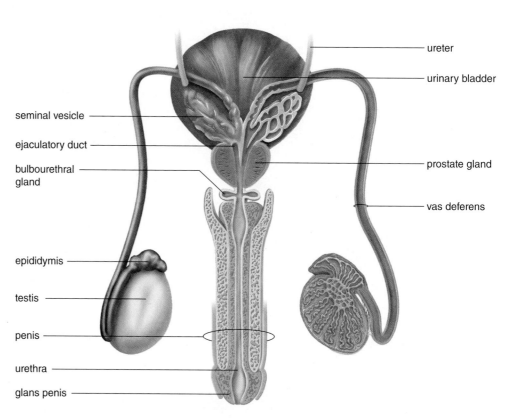

Table 27.3	Location of Penis in Male Fetal Pig and Human Male	
	Fetal Pig	**Human**
Penis		

27.4   Female Reproductive System

The **female reproductive system** (Table 27.4) consists of the **ovaries,** which produce eggs, and the **oviducts,** which transport eggs to the **uterus,** where development occurs. In the fetal pig, the uterus does not form a single organ, as in humans, but is partially divided into external structures called **uterine horns,** which connect with the oviduct. The **vagina** is the birth canal and the female organ of sexual intercourse.

Table 27.4 Female Reproductive Organs and Functions	
Organ	**Function**
Ovary	Produces egg and sex hormones
Oviduct (fallopian tube)	Conducts egg toward uterus
Uterus	Houses developing offspring
Vagina	Receives penis during copulation and serves as birth canal

Observation: Female Reproductive System in Pigs

While doing this dissection, consult Table 27.4 for the function of the female reproductive organs.

Ovaries and Oviducts

1. Locate the paired ovaries, small bodies suspended from the peritoneal wall in mesenteries, posterior to the kidneys (Figs. 27.10 and 27.11).
2. Closely examine one ovary. Note the small, short, coiled **oviduct**, sometimes called the fallopian tube. The oviduct does not attach directly to the ovary but ends in a funnel-shaped structure with fingerlike processes (fimbriae) that partially encloses the ovary. The egg produced by an ovary enters an oviduct where it is fertilized by a sperm, if reproduction will occur. Any resulting embryo passes to the uterus.

Uterine Horns

1. Locate the **uterine horns.** (Do not confuse the uterine horns with the oviducts; the latter are much smaller and are found very close to the ovaries.)
2. Find the body of the uterus located where the uterine horns join.

Vagina

1. Separate the hindlimbs of your specimen, and cut down along the midventral line. The cut will pass through muscle and the cartilaginous pelvic girdle. With your fingers, spread the cut edges apart, and use blunt dissecting instruments to separate connective tissue.
2. Now find the vagina, which passes from the uterus to the **urogenital sinus.** The vagina is dorsal to the urethra, which also enters the urogenital sinus, and ventral to the **rectum,** which exits at the **anus.** The urogenital sinus opens at the urogenital papilla (Figs. 27.10 and 27.11).
3. The vagina is the organ of copulation and is the birth canal. The receptacle for the vagina in a pig, the urogenital sinus, is absent in adult humans and several other adult female mammals in which both the urethra and vagina have their own openings.
4. The vagina plays a critical role in reproduction even though development of the offspring occurs in the uterus. Explain. _____

_____

Figure 27.10 Female reproductive system of the fetal pig.

In a pig, both the vagina and the urethra enter the urogenital sinus, which opens at the urogenital papilla.

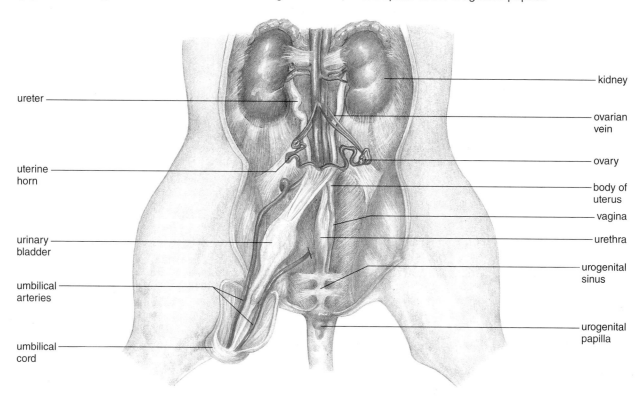

ureter

uterine horn

urinary bladder

umbilical arteries

umbilical cord

kidney

ovarian vein

ovary

body of uterus

vagina

urethra

urogenital sinus

urogenital papilla

Figure 27.11 Photograph of the female reproductive system of the fetal pig.

Compare the diagram in Figure 27.10 with this photograph to help identify the structures of the female urinary and reproductive systems.

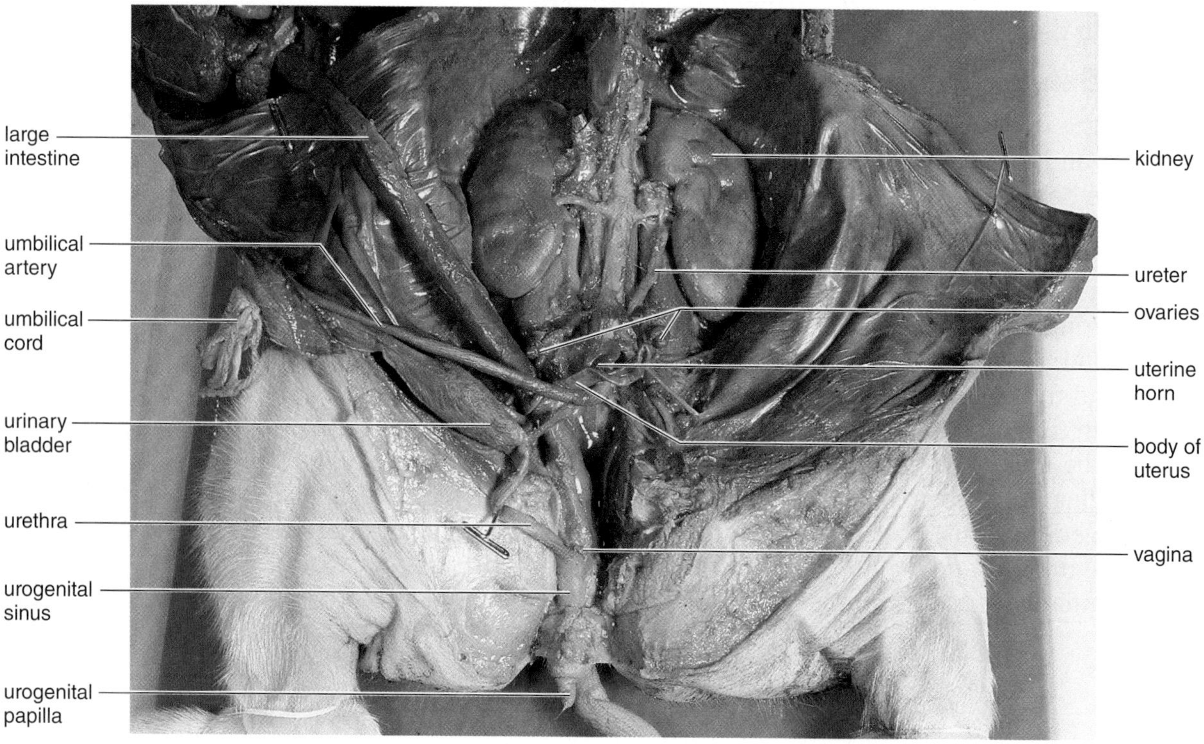

large intestine

umbilical artery

umbilical cord

urinary bladder

urethra

urogenital sinus

urogenital papilla

kidney

ureter

ovaries

uterine horn

body of uterus

vagina

Comparison of Female Fetal Pig with Human Female

Use Figure 27.12 to compare the female pig reproductive system with the human female reproductive system. Complete Table 27.5, which compares the appearance of the oviducts and the uterus, as well as the presence or absence of a urogenital sinus in these two mammals.

Figure 27.12 Human female reproductive system.
Especially compare the anatomy of the oviducts in humans with that of the uterine horns in a pig. In a pig, the fetuses develop in the uterine horns; in a human female, the fetus develops in the body of the uterus.

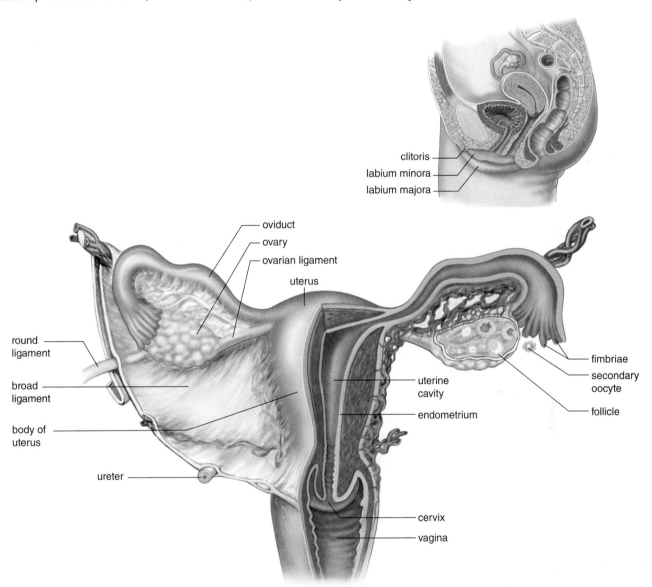

Table 27.5 Comparison of Female Fetal Pig with Human Female		
	Fetal Pig	**Human**
Oviducts		
Uterus		
Urogenital sinus		

Storage of Pigs

1. Before leaving the laboratory, place your pig in the plastic bag provided.
2. Expel excess air from the bag, and tie it shut.
3. Write your *name* and *section* on the tag provided, and attach it to the bag. Your instructor will indicate where the bags are to be stored until the next laboratory period.
4. Clean the dissecting tray and tools, and return them to their proper location.
5. Wipe off your goggles.
6. Wash your hands.

Laboratory Review 27

1. What are the four chambers of the mammalian heart? _____

   _____

2. Contrast the pumping function of the right and left sides of the heart. _____

   _____

3. Sequence the blood vessels in the systemic circuit to trace the path of blood from the left ventricle to the

   kidneys and back to the right atrium. _____

   _____

4. Sequence the organs of the respiratory system from the glottis to the lungs. _____

   _____

5. What's the difference between the ureters and the urethra in the urinary system? _____

   _____

6. Sequence the following organs: stomach, large intestine, small intestine, pharynx, mouth, esophagus,

   anus. _____

   _____

7. Sequence the path of sperm from the testes to the urogenital opening. _____

   _____

8. What organs enter the urogenital sinus in female pigs? _____

9. Which organ in males produces sperm, and which organ in females produces eggs? _____

   _____

10. How and when do sperm acquire access to an egg in mammals? _____

   _____

28

Chemical Aspects of Digestion

Learning Outcomes

Introduction
- Sequence the organs of the digestive tract from the mouth to the anus.   400
- State the contribution of each organ, if any, to the process of chemical digestion.   400

28.1 Protein Digestion by Pepsin
- Associate the enzyme pepsin with the ability of the stomach to digest protein.   401–2
- Explain why stomach contents are acidic and how a warm body temperature aids digestion.   401–2

28.2 Fat Digestion by Pancreatic Lipase
- Associate the enzyme lipase with the ability of the small intestine to digest fat.   402–3
- Explain why the emulsification process assists the action of lipase.   402
- Explain why a change in pH indicates that fat digestion has occurred.   403–4
- Explain the relationship between time and enzyme activity.   403–4

28.3 Starch Digestion by Pancreatic Amylase
- Associate the enzyme pancreatic amylase with the ability of the small intestine to digest starch.   405
- Explain the effect of boiling on the digestive action of amylase.   405

28.4 Requirements for Digestion
- Assuming a specific enzyme, list four factors that can affect the activity of all enzymes.   407
- Explain why the operative procedure called duodenal switch causes an individual to lose weight.   407

Introduction

In Laboratories 26 and 27, we examined the organs of digestion in the fetal pig. Now we wish to further our knowledge of the digestive process by associating certain digestive enyzmes with particular organs, as shown in Figure 28.1. This laboratory will also give us an opportunity to study the action of enzymes, much as William Beaumont did when he removed food samples through a hole in the stomach wall of his patient, Alexis St. Martin. Every few hours, Beaumont would see how well the food had been digested.

In Laboratory 5 we learned that enzymes are very specific and usually participate in only one type of reaction. The active site of an enzyme has a shape that accommodates its substrate, and if an environmental factor such as a boiling temperature or a wrong pH alters this shape, the enzyme loses its ability to function well, if at all. We will have an opportunity to make these observations with controlled experiments. The box on the next page reviews what is meant by a controlled experiment.

> **Planning Ahead** Be advised that protein digestion (page 401) requires 1 1/2 hours and fat digestion (page 403) requires 1 hour. Also a boiling water bath is required for starch digestion (page 405).

Figure 28.1 Organs of the digestive tract (right) and accessory organs (left).

Virtual Lab Digestive and Endocrine System Despite the title, this virtual laboratory available on the *Biology* website, **www.mhhe.com/mader biology11**, is about nutrition. An opportunity is provided to create a diet of varied foods (be sure to click on the forward arrow for choices from all food groups); keep within the allotted calories for your build; fulfill but do not exceed the percent daily requirements for nutrients.

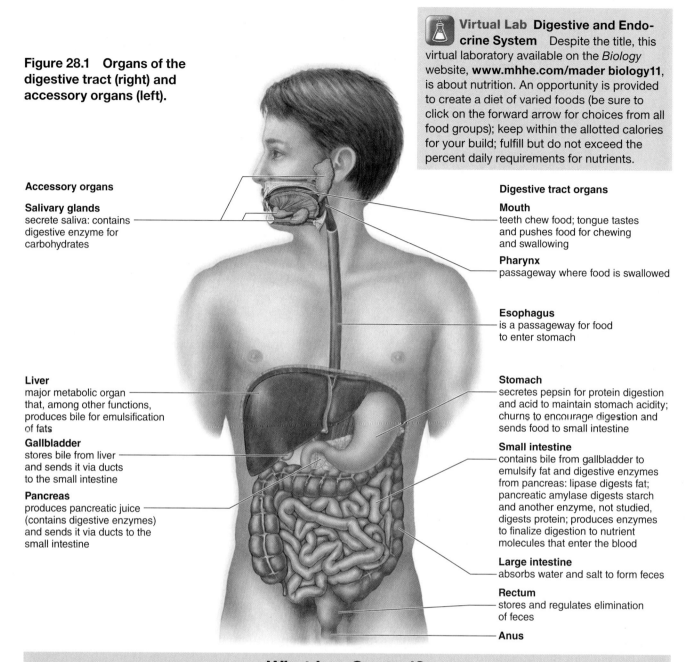

Accessory organs

Salivary glands
secrete saliva: contains digestive enzyme for carbohydrates

Liver
major metabolic organ that, among other functions, produces bile for emulsification of fats

Gallbladder
stores bile from liver and sends it via ducts to the small intestine

Pancreas
produces pancreatic juice (contains digestive enzymes) and sends it via ducts to the small intestine

Digestive tract organs

Mouth
teeth chew food; tongue tastes and pushes food for chewing and swallowing

Pharynx
passageway where food is swallowed

Esophagus
is a passageway for food to enter stomach

Stomach
secretes pepsin for protein digestion and acid to maintain stomach acidity; churns to encourage digestion and sends food to small intestine

Small intestine
contains bile from gallbladder to emulsify fat and digestive enzymes from pancreas: lipase digests fat; pancreatic amylase digests starch and another enzyme, not studied, digests protein; produces enzymes to finalize digestion to nutrient molecules that enter the blood

Large intestine
absorbs water and salt to form feces

Rectum
stores and regulates elimination of feces

Anus

What Is a Control?

The experiments in today's laboratory have both a positive control and a negative control, *which should be saved for comparison purposes until the experiment is complete*. The **positive control** goes through all the steps of the experiment and does contain the substance being tested. Therefore, positive results are expected. The **negative control** goes through all the steps of the experiment except it does not contain the substance being tested. Therefore, negative results are expected.

For example, if a test tube contains glucose (the substance being tested) and Benedict's reagent (blue) is added, a red color develops upon heating. This test tube is the positive control; it tests positive for gluclose. If a test tube does not contain glucose and Benedict's reagent is added, Benedict's is expected to remain blue. This test tube is the negative control; it tests negative for glucose.

What benefit is a positive control? Positive controls give you a standard by which to tell if the substance being tested is present (or acting properly) in an unknown sample. Negative controls ensure that the experiment is giving reliable results; after all, if a negative control should happen to give a positive result, then the entire experiment may be faulty and unreliable.

28.1   Protein Digestion by Pepsin

Certain foods, such as meat and egg whites, are rich in protein. Egg whites contain albumin, which is the protein used in this Experimental Procedure. Protein is digested by **pepsin** in the stomach (Fig. 28.2), a process described by the following reaction:

$$\text{protein} + \text{water} \xrightarrow{\text{pepsin (enzyme)}} \text{peptides}$$

The stomach has a very low pH. Does this indicate that pepsin works effectively in an acidic or

basic environment? _____

The acidity of the stomach can be troublesome to humans who often have to take antacids or even prescription medications to treat heartburn.

Test for Protein Digestion

Biuret reagent is used to test for protein digestion. If digestion has not occurred, biuret reagent turns purple, indicating that protein is present. If digestion has occurred, biuret reagent turns pinkish-purple, indicating that peptides are present.

> ⚠ **Biuret reagent** is highly corrosive. Exercise care in using this chemical. If any should spill on your skin, wash the area with mild soap and water. Follow your instructor's directions for its disposal.

Experimental Procedure: Protein Digestion

With a wax pencil, number four test tubes 1 through 4, and mark at the 2 cm, 4 cm, 6 cm, and 8 cm levels. Fill all tubes to the 2 cm mark with *albumin solution.* Albumin is a protein.

Tube 1   Fill to the 4 cm mark with *pepsin solution,* and to the 6 cm mark with *0.2% HCl. HCl* simulates the acidic conditions of the stomach. Swirl to mix, and incubate at 37°C. After 1½ hours, fill to the 8 cm mark with *biuret reagent.* Record the temperature and your results in Table 28.1.

Tube 2   Fill to the 4 cm mark with *pepsin solution,* and to the 6 cm mark with *0.2% HCl.* Swirl to mix, and keep at room temperature. After 1½ hours, fill to the 8 cm mark with *biuret reagent.* Record the temperature and your results in Table 28.1.

Tube 3   Fill to the 4 cm mark with *pepsin solution,* and to the 6 cm mark with *water.* Swirl to mix, and incubate at 37°C. After 1½ hours, fill to the 8 cm mark with *biuret reagent.* Record the temperature and your results in Table 28.1.

Tube 4   Fill to the 6 cm mark with *water.* Swirl to mix, and incubate at 37°C. After 1½ hours, fill to the 8 cm mark with *biuret reagent.* Record the temperature and your results in Table 28.1.

Figure 28.2   Digestion of protein.
Pepsin, produced by the gastric glands of the stomach, helps digest protein.

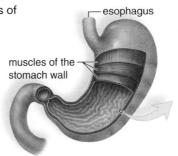

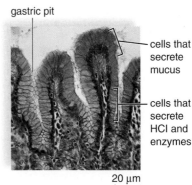

Table 28.1   Protein Digestion by Pepsin

Tube	Contents	Temperature	Results	Explanation
1	Albumin Pepsin HCl Biuret reagent			
2	Albumin Pepsin HCl Biuret reagent			
3	Albumin Pepsin Water Biuret reagent			
4	Albumin Water Biuret reagent			

Conclusions: Protein Digestion

- Explain your results in Table 28.1 by giving an explanation why digestion did or did not occur. To be complete, consider all the requirements for an enzymatic reaction as per listed in Table 28.4. Now show here that Tube 1 met all the requirements for digestion:

Pepsin is the correct _____

Albumin is the correct _____

37°C is the optimum _____

HCL provides the optimum _____

1½ hours provides _____ for the reaction to occur.

- Review "What Is a Control?" on page 400. Which tube was the negative control? _____

  Explain. _____

- If this control tube had given a positive result for protein digestion, what could you conclude about this experiment? _____

28.2   Fat Digestion by Pancreatic Lipase

Lipids include fats (e.g., butterfat) and oils (e.g., sunflower, corn, olive, and canola). Lipids are digested by **pancreatic lipase** in the small intestine (Fig. 28.3).

Figure 28.3   Emulsification and digestion of fat.
Bile from the liver (stored in the gallbladder) enters the small intestine, where lipase in pancreatic juice from the pancreas digests fat.

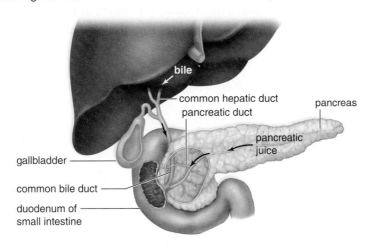

The following two reactions describe fat digestion:

1. $$\text{fat} \xrightarrow{\text{bile (emulsifier)}} \text{fat droplets}$$

2. $$\text{fat droplets} + \text{water} \xrightarrow{\text{lipase (enzyme)}} \text{glycerol} + \text{fatty acids}$$

With regard to the first step, consider that fat is not soluble in water; yet, lipase makes use of water when it digests fat. Therefore, bile is needed to emulsify fat—cause it to break up into fat droplets that disperse in water. The reason for dispersal is that bile contains molecules with two ends. One end is soluble in fat, and the other end is soluble in water. Bile can emulsify fat because of this.

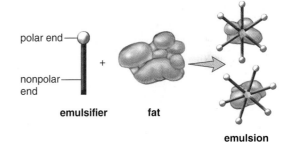

With regard to the second step, would the pH of the solution be lower before or after the enzymatic reaction? (*Hint:* Remember that an acid decreases pH and a base increases pH.) _____

_____

Test for Fat Digestion

In the test for fat digestion, you will be using a pH indicator, which changes color as the solution in the test tube goes from basic conditions to acidic conditions. Phenol red is a pH indicator that is red in basic solutions and yellow in acidic solutions.

Experimental Procedure: Fat Digestion

With a wax pencil, number three clean test tubes 1 through 3, and mark at the 1 cm, 3 cm, and 5 cm levels. Fill all the tubes to the 1 cm mark with *vegetable oil,* and to the 3 cm mark with *phenol red.*

Tube 1   Fill to the 5 cm mark with *pancreatin solution* (pancreatic lipase). Add a pinch of *bile salts,* the emulsifier. Invert gently to mix, and record the initial color in Table 28.2. Incubate at 37°C, and check every 20 minutes. Record any color change and the time taken for the change.

Tube 2   Fill to the 5 cm mark with *pancreatin solution.* Invert gently to mix, and record the initial color in Table 28.2. Incubate at 37°C, and check every 20 minutes. Record any color change and the time taken for the change.

Tube 3   Fill to the 5 cm mark with *water.* Invert gently to mix, and record the initial color in Table 28.2. Incubate at 37°C, and check every 20 minutes. Record any color change and the time taken for the change.

Table 28.2	Fat Digestion by Pancreatic Lipase				
Tube	Contents	Color	Time Taken		Explanation
			Initial	*Final*	
1	Vegetable oil Phenol red Pancreatin Bile salts				
2	Vegetable oil Phenol red Pancreatin				
3	Vegetable oil Phenol red Water				

Conclusions: Fat Digestion

- Explain your results in Table 28.2 by giving an explanation why digestion did or did not occur.
- What role did bile salts play in this experiment? _____

  _____
- What role did phenol red play in this experiment? _____
- Review "What Is a Control?" on page 400. Which test tube in this experiment could be considered a negative control? _____

  _____

28.3 Starch Digestion by Pancreatic Amylase

Starch is present in bakery products and in potatoes, rice, and corn. Starch is digested by **pancreatic amylase** in the small intestine, a process described by the following reaction:

$$\text{starch} + \text{water} \xrightarrow{\text{amylase (enzyme)}} \text{maltose}$$

1. If digestion *does not* occur, which will be present—starch or maltose? _____

2. If digestion *does* occur, which will be present—starch or maltose? _____

3. What happens to enzymes when they are boiled? _____

   Will digestion occur? _____

Tests for Starch Digestion

You will be using two tests for starch digestion:

1. If digestion has not taken place, the iodine test for starch will be positive (+). If digestion has occurred, the iodine test for starch will be negative (−).
2. If digestion has taken place, the Benedict's test for sugar (maltose) will be positive (+). If digestion has not taken place, the Benedict's test for sugar will be negative (−). To test for sugar, add an equal amount of Benedict's reagent to each test tube. Place the tube in a boiling water bath for 2 to 5 minutes, and note any color changes (see Table 3.3 on page 31). Boiling the test tube is necessary for the Benedict's reagent to react.

> ⚠ **Benedict's reagent** is highly corrosive. Use protective eyewear when performing this experiment. Exercise care in using this chemical. If any should spill on your skin, wash the area with mild soap and water. Follow your instructor's directions for disposal of this chemical.

Experimental Procedure: Starch Digestion

Preparation

1. With a wax pencil, number eight clean test tubes 1 through 8 above the level of boiling water bath, and mark at the 1 cm and 2 cm levels.
2. Fill tubes 1 through 6 to the 1 cm mark with *pancreatic-amylase solution*. Fill tubes 7 and 8 to the 1 cm mark with *water*. See Testing for tubes 1 and 2.
3. Shake the *starch suspension* well each time before dispensing. After tubes 3 through 8 have received the starch suspension, at least 30 minutes will lapse before testing occurs.
4. Fill tubes 3 and 4 to the 2 cm mark with *starch suspension,* and allow them to stand at room temperature for 30 minutes.
5. Place tubes 5 and 6 in a boiling water bath for 10 minutes. After boiling, fill to the 2 cm mark with *starch suspension,* and allow the tubes to stand for 20 minutes more.
6. Fill tubes 7 and 8 to the 2 cm mark with *starch suspension.* Allow the tubes to stand for 30 minutes.

Testing

Tube 1 Fill to the 2 cm mark with *starch suspension,* and test for starch *immediately,* using the iodine test described previously. As an example, all but explanation has been completed for you for tube 1 in Table 28.3.

Tube 2 Fill to the 2 cm mark with *starch suspension,* and test for sugar immediately, using *Benedict's reagent,* described earlier, which requires boiling. Complete all but explanation in Table 28.3.

Why do you expect tube 1 to have a positive test for starch and tube 2 to have a negative test for

sugar? _____

Record your explanation for tubes 1 and 2 in Table 28.3.

Tubes 3, 5, and 7 After 30 minutes, test for starch using the iodine test. Complete all but explanation in Table 28.3.

Tubes 4, 6, and 8 After 30 minutes, test for sugar using the Benedict's test. Complete all but explanation in Table 28.3.

Why do you expect tube 3 to have a negative test for starch and tube 4 to have a positive test for sugar? _____ .

Why do you expect tube 5 to have a positive test for starch and tube 6 to have a negative test for sugar? _____ .

Why do you expect tube 7 to have a positive test for starch and tube 8 to have a negative test for sugar? _____ .

Record your explanations for tubes 3–8 in Table 28.3.

Table 28.3	Starch Digestion by Amylase				
Tube	Contents	Time*	Type of Test	Results	Explanation
1	Pancreatic amylase Starch	0	Iodine	+	
2	Pancreatic amylase Starch				
3	Pancreatic amylase Starch				
4	Pancreatic amylase Starch				
5	Pancreatic amylase, boiled Starch				
6	Pancreatic amylase, boiled Starch				
7	Water Starch				
8	Water Starch				

* Enter either 0 for Immediately or T for after 30 minutes.

Conclusions: Starch Digestion

- This experiment demonstrated that, for an enzymatic reaction to occur, an active _____ must be present, and _____ must pass to allow the reaction to occur.
- Which test tubes served as a negative control in this experiment? _____
  Explain. _____

Absorption of Sugars and Other Nutrients

Figure 28.4 shows that the folded lining of the small intestine has many fingerlike projections called villi. The small intestine not only digests food; it also absorbs the products of digestion, such as sugars from carbohydrate digestion, amino acids from protein digestion, and glycerol and fatty acids from fat digestion at the villi.

Figure 28.4   Anatomy of the small intestine.
Nutrients enter the bloodstream across the much convoluted walls of the small intestine.

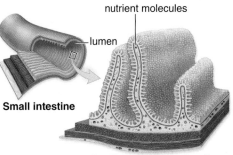

28.4   Requirements for Digestion

Explain in Table 28.4 how each of the requirements listed influences effective digestion.

Table 28.4 Requirements for Digestion	
Requirement	**Explanation**
Specific enzyme	
Specific substrate	
Warm temperature	
Specific pH	
Time	
Fat emulsifier	

To lose weight, some obese individuals undergo an operation in which (1) the stomach is reduced to the size of a golf ball, and (2) food bypasses the duodenum (first 2 feet) of the intestine. Answer these questions to explain how this operation would affect the requirements for digestion.

1. How is the amount of substrate reduced? _____

2. How is the amount of digestive enzymes reduced? _____

3. How is time reduced? _____

4. What makes the pH of the small intestine higher than before? _____

5. How is fat emulsification reduced? _____

6. How does surgery to reduce obesity sometimes result in malnutrition? _____

1. Give a reason why amylase will not digest protein.

   _____

2. Enzymes perform better at room temperature than when they are boiled. Explain.

   _____

3. Relate the expectation of more product per length of time to the fact that enzymes are used over and over.

   _____

   _____

4. Why do enzymes work better at their optimum pH?

   _____

5. Why is an emulsifier needed for the lipase experiment but not for the pepsin and amylase experiments?

   _____

6. Which of the following two combinations is most likely to result in digestion?

   a. Pepsin, protein, water, body temperature

   b. Pepsin, protein, hydrochloric acid (HCl), body temperature

   Explain. _____

7. Which of the following two combinations is most likely to result in digestion?

   a. Amylase, starch, water, body temperature, testing immediately

   b. Amylase, starch, water, body temperature, waiting 30 minutes

   Explain. _____

   _____

8. The test used for fat digestion was the presence of acids. Explain why this works.

   _____

   _____

9. Given that, in this laboratory, you tested for the action of digestive enzymes on their substrates, what substance would be missing from a negative control sample?

   _____

10. Knowing that blood clotting is an enzymatic reaction, would you place a cut finger under warm or cold running water? Explain. _____

    _____

Biology **Website**

Enhance your study of the text and laboratory manual with study tools, practice tests, and virtual labs. Also ask your instructor about the resources available through ConnectPlus, including the media-rich eBook, interactive learning tools, and animations.

www.mhhe.com/maderbiology11

McGraw-Hill Access Science Website

An Encyclopedia of Science and Technology Online which provides more information including videos that can enhance the laboratory experience.

www.accessscience.com

Learning Outcomes

Introduction
- Define homeostasis and the internal environment of vertebrates.   409

29.1 Heartbeat and Blood Flow
- Describe the cardiovascular system and relate the heartbeat cycle to blood pressure and blood flow.   410–12
- Measure blood pressure, and explain the relationship between blood pressure and heart rate.   412–13

29.2 Blood Flow and Systemic Capillary Exchange
- Describe the exchange of molecules across a capillary wall and how this exchange relates to blood pressure and osmotic pressure.   413–14

29.3 Lung Structure and Human Respiratory Volumes
- Describe the mechanics of breathing and the role of the alveoli in gas exchange.   414–15
- Measure respiratory volumes (e.g., tidal volume) and explain their relationship to homeostasis.   415–17

29.4 Kidneys
- Understand kidney and nephron structure and blood supply.   418–19
- State the three steps in urine formation and how they relate to the parts of a nephron.   420–22
- Perform a urinalysis and explain how the results are related to kidney functions.   422–23

Introduction

Homeostasis refers to the dynamic equilibrium of the body's internal environment. The **internal environment** of vertebrates, including humans, consists of blood and tissue fluid. The body's cells take nutrients from tissue fluid and return their waste molecules to it. Tissue fluid, in turn, exchanges molecules with the blood. This is called capillary exchange. All internal organs contribute to homeostasis, but this laboratory specifically examines the contributions of the blood, lungs, and kidneys (Fig. 29.1).

Figure 29.1   Contributions of organs to homeostasis.
The lungs exchange gases with blood; the kidneys remove nitrogenous wastes from blood; and the intestinal tract adds nutrients as regulated by the liver to blood.

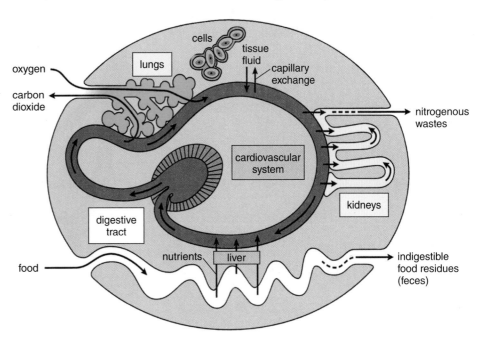

29.1 Heartbeat and Blood Flow

The cardiovascular system consists of the heart, blood vessels, and blood (Fig. 29.2). Arteries carry blood away from the heart while veins transport blood toward the heart. Arteries branch into smaller vessels called arterioles that enter capillary beds. Capillary beds are present throughout the organs and tissues of the body. An exchange of gases takes place across the thin walls of **pulmonary capillaries**. In the lungs, CO_2 leaves the blood and O_2 enters the blood. An exchange of gases and nutrients for metabolic wastes takes place across the thin walls of **systemic capillaries**. In the body tissues, O_2 and nutrients exit the blood; while CO_2 and metabolic wastes enter the blood.

We will see that the heart is vital to homeostasis because its contraction (called the **heartbeat**) keeps the blood moving in the arteries and arterioles which take blood to the capillaries. The exchanges that take place across capillaries help maintain homeostasis.

Liver

The hepatic portal vein lies between _____ and the _____ . This placement allows the liver to regulate what molecules enter the blood from the digestive tract. For example, if the hormone insulin is present, the liver removes excess glucose and stores it as glycogen. Later, the liver breaks down glycogen to glucose to keep the glucose concentration constant.

Figure 29.2 The circulatory system.
The heart provides the pumping action that transports the blood though the arteries, capillary beds, and veins.

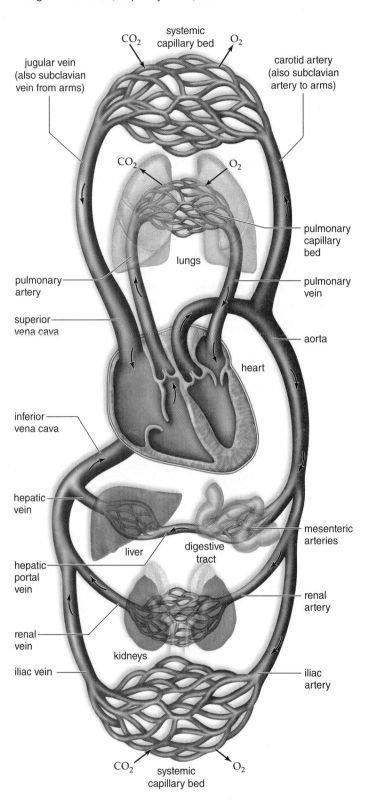

Heartbeat

During a heartbeat, first the atria contract and then the ventricles contract. When a chamber contracts, it is called **systole**, and when a chamber relaxes, it is called **diastole**. The atria and ventricles take turns being in systole.

Time	Atria	Ventricles
0.15 sec	Systole	Diastole
0.30 sec	Diastole	Systole
0.40 sec	Diastole	Diastole

Usually, there are two heart sounds with each heartbeat (Fig. 29.3). The first sound (*lub*) is low and dull and lasts longer than the second sound. It is caused by the closure of valves following atrial systole. The second sound (*dub*) follows the first sound after a brief pause. The sound has a snapping quality of higher pitch and shorter duration. The *dub* sound is caused by the closure of valves following ventricle systole.

Figure 29.3 The heartbeat sounds.
a. When the atria contract (are in systole), the ventricles fill with blood. **b.** Closure of the valves between atria and ventricles results in a *lub* sound. When the ventricles contract, blood enters the attached arteries (aorta and pulmonary trunk).
c. Closure of valves between ventricles and arteries results in a *dub* sound. Blood enters the heart from the attached veins (vena cavas and pulmonary veins) once more.

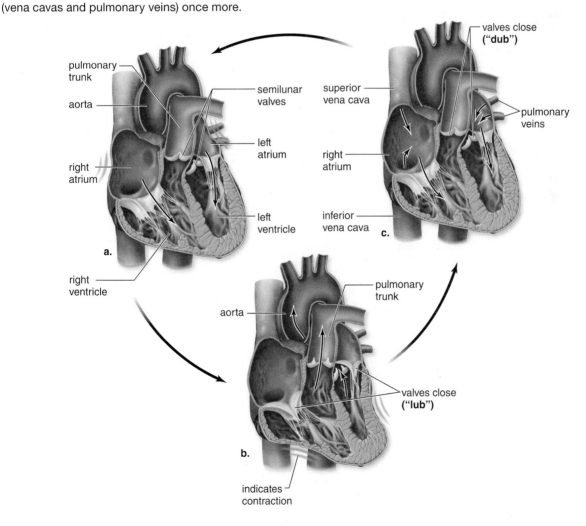

In the following procedure, you will work with a partner and use a stethoscope to listen to the heartbeat. It will not be necessary for you to count the number of beats per minute.

1. Obtain a stethoscope, and properly position the earpieces. They should point forward. Place the bell of the stethoscope on the left side of your partner's chest between the fourth and fifth ribs. This is where the apex (tip) of the heart is closest to the body wall.

2. Which of the two sounds (*lub* or *dub*) is louder? _____

3. Now switch, and your partner will determine your heartbeat.

Blood Pressure

Blood pressure is highest just after ventricular systole (contraction) and it is lowest during ventricular

diastole (relaxation). Why? _____

We would expect a person to have lower blood pressure readings at rest than after exercise. Why?

_____

Experimental Procedure: Blood Pressure at Rest and After Exercise

A number of different types of digital blood pressure monitors are available, and your instructor will instruct you on how to use the type you will be using for this Experimental Procedure. The resting blood pressure readings for an individual are displayed on the monitor shown in Figure 29.4. A blood pressure reading of 120/80 (systolic/diastolic) is considered normal.

You may work with a partner or by yourself. If working with a partner, each of you will assist the other in taking blood pressure readings. After you have noted the blood pressure readings, also note the pulse reading.

Figure 29.4 Measurement of blood pressure and pulse.
There are many different types of digital blood pressure/pulse monitors now available. The one shown here uses a cuff to be placed on the arm. Others use a cuff for the wrist.

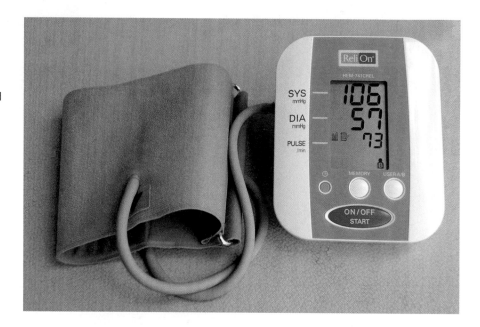

Blood Pressure at Rest

1. Reduce your activity as much as possible.
2. Use the blood pressure monitor to obtain several blood pressure readings, average them, and record your results in Table 29.1.

Blood Pressure After Exercise

1. Run in place for 1 minute.
2. Immediately use the blood pressure monitor to obtain a blood pressure reading, and record it in Table 29.1.

Table 29.1 Blood Pressure		
	Blood Pressure at Rest	**Blood Pressure After Exercise**
Partner		
Yourself		

Conclusions: Blood Pressure

- Knowing that exercise increases the heartrate offer an explanation for your results. _____
  _____
  _____

- Under what conditions in everyday life would you expect the heartrate and the blood pressure to increase, even though you are not exercising? _____
  _____

  When might this be an advantage? _____

  A disadvantage? _____

29.2   Blood Flow and Systemic Capillary Exchange

We associate death with lack of a heartbeat but the real problem is lack of blood flow to the capillaries.

Blood Flow

The beat of the heart moves blood into the aorta, which divides into arterioles and then arterioles divide into capillaries. Venules which receive blood from capillaries, combine to form veins which take blood back to the heart.

Experimental Procedure: Blood Flow

1. Observe blood flow through arterioles, capillaries, and venules, either in the tail of a goldfish or in the webbed skin between the toes of a frog, as prepared by your instructor.
2. Examine under low and high power of the microscope.
3. Watch the pulse and the swiftly moving blood in the arterioles.
4. Contrast this with the more slowly moving blood that circulates in the opposite direction in the venules. Many criss-crossing capillaries are visible.
5. Look for blood cells floating in the bloodstream (see Fig. 25.2). Don't confuse blood cells with chomatophores, irregular, black patches of pigment that may be visible in the skin.

Systemic Capillary Exchange

The beat of the heart creates blood pressure, and blood pressure is necessary to capillary exchange. Blood pressure acts to move water and substances out of a capillary, osmotic pressure acts to move water and substances into a capillary (Fig. 29.5). Blood pressure is higher than osmotic pressure at the arteriole end of a capillary, but blood pressure lessens as the blood moves through a capillary bed. This means that osmotic pressure (created by the presence of proteins in the blood) is higher than blood pressure at the venule end of a capillary.

Figure 29.5 Systemic capillary exchange.
At a systemic capillary, an exchange takes place across the capillary wall. In between the arterial end and the venule end molecules follow their concentration gradient.

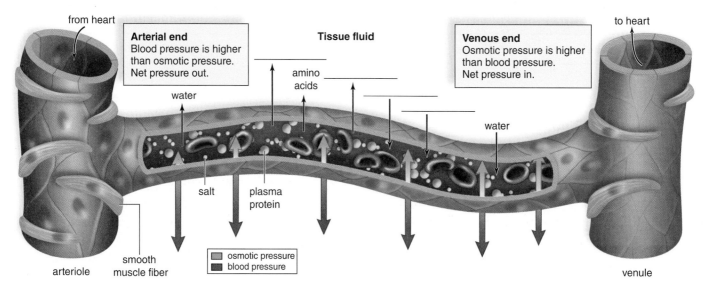

Conclusions: Systemic Capillary Exchange

- What generates blood pressure? _____
  _____

- Why are tissue cells always in need of glucose and oxygen? _____
  Add glucose at the end of an appropriate arrow in Figure 29.5. Do the same for oxygen.

- Why are tissue cells always producing carbon dioxide? _____
  Add carbon dioxide at the start of an appropriate arrow in Figure 29.5. Do the same for metabolic wastes.

29.3 Lung Structure and Human Respiratory Volumes

Air moves from the nasal passages to the trachea, bronchi, bronchioles, and finally, lungs. The right and left lungs lie in the thoracic cavity on either side of the heart (see Fig. 27.4).

Lung Structure

A **lung** is a spongy organ consisting of irregularly shaped air spaces called **alveoli** (sing., alveolus). The alveoli are lined with a single layer of squamous epithelium and are supported by a mesh of fine, elastic fibers. The alveoli are surrounded by a rich network of tiny blood vessels called pulmonary capillaries.

1. Observe a prepared slide of a stained section of a lung. In stained slides, the nuclei of the cells forming the thin alveolar walls appear purple or dark blue.
2. Look for areas that show red or orange disc-shaped **erythrocytes** (red blood cells). When these appear in strings, you are looking at capillary vessels in side view.
3. In some part of the slide, you may even observe an artery. Thicker, circular or oval structures with a lumen (cavity) are cross sections of **bronchioles,** tubular pathways through which air reaches the air spaces.
4. In Figure 29.6b, *add* arrows to show gas exchange between normal alveoli and pulmonary capillaries. In Figure 29.6c, note that in emphysema, alveoli have burst. In smokers, small bronchioles collapse and trapped air in alveoli cause them to burst.

Figure 29.6   Healthy lung tissue versus emphysema.
a. The lungs normally contain many air sacs called alveoli where gas exchange occurs. **b.** Micrograph of normal lung tissue. **c.** In smokers, emphysema can occur; the alveoli burst and gas exchange is inadequate.

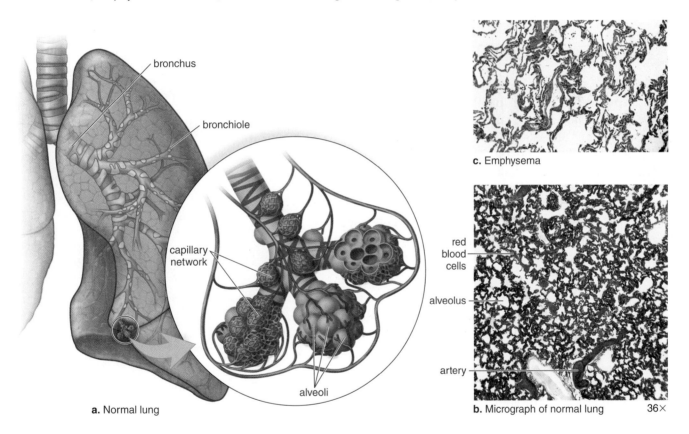

c. Emphysema

a. Normal lung

b. Micrograph of normal lung   36×

Human Respiratory Volumes

Breathing in, called **inspiration** or inhalation, is the active part of breathing because that's when contraction of rib cage muscles causes the rib cage to move up and out, and contraction of the diaphragm causes the diaphragm to lower. Due to an enlarged thoracic cavity, air is drawn into the lungs. Breathing out, called **expiration** or exhalation, occurs when relaxation of these same muscles causes the thoracic cavity to resume its original capacity. Now air is pushed out of the lungs (Fig. 29.7).

Figure 29.7 Inspiration and expiration.

a. Inspiration occurs after the rib cage moves up and out and the diaphragm moves down. Air rushes in because of the expanded thoracic cavity. **b.** Expiration occurs as the rib cage moves down and in and the diaphragm moves up. As the thoracic cavity gets smaller, air is pushed out.

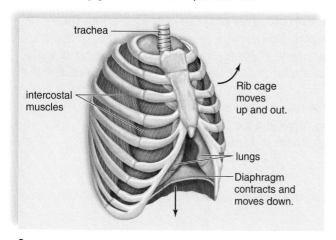

trachea

intercostal muscles

Rib cage moves up and out.

lungs

Diaphragm contracts and moves down.

a.

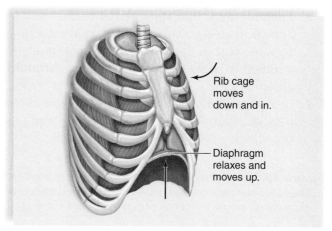

Rib cage moves down and in.

Diaphragm relaxes and moves up.

b.

Experimental Procedure: Human Respiratory Volumes

During this Experimental Procedure you will be working with a spirometer, an instrument that measures the amount of inhaled and exhaled air (Fig. 29.8). Normally, about 500–600 mL of air move into and out of the lungs with each breath. This is called the **tidal volume** (TV). You can inhale deeply after a normal breath and more air will enter the lungs; this is the **inspiratory reserve volume** (IRV). You can also force more air out of your lungs after a normal breath; this is the **expiratory reserve volume** (ERV). **Vital capacity** is the volume of air that can be forcibly exhaled after forcibly inhaling.

Figure 29.8 Nine-liter student wet spirometer.

Tidal Volume (TV)

1. When it's your turn to use the spirometer, install a new disposable mouthpiece and set the spirometer to zero.
2. Inhale normally, then exhale normally (with *no* extra effort) through the mouthpiece of the spirometer. Record your measurement in Table 29.2.
3. Three readings are needed, so twice more set the spirometer to zero and repeat the same procedure. Record your measurements in Table 29.2.
4. Later, if necessary, change your readings to milliliters (mL), and calculate your average TV in mL.

   In your own words, what is tidal volume?

   _____

   _____

Expiratory Reserve Volume (ERV)

1. Make sure the spirometer is set to zero.
2. Inhale and exhale normally and then force as much air out as possible into the spirometer. Record your measurement in Table 29.2.

3. Three readings are needed, so twice more set the spirometer to zero and repeat the same procedure. Record your measurements in Table 29.2.
4. Later, if necessary, change your readings to mL, and calculate your average ERV.

In your own words, what is expiratory reserve volume? _____

_____

Vital Capacity (VC)

1. Make sure the spirometer is set to zero.
2. Inhale as much as possible and then exhale as much as possible into the spirometer.
3. Three readings are needed, so twice more set the spirometer to zero and repeat the same procedure. Record your measurements in Table 29.2.
4. Later, if necessary, change your readings to mL, and calculate your average VC.

In your own words, what is vital capacity? _____

_____

Inspiratory Reserve Volume (IRV)

It will be necessary for us to calculate IRV because a spirometer only measures exhaled air, not inhaled air. Explain. _____

From having measured vital capacity (VC) you can see that VC = TV + IRV + ERV. To calculate IRV, simply subtract the average TV + the average ERV from the value you recorded for the average VC:

$$IRV = VC - (TV + ERV) = \text{_____} \text{ mL. Record your IRV in Table 29.2.}$$

Table 29.2 Measurements of Lung Volumes			
Tidal Volume (TV)	Expiratory Reserve Volume (ERV)	Vital Capacity (VC)	Inspiratory Reserve Volume (IRV)
1st	1st	1st	———
2nd	2nd	2nd	———
3rd	3rd	3rd	———
Average mL	Average mL	Average mL	Calculated value = mL

Conclusions: Human Respiratory Volumes

- Vital capacity varies with age, sex, and height; however, typically for men vital capacity is about 5,200 mL and for women it is about 4,000 mL. How does your vital capacity compare to the typical values for your gender? _____ If smaller than normal, are you a smoker or is there any health reason why it would be smaller? If larger than normal, are you a sports enthusiast or do you play a musical instrument that involves inhaling and exhaling deeply? _____

- Diffusion alone accounts for pulmonary gas exchange. Therefore, how does good lung ventilation assist gas exchange? _____

_____

29.4 Kidneys

The **kidneys** are bean-shaped organs that lie along the dorsal wall of the abdominal cavity.

Kidney Structure

Figure 29.9 shows the structure of a kidney, macroscopic and microscopic. The macroscopic structure of a kidney is due to the placement of over 1 million **nephrons.** Nephrons are tubules that do the work of producing urine.

Figure 29.9 Longitudinal section of a kidney.
a. The kidneys are served by the renal artery and renal vein. **b.** Macroscopically, a kidney has three parts: renal cortex, renal medulla, and renal pelvis. **c.** Microscopically, each kidney contains over a million nephrons.

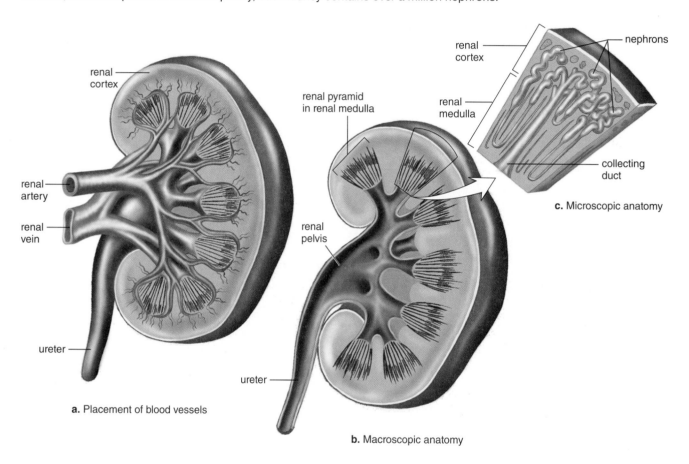

Observation: Kidney Model

Study a model of a kidney, and with the help of Figure 29.9, locate the following:

1. **Renal cortex:** a granular region
2. **Renal medulla:** contains the renal pyramids
3. **Renal pelvis:** where urine collects

Observation: Nephron Structure

Study a nephron model and, with the help of Figure 29.10, identify the following parts of a nephron:

1. **Glomerular capsule:** (Bowman's capsule): closed end of the nephron pushed in on itself to form a cuplike structure; the inner layer has pores that allow **glomerular filtration** to occur; substances move from the blood to inside the nephron.
2. **Proximal convoluted tubule:** The inner layer of this region has many microvilli that allow tubular reabsorption to occur; substances move from inside the nephron to the blood.
3. **Loop of the nephron:** Nephron narrows to form a U-shaped portion. Functions in water reabsorption.
4. **Distal convoluted tubule:** second convoluted section that lacks microvilli and functions in **tubular secretion**; substances move from blood to inside nephron.

Several nephrons enter one collecting duct. The **collecting ducts** also function in water reabsorption, and they conduct urine to the pelvis of a kidney.

Observation: Circulation About a Nephron

Study a nephron model and, with the help of Figure 29.10 and Table 29.3, trace the path of blood from the renal artery to the renal vein:

1. **Afferent arteriole:** small vessel that conducts blood from the renal artery to a nephron.
2. **Glomerulus:** capillary network that exists inside the glomerular capsule; small molecules move from inside the capillary to the inside of the glomerulus during glomerular filtration.
3. **Efferent arteriole:** small vessel that conducts blood from the glomerulus to the peritubular capillary network.
4. **Peritubular capillary network:** surrounds the proximal convoluted tubule, the loop of the nephron, and the distal convoluted tubule.
5. **Venule:** takes blood from the peritubular capillary network to the renal vein.

Table 29.3 Blood Vessels Serving the Nephron	
Name of Structure	**Significance**
Afferent arteriole	Brings arteriolar blood to the glomerulus
Glomerulus	Capillary tuft enveloped by glomerular capsule
Efferent arteriole	Takes arteriolar blood away from the glomerulus
Peritubular capillary network	Capillary bed that envelops the rest of the nephron
Venule	Takes venous blood away from the peritubular capillary network

Kidney Function

The kidneys produce urine and in doing so help maintain homeostasis in several ways. Urine formation requires three steps: **glomerular filtration**, **tubular reabsorption**, and **tubular secretion** (see Fig. 29.10).

Figure 29.10 Nephron structure and blood supply.

The three main processes in urine formation are described in boxes and color coded to arrows that show the movement of molecules out of or into the nephron at specific locations. In the end, urine is composed of the substances within the collecting duct (see brown arrow).

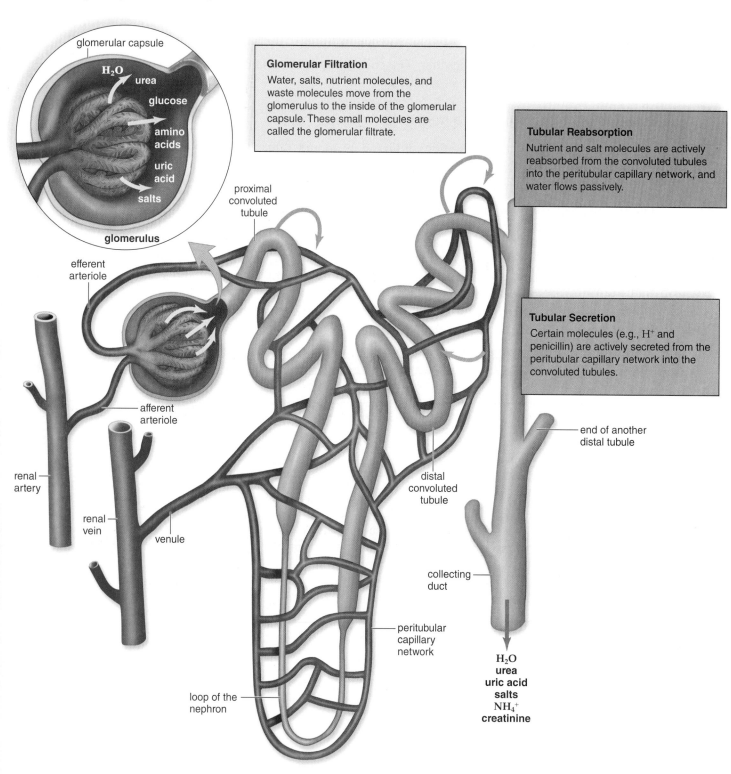

Glomerular Filtration

Water, salts, nutrient molecules, and waste molecules move from the glomerulus to the inside of the glomerular capsule. These small molecules are called the glomerular filtrate.

Tubular Reabsorption

Nutrient and salt molecules are actively reabsorbed from the convoluted tubules into the peritubular capillary network, and water flows passively.

Tubular Secretion

Certain molecules (e.g., H^+ and penicillin) are actively secreted from the peritubular capillary network into the convoluted tubules.

glomerular capsule

H_2O
urea
glucose
amino acids
uric acid
salts

glomerulus

efferent arteriole

afferent arteriole

renal artery

renal vein

venule

loop of the nephron

proximal convoluted tubule

peritubular capillary network

distal convoluted tubule

end of another distal tubule

collecting duct

H_2O
urea
uric acid
salts
NH_4^+
creatinine

Glomerular Filtration

1. Blood entering the glomerulus contains cells, proteins, glucose, amino acids, salts, urea, and water. Use this information to complete the first column of Table 29.4. In this table use an X to indicate that the substance is at the locations noted.

2. Blood pressure causes small molecules of glucose, amino acids, salts, urea, and water to exit the blood and enter the glomerular capsule. The fluid in the glomerular capsule is called the **filtrate**. In the list that follows, draw an arrow from left to right for the small substances that leave the glomerulus and become a part of the filtrate. Use this information to complete the second column of Table 29.4.

Glomerulus	**Glomerular Capsule (Filtrate)**
Cells	
Proteins	
Glucose	
Amino acids	
Urea	
Water and salts	

3. What substances are too large to leave the glomerulus and enter the glomerular capsule?

_____

These substances remain in the blood.

Table 29.4 Urine Constituents		
In Glomerulus	**In Filtrate**	**In Urine**
Cells		
Proteins		
Glucose		
Amino Acids		
Urea		
Water and salts		

Tubular Reabsorption

1. When the filtrate enters the proximal convoluted tubule, it contains glucose, amino acids, urea, water, and salts. Some water and salts remain in the nephron but enough are *passively* reabsorbed into the peritubular capillary to maintain blood volume and blood pressure. Use this information to state a way the kidneys help maintain homeostasis: _____

2. The cells that line the proximal convoluted tubule are also engaged in active transport and usually completely reabsorb nutrients (glucose and amino acids) into the peritubular capillary. What would

   happen to cells if the body lost all its nutrients by way of the kidneys? _____

_____

3. In the list that follows, draw an arrow from left to right for all those molecules passively reabsorbed into the blood. Use darker arrows for those actively reabsorbed.

Proximal Convoluted Tubule   **Peritubular Capillary**
Water and salts
Glucose
Amino acids
Urea

4. Which of these substances is not reabsorbed and will become a part of urine? _____ Urea is a nitrogenous waste. State here a way that kidneys contribute to homeostasis. _____

_____

Tubular Secretion

1. During tubular secretion, certain substances—for example, penicillin and histamine—are actively secreted from the peritubular capillary into the fluid of the tubule. Also, hydrogen ions (H^+) and ammonia (NH_3) are secreted as NH_4^+ as necessary. Complete the last column of Table 29.4. Check your entries against Fig. 29.10.
2. The blood is buffered but only the kidneys can excrete H^+. The excretion of H^+ by the kidneys raises the pH of the blood. Use this information to state a third way the kidneys contribute to homeostasis.

_____

Kidney Function

• The presence of urea in urine illustrates which of the kidney's functions? _____
  The liver makes urea. What do the kidneys produce? _____

• The presence of NH_4^+ in the urine illustrates which of the kidney's functions? _____

_____

• Water and salts are in the glomerulus, the filtrate and the urine but much is reabsorbed. Regulation of the blood's water and salt content by the kidneys helps maintain _____

_____ within normal limits.

Urinalysis: A Diagnostic Tool

Urinalysis can indicate whether the kidneys are functioning properly or whether an illness such as diabetes mellitus is present. The procedure is easily performed with a Chemstrip test strip, which has indicator spots that produce specific color reactions when certain substances are present in urine.

Experimental Procedure: Urinalysis

A urinalysis has been ordered, and you are to test the urine for a possible illness. (In this laboratory, you will be testing simulated urine.)

Assemble Supplies

1. Obtain three Chemstrip urine test strips each of which tests for leukocytes, pH, protein, glucose, ketones, and blood, as noted in Figure 29.11.
2. The color key on the diagnostic color chart or on the Chemstrip vial label will explain what any color changes mean in terms of the pH level and amount of each substance present in the urine sample. You will use these color blocks to read the results of your test.
3. Obtain three "specimen containers of urine" marked 1 through 3. Among them is a normal specimen and two that indicate the patient has an illness.

Test the Specimen

1. Be sure the chemically treated patches on the test strip are totally immersed. Briefly (no longer than 1 second) dip a test strip into the first specimen of urine.
2. Draw the edge of the strip along the rim of the specimen container to remove excess urine.
3. Turn the test strip on its side, and tap once on a piece of absorbent paper to remove any remaining urine and to prevent the possible mixing of chemicals.
4. After 60 seconds, read the results as follows: Hold the strip close to the color blocks on the diagnostic color chart (Fig. 29.11) or vial label, and match carefully, ensuring that the strip is properly oriented to the color chart. Enter the test results in Figure 29.11. Use a negative symbol (−) for items that are not present in the urine, a plus symbol (+) for those that are present, and a number for the pH.
5. Test the other two specimens.

Figure 29.11 Urinalysis test.
A Chemstrip test strip can help determine illness in a patient by detecting substances in the urine. If leukocytes (white blood cells), protein, or blood are in the urine, the kidneys are not functioning properly. If glucose and ketones are in the urine, the patient has diabetes mellitus (type 1 or type 2).

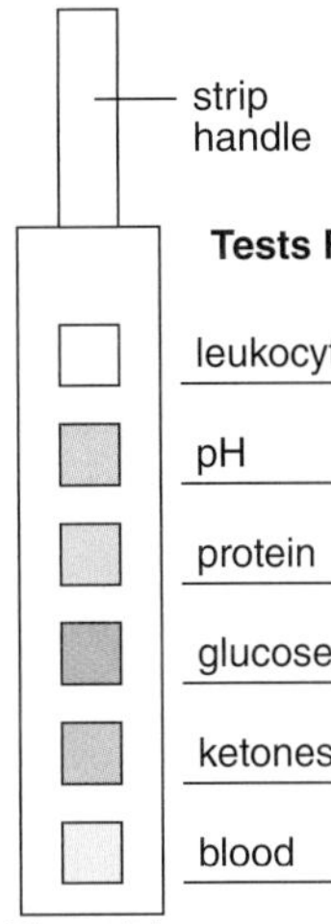

Tests For:	Normal	Test 1	Test 2	Test 3
leukocytes	−			
pH	pH 5			
protein	−			
glucose	−			
ketones	−			
blood	−			

Chemstrip before urine test

Conclusion: Urinalysis

- State below if the urinalysis is normal or indicates a urinary tract infection (leukocytes, blood, and possibly protein in the urine) or the patient has diabetes mellitus.

  Test strip 1 _____

  Test strip 2 _____

  Test strip 3 _____

- The hormone insulin promotes the uptake of glucose by cells. When glucose is in the urine, either the pancreas is not producing insulin (diabetes mellitus type 1) or cells are resistant to insulin (diabetes mellitus type 2). Ketones (acids) are also in the urine because the cells are metabolizing fat instead of glucose. Explain why. _____

  _____

  Why is the pH of urine lower than normal? _____

  _____

- If urinalysis shows that proteins are excreted instead of retained in the blood, would capillary exchange in the tissues (see Fig. 29.5) be normal? _____ Why or why not? _____

  _____

1. In your own words, what is homeostasis? _____

2. Explain how the systemic capillaries help maintain homeostasis. _____

   _____

3. Relate a blood pressure of 120/80 to systole and diastole of the ventricles during a heartbeat. When

   would blood pressure be 120? _____ When would blood pressure be 80? _____

   _____

4. If a smoker has a low tidal volume, why might he feel tired and run down? _____

   _____

5. How would you measure the effects of exercise on vital capacity? _____

   _____

6. What role do the kidneys have in maintaining blood pressure and volume? _____

   _____

7. List the three steps in urine formation and define.

   a. _____

   b. _____

   c. _____

8. With regard to urine formation, name a substance found in both the filtrate and the urine. _____

   Explain why the substance is in both places. _____

   _____

9. With regard to urine formation, name a substance found in the filtrate and not in the urine. _____

   Explain why the substance is only in the filtrate and not in the urine. _____

   _____

10. After a urinalysis test, what medical condition would be indicated by a positive test for glucose in the

    urine? _____

    _____

Biology Website

Enhance your study of the text and laboratory manual with study tools, practice tests, and virtual labs. Also ask your instructor about the resources available through ConnectPlus, including the media-rich eBook, interactive learning tools, and animations.

www.mhhe.com/maderbiology11

McGraw-Hill Access Science Website

An Encyclopedia of Science and Technology Online which provides more information including videos that can enhance the laboratory experience.

www.accessscience.com

30

Nervous System and Senses

Introduction

An animal's moment-to-moment survival relies on its ability to detect and respond to stimuli in the external environment. This ability is dependent on the nervous system and the sense organs. In vertebrates, the evolution of a variety of strategies for coping with life challenges has resulted in a variety of modifications of both the brain and sense organs.

The vertebrate nervous system consists of the brain, spinal cord, and nerves. Sensory receptors detect changes in environmental stimuli, and nerve impulses move along sensory nerve fibers to the brain and the spinal cord. The brain and spinal cord sum up the data before sending impulses via motor neurons to effectors (muscles and glands) so a response to stimuli is possible (Fig. 30.1). Nervous tissue consists of neurons; whereas the brain and spinal cord contain all parts of neurons, nerves contain only axons.

Figure 30.1   Motor neuron anatomy.
Neurons are cells specialized to conduct nerve impulses.

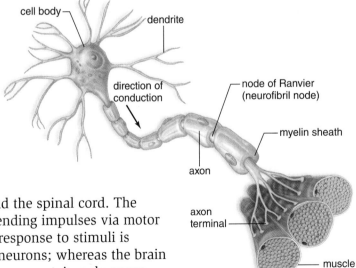

cell body
dendrite
direction of conduction
node of Ranvier (neurofibril node)
myelin sheath
axon
axon terminal
muscle

30.1 Animal Nervous Systems

The brain is the enlarged, anterior end of the nerve cord. In vertebrates, the nerve cord is called the spinal cord. The brain contains parts and centers that receive input from, and can command other regions of, the nervous system.

> ⚠ **Latex gloves** Wear protective latex gloves when handling preserved animal organs. Use protective eyewear and exercise caution when using sharp instruments during this laboratory. Wash hands thoroughly upon completion of this laboratory.

Sheep Brain

The mammalian brain has many parts, and the sheep brain is often used to study the mammalian brain.

Observation: Preserved Sheep Brain

Examine the exterior and a midsaggital (longitudinal) section of a preserved sheep brain or a model of the human brain, and with the help of Figure 30.2 identify the following:

1. **Ventricles:** interconnecting spaces that produce and serve as a reservoir for cerebrospinal fluid, which cushions the brain. Toward the anterior, note the lateral ventricle (on one longitudinal section) and similarly a lateral ventricle (on the other longitudinal section). Trace the second ventricle to the third and then the fourth ventricles.
2. **Cerebrum:** most developed area of the brain; responsible for higher mental capabilities. The cerebrum is divided into the right and left **cerebral hemispheres**, joined by the **corpus callosum**, a broad sheet of white matter. The outer portion of the cerebrum is highly convoluted and divided into the following surface lobes.
   a. **Frontal lobe:** controls motor functions and permits voluntary muscle control; it also is responsible for abilities to think, problem solve, speak, and smell.
   b. **Parietal lobe:** receives information from sensory receptors located in the skin and also the taste receptors in the mouth. A groove called the **central sulcus** separates the frontal lobe from the parietal lobe.
   c. **Occipital lobe:** interprets visual input and combines visual images with other sensory experiences. The optic nerves split and enter opposite sides of the brain at the optic chiasma, located in the diencephalons.
   d. **Temporal lobe:** has sensory areas for hearing and smelling. The olfactory bulb contains nerve fibers that communicate with the olfactory cells in the nasal passages and take nerve impulses to the temporal lobe.
3. **Diencephalon:** portion of the brain where the third ventricle is located. The hypothalamus and thalamus are also located here.
   a. **Thalamus:** two connected lobes located in the roof of the third ventricle. The thalamus is the highest portion of the brain to receive sensory impulses before the cerebrum. It is believed to control which received impulses are passed on to the cerebrum. For this reason, the thalamus sometimes is called the "gatekeeper to the cerebrum."
   b. **Hypothalamus:** forms the floor of the third ventricle and contains control centers for appetite, body temperature, and water balance. Its primary function is homeostasis. The hypothalamus also has centers for pleasure, reproductive behavior, hostility, and pain.
4. **Cerebellum:** located just posterior to the cerebrum as you observe the brain dorsally, the cerebellum's two lobes make it appear rather like a butterfly. In cross section, the cerebellum has an internal pattern that looks like a tree. The cerebellum coordinates equilibrium and motor activity to produce smooth movements.
5. **Brain stem:** part of the brain that connects with the spinal cord. Because it includes the pons and medulla oblongata, it contains centers for the functioning of internal organs; because of its location, it serves as a relay station for nerve impulses passing from the cord to the brain. Therefore, it helps keep the rest of the brain alert and functioning.

Figure 30.2   The sheep brain.

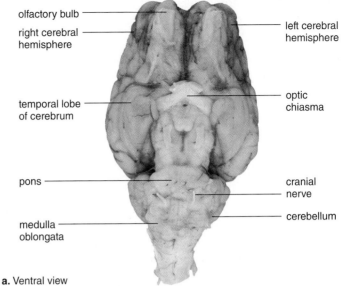

olfactory bulb

right cerebral hemisphere

left cerebral hemisphere

temporal lobe of cerebrum

optic chiasma

pons

cranial nerve

cerebellum

medulla oblongata

a. Ventral view

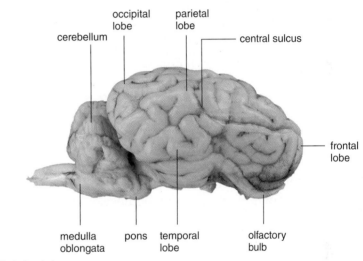

occipital lobe

parietal lobe

cerebellum

central sulcus

frontal lobe

medulla oblongata

pons

temporal lobe

olfactory bulb

b. Lateral view

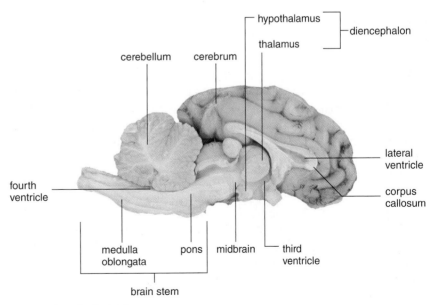

hypothalamus

diencephalon

thalamus

cerebellum

cerebrum

lateral ventricle

corpus callosum

fourth ventricle

medulla oblongata

pons

midbrain

third ventricle

brain stem

c. Longitudinal cut

a. **Midbrain:** anterior to the pons, the midbrain serves as a relay station for sensory input and motor output. It also contains a reflex center for eye muscles.

b. **Pons:** the ventral, bulblike enlargement on the brain stem. It serves as a passageway for nerve impulses running between the medulla and the higher brain regions.

c. **Medulla oblongata** (or simply **medulla**): the most posterior portion of the brain stem. It controls internal organs; for example, cardiac and breathing control centers are present in the medulla. Nerve impulses pass from the spinal cord through the medulla to higher brain regions.

Comparison of Vertebrate Brains

The vertebrate brain has a forebrain, midbrain, and hindbrain. In the earliest vertebrates, the forebrain was largely a center for sense of smell, the midbrain was a center for the sense of vision, and the hindbrain was a center for the sense of hearing and balance. How the functions of these parts changed to accommodate the lifestyles of different vertebrates can be traced.

Observation: Comparison of Vertebrate Brains

1. Examine the brain of a fish, amphibian, bird, and mammal (Fig. 30.3).
2. Identify the following three areas:
   Forebrain: Forebrain contains the olfactory bulb (not shown) and the **cerebrum,** which function in the sense of smell; the cerebrum is also responsible for memory and intelligence. In animals in which the cerebrum is less developed, the cerebrum's main function is olfactory. In animals with a highly developed, complex cerebrum, the areas involved in thought and reasoning are more evolved.
   Midbrain: Midbrain contains the **optic lobes** and other structures. The size of the optic lobes increases with the importance of vision to the animal's lifestyle. In fishes and amphibians, midbrain is often the most prominent region of the brain because it contains the nervous center, responsible for controlling most of the body's activities.
   Hindbrain: The **cerebellum** is located here. It is responsible for the coordination of complex muscular movement. A well-developed cerebellum is associated with agility.
3. Answer these questions.

   a. Based on the relative size of the optic lobe, in which animal is vision particularly important to survival? _____

Figure 30.3   Comparative vertebrate brains.
Vertebrate brains differ in particular by the comparative sizes of the cerebrum, optic lobes, and cerebellum.

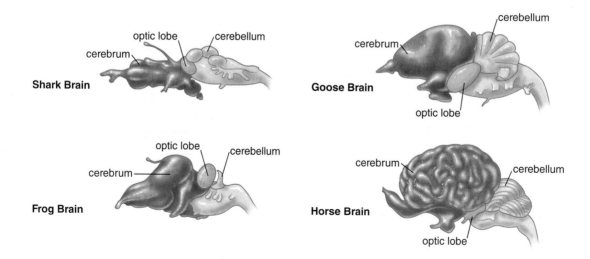

b. Which animal would you predict to be the least agile? _____

Why? _____

c. Which two of the animals would be expected to exhibit more complex behaviors than the

others? _____ Explain. _____

_____

4. Which of the labeled structures appears to have undergone the greatest change in the horse brain

compared to the frog brain? _____

Relate this to each animal's lifestyle. _____

_____

5. Which animal has the proportionately largest cerebellum? _____

How might this be explained? _____

_____

6. Which areas of the human brain would you expect to be particularly large or small? _____

_____

Using Figure 30.2 as a guide, label Figure 30.4. The optic lobe, a part of the midbrain, is not called out in Figure 30.4.

Figure 30.4 The human brain (longitudinal section).

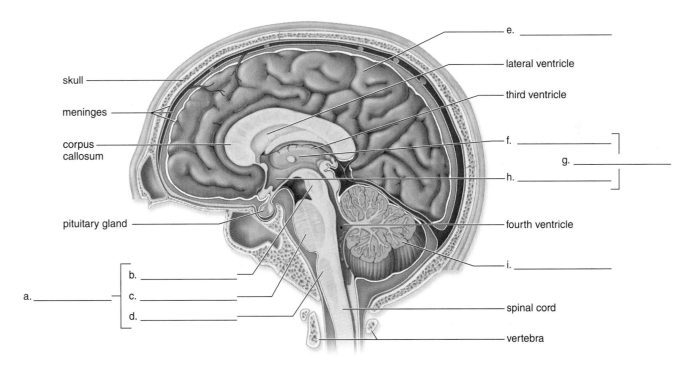

The Spinal Cord

The brain and spinal cord comprise the **central nervous system (CNS)** because they are in the center of the body. In humans, the brain is protected by the skull and the spinal cord is protected by the vertebrae (see Fig. 30.4). The spinal cord functions as a relay station: it receives out-going nerve impulses from the brain which are transferred to the body and receives in-coming nerve impulses from the body many of which are transferred to the brain. Each part of the brain then carries out the functions already studied.

Observation: Spinal Cord

1. Examine a prepared slide of a cross section of the spinal cord under the lowest magnification possible. For example, some microscopes are equipped with a very short scanning objective that enlarges about 3.5×, with a total magnification of 35×. If a scanning objective is not available, observe the slide against a white background with the naked eye.

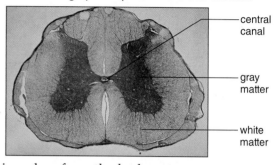

Figure 30.5 The spinal cord.
Photomicrograph of spinal cord cross section.

central canal

gray matter

white matter

2. Identify the following with the help of Figure 30.5:
   a. **Gray matter:** A central, butterfly-shaped area composed of masses of short nerve fibers (see Fig. 30.1), certain neurons and cell bodies. This is the part of the spinal cord that receives in-coming nerve impulses from the body.
   b. **White matter:** Masses of long fibers that lie outside the gray matter. White matter appears white because an insulating myelin sheath surrounds long fibers (see Fig. 30.1). This is the part of the spinal cord that sends nerve impulses to the brain and receives them from the brain.

   Considering that the spinal cord runs the length of the body proper, why is it expected that it would

   have white matter?_____

Spinal Nerves

The spinal nerves extend outward from the cord and therefore they are in the **peripheral nervous system (PNS)**. The peripheral nervous system carries nerve impulses from the body to the CNS and from the CNS to the body because each spinal nerve contains long fibers of sensory neurons and long fibers of motor neurons (see page 425). In Figure 30.6, identify the following:

1. **Sensory neuron** takes nerve impulses from a sensory receptor to the spinal cord. The cell body of a sensory neuron is in the dorsal root ganglion instead of in the spinal cord.
2. **Interneuron** lies completely within the spinal cord. Some interneurons are in the white matter of the cord and take nerve impulses to and from the brain. The interneurons that remain wholly in gray matter are short because they take nerve impulses from a sensory neuron to a motor neuron.
3. **Motor neuron** takes nerve impulses from the spinal cord to an effector—in this case, a muscle. Muscle contraction is one type of response to stimuli.

Suppose you were walking barefoot and stepped on a prickly sandbur. Describe the pathway of nerve impulses, starting with a pain receptor in your foot, that would allow you to respond to this unwelcome stimulus. _____

_____

_____

How would the brain become aware of your pain?_____

Figure 30.6 Pathway of a spinal reflex.

The arrows mark the path of nerve impulses from a sensory receptor to an effector by way of the spinal cord.

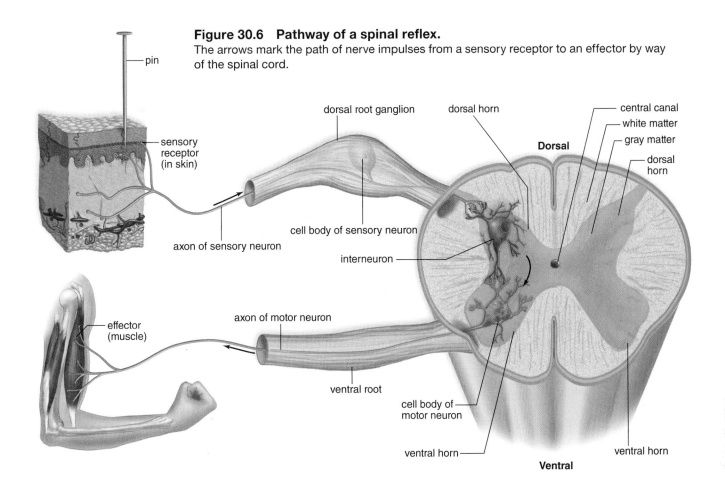

Spinal Reflexes

A **spinal reflex** is an involuntary and predictable response to a given stimulus. When you touch a sharp tack, you immediately withdraw your hand (Fig. 30.6). When a spinal reflex occurs, a sensory receptor is stimulated and generates nerve impulses that pass along the three neurons mentioned previously—a sensory neuron, interneuron, and motor neuron—until the effector responds. Notice that spinal reflexes do not involve the brain. The brain can also carry out reflexes as when a flying object causes you to blink your eyes. In that case the reflex involves one of the twelve pairs of cranial nerves that project from the inferior surface of the brain. Before we proceed, distinguish between the central nervous

system and the peripheral nervous system. _____

_____

Do the two systems work together? _____ How so? _____

_____

We will now study certain types of spinal reflexes. The spinal reflexes we will study begin with stretch receptors that detect a tap before sensory neurons conduct nerve impulses to interneurons in the spinal cord. The interneurons send a message via motor neurons to the effectors (muscles in the leg or foot). These reflexes are involuntary because the brain is not involved in formulating the response. _Consciousness_ of the stimulus lags behind the response because nerve impulses must be sent up the spinal cord to the brain before you can become aware of the tap.

Although many reflexes occur in the body, only tendon reflexes are investigated in this Experimental Procedure. Two easily tested tendon reflexes involve the **Achilles** and **patellar tendons.** When these tendons are mechanically stimulated by tapping with a reflex hammer (Fig. 30.7), or in this case, with a meterstick, sensory neurons in the tendon transmit impulses through the reflex arc to the spinal cord and then back through motor neurons to the muscle attached to the stimulated tendon. The muscle contracts and tugs on the tendon, causing movement of a bone opposite the joint.

Ankle (Achilles) Reflex

1. Have the subject sit on the table so that the leg hangs free.
2. Tap the subject's Achilles tendon at the ankle with a meterstick.
3. Which way does the foot move? Does it extend (move away from the knee) or flex (move toward

   the knee)? _____

Knee-Jerk (Patellar) Reflex

1. Have the subject sit on the table so that the leg hangs free.
2. Sharply tap the patellar tendon just below the patella (kneecap) with a meterstick.
3. In this relaxed state, does the leg flex (move toward the buttocks) or extend (move away from
   the buttocks)? _____
   Ordinarily, the cerebrum initiates the nerve impulses that cause a body part to move. Of what
   benefit are reflexes which involve only the spinal cord and not the brain?_____

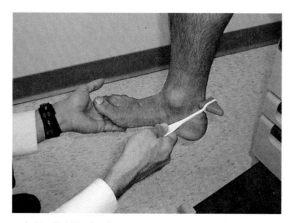

a. Ankle (Achilles) reflex

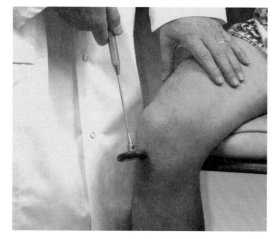

b. Knee-jerk (patellar) reflex

Figure 30.7 Two human reflexes.
The quick response when either **a.** the Achilles tendon or **b.** the patellar tendon is stimulated by tapping with a rubber hammer indicates that a reflex has occurred.

30.2 Animal Eyes

The eye is a special sense organ for detecting light rays in the environment. Most animals have some ability to detect light and the basic mechanisms for vision are the same for both invertebrates and vertebrates: photoreceptors are stimulated by light, generating nerve impulses that are interpreted by the brain.

Anatomy of Invertebrate Eyes

Some simple invertebrates, like planarians, have light detection without image formation; their eyespots allow them to detect and avoid light without actually "seeing." The eyes of arthropods and squid provide examples of the two major types of invertebrates' image-forming eyes.

Arthropods have **compound eyes** composed of many independent visual units, called ommatidia, each of which has its own photoreceptor cells (Fig. 30.8). Each unit "sees" a separate portion of the object. How well the brain combines this information is not known.

A squid has a camera-type eye. A single lens focuses an image of the visual field on the photoreceptors, packed closely together (Fig. 30.9).

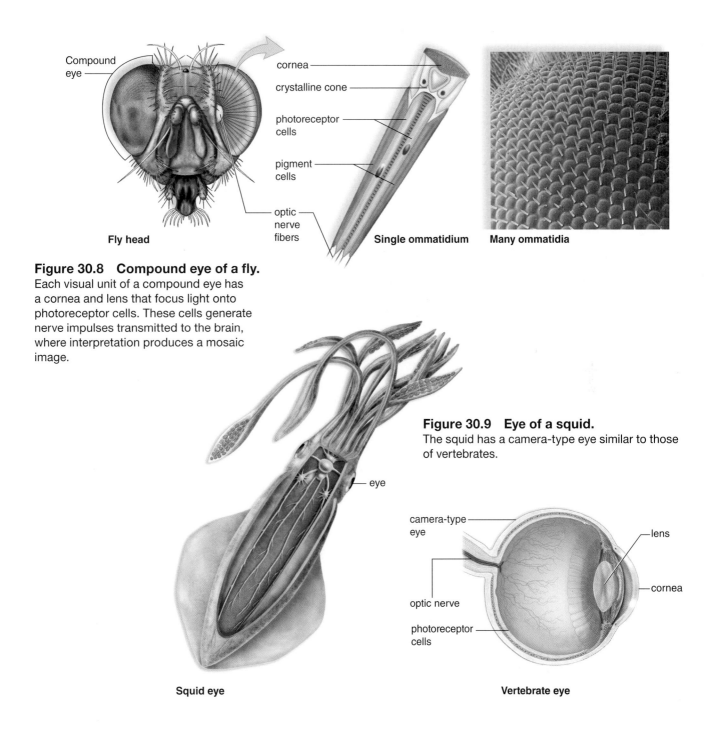

Fly head

cornea
crystalline cone
photoreceptor cells
pigment cells
optic nerve fibers

Single ommatidium　**Many ommatidia**

Figure 30.8　Compound eye of a fly.
Each visual unit of a compound eye has a cornea and lens that focus light onto photoreceptor cells. These cells generate nerve impulses transmitted to the brain, where interpretation produces a mosaic image.

eye

Squid eye

Figure 30.9　Eye of a squid.
The squid has a camera-type eye similar to those of vertebrates.

camera-type eye
lens
cornea
optic nerve
photoreceptor cells

Vertebrate eye

Observation: Invertebrate Eyes

1. Examine the demonstration slide of a compound eye set up under a stereomicroscope or examine a model of a compound eye.
2. Examine the eyes of any other invertebrates on display.

Anatomy of the Human Eye

The anatomy of the human eye allows light rays to enter the eye and strike the **rod cells** and **cone cells,** the photoreceptors for sight. The rods and cones generate nerve impulses that go to the brain via the optic nerve.

Observation: Human Eye

1. Examine a human eye model, and identify the structures listed in Table 30.1 and depicted in Figure 30.10.
2. Trace the path of light from outside the eye to the retina.

_____

_____

3. Specifically, what are the receptors for sight, and where are they located in the eye?

_____

4. What structure takes nerve impulses to the brain from the rod cells and cone cells?

_____

Table 30.1 Parts of the Human Eye

Part	Location	Function
Sclera	Outer layer of the eye	Protects and supports eyeball
Cornea	Transparent portion of sclera	Refracts light rays
Choroid	Middle layer of the eye	Absorbs stray light rays
Retina	Inner layer of the eye	Contains receptors for sight
Rod cells	In retina	Make black-and-white vision possible
Cone cells	Concentrated in fovea centralis	Make color vision possible
Fovea centralis	Special region of retina	Makes acute vision possible
Lens	Interior of eye between cavities	Refracts and focuses light rays
Ciliary body	Extension from choroid	Holds lens in place; functions in accommodation
Iris	More anterior extension of choroid	Regulates light entrance
Pupil	Opening in middle of iris	Admits light
Humors (aqueous and vitreous)	Fluid media in anterior and posterior compartments, respectively, of eye	Transmit and refract light rays; support the eyeball
Optic nerve	Extension from posterior of eye	Transmits impulses to brain

Figure 30.10 Anatomy of the human eye.

The sensory receptors for vision are the rod cells and cone cells present in the retina of the eye.

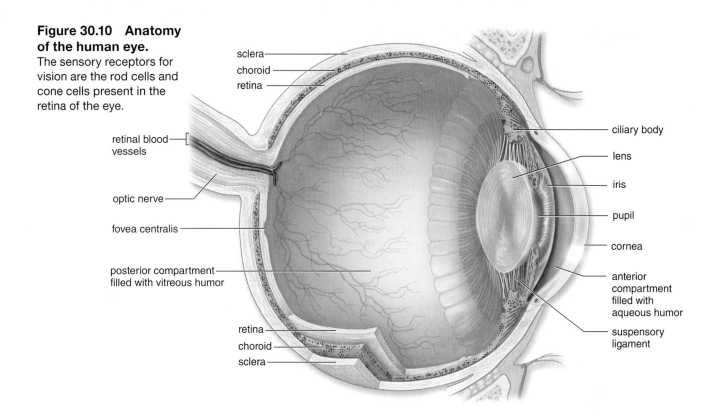

sclera
choroid
retina

ciliary body

lens

iris

pupil

cornea

retinal blood vessels

optic nerve

fovea centralis

posterior compartment filled with vitreous humor

anterior compartment filled with aqueous humor

suspensory ligament

retina
choroid
sclera

Physiology of the Human Eye

When we look at an object, light rays pass through the cornea, and as they do, they bend and converge. Further bending occurs as the rays pass through the lens. The shape of the lens changes according to whether we are looking at a distant or near object (Fig. 30.11). When looking at a near object, the lens rounds so that the light rays are bent a great deal to bring all rays to a focus on the retina. You will be measuring the ability of your lens to round up (accommodate) for viewing a near object.

Where the optic nerve pierces the wall of the eyeball and, therefore, the retina, there are no rod cells and cone cells, and therefore vision is impossible. You will be finding the blind spot in an Experimental Procedure.

Figure 30.11 Focusing of the human eye.

a. When focusing on a distant object, the ciliary muscle (in the ciliary body) is relaxed and the lens is flattened. **b.** When focusing on a near object, the ciliary muscle contracts and the lens rounds up. This is called accommodation.

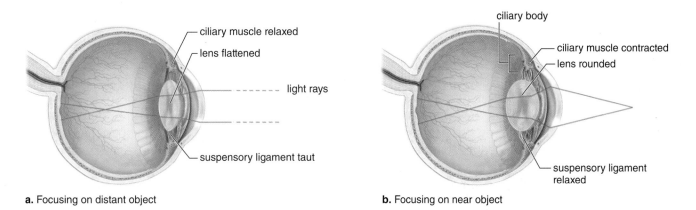

ciliary muscle relaxed

lens flattened

light rays

suspensory ligament taut

a. Focusing on distant object

ciliary body

ciliary muscle contracted

lens rounded

suspensory ligament relaxed

b. Focusing on near object

Experimental Procedure: Accommodation of the Eye

When the eye accommodates to see objects at different distances, the shape of the lens changes. The lens shape is controlled by the ciliary muscles attached to it. When you are looking at a distant object, the lens is in a flattened state. When you are looking at a closer object, the lens becomes more rounded. The elasticity of the lens determines how well the eye can accommodate. Lens elasticity decreases with increasing age, a condition called **presbyopia.** Because of presbyopia, many older people need bifocals to see near objects.

This Experimental Procedure requires a laboratory partner. It tests accommodation of either your left or right eye.

Figure 30.12 Accommodation.
When testing the ability of your eyes to accommodate in order to see a near object, always keep the pencil in this position.

1. Hold a pencil upright by the eraser and at arm's length in front of whichever of your eyes you are testing (Fig. 30.12).
2. Close the opposite eye.
3. Move the pencil from arm's length toward your eye.
4. Focus on the end of the pencil.
5. Move the pencil toward you until the end is out of focus. Measure the distance (in centimeters)

    between the pencil and your eye. _____ cm

6. At what distance can your eye no longer

    accommodate for distance? _____ cm

7. If you wear glasses, repeat this experiment

    without your glasses, and note the

    accommodation distance of your eye without glasses. _____ cm (Contact lens wearers need

    not make these determinations, and they should write the words *contact lens* in this blank.)

8. The "younger" lens can easily accommodate for closer distances. The nearest point at which the end of the pencil can be clearly seen is called the **near point.** The more elastic the lens, the

    "younger" the eye (Table 30.2). How "old" is the eye you tested? _____

Table 30.2 Near Point and Age Correlation

Age (Years)	10	20	30	40	50	60
Near Point (cm)	9	10	13	18	50	83

Experimental Procedure: Blind Spot of the Eye

The **blind spot** occurs where the optic nerves exit the retina (see Fig. 30.10). No vision is possible at this location because of the absence of rods and cones. Why are you unaware of a blind spot under normal conditions? Because although the eye detects patterns of light and color, it is the brain that determines what we perceive visually. The brain interprets the visual input based in part on past experiences.

This Experimental Procedure requires a laboratory partner. Figure 30.13 shows a small circle and a cross several centimeters apart.

Left Eye

1. Hold Figure 30.13 approximately 30 cm from your eyes. The cross should be directly in front of your left eye. If you wear glasses, keep them on.
2. Close your right eye.
3. Stare only at the cross with your left eye. You should also be able to see the dark circle in the same field of vision. Slowly move the paper toward you until the circle disappears.
4. Repeat the procedure as many times as needed to find the blind spot.
5. Then slowly move the paper closer to your eyes until the circle reappears. Because only your left eye is open, you have found the blind spot of your left eye.
6. Measure the distance (with your partner's help) from your eye to the paper when the circle first

   disappeared. _____ cm, left eye

Right Eye

1. Hold Figure 30.13 approximately 30 cm from your eyes. The circle should be directly in front of your right eye. If you wear glasses, keep them on.
2. Close your left eye.
3. Stare only at the circle with your right eye. You should also be able to see the cross in the same field of vision. Slowly move the paper toward you until the cross disappears.
4. Repeat the procedure as many times as needed to find the blind spot.
5. Then slowly move the paper closer to your eyes until the cross reappears. Because only your right eye is open, you have found the blind spot of your right eye.
6. Measure the distance (with your partner's help) from your eye to the paper when the cross first

   disappeared. _____ cm, right eye

Figure 30.13   Blind spot.
This dark circle (or cross) will disappear at one location because there are no rod cells or cone cells at each eye's blind spot, where vision does not occur.

Experimental Procedure: An Optical Illusion

This exercise reveals how brain anatomy relates to vision. The optical illusion creates an artificial situation that "confuses" certain neurons in the visual cortex that are responsible for two functions, the interpretation of both the orientation and the direction of movement of visual lines and curves.

1. Focus on the dot in the center of Figure 30.14.
2. Slowly move your head, first toward the image, and then away from it several times.
3. What do you see? _____

   _____

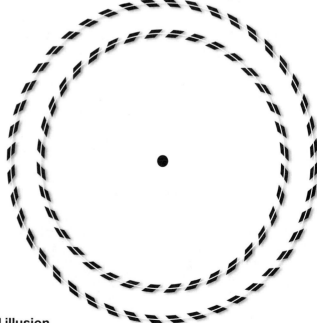

Figure 30.14   An optical illusion.

30.3 Animal Ears

Ears contain specialized receptors for detecting sound waves in the environment. They also often function as organs of balance.

Anatomy of Invertebrate Ears

Among invertebrates, only certain arthropod groups—crustaceans, spiders, and insects—have receptors for detecting sound waves. The invertebrate ear usually has a simple design: a pair of air pockets enclosed by a tympanum that passes sound waves to sensory neurons.

Observation: An Invertebrate Ear

Examine the preserved grasshopper on display, and with the help of Figure 30.15*a,* locate the tympanum. The tympanum covers an internal air sac that allows the tympanum to vibrate when struck by sound waves. Sensory neurons attached to the tympanum are stimulated directly by the vibration.

Figure 30.15 Evolution of the human ear.
a. A few invertebrates have ears such as that of the grasshopper. **b.** The lateral line of fishes contains hair cells with cilia embedded in a gelatinous cupula. **c.** Hair cells in the human ear have stereocilia embedded in the gelatinous tectorial membrane. When the basal membrane vibrates, bending of the stereocilia generates nerve impulses.

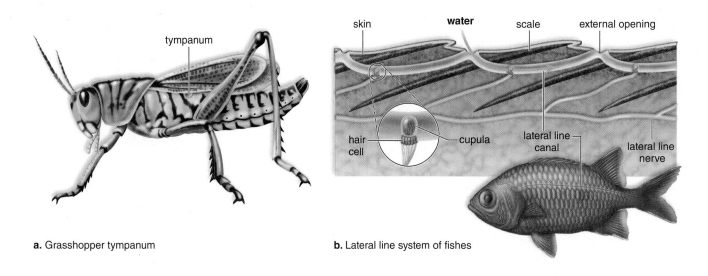

a. Grasshopper tympanum

b. Lateral line system of fishes

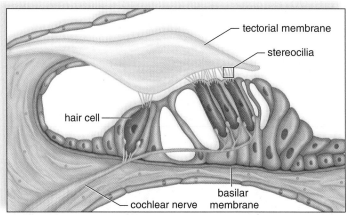

c. Cochlear cross section of the human ear

Anatomy of the Human Ear

The tetrapod vertebrate ear may have evolved from the fishes. Fishes have a lateral line, a series of hair cells with cilia embedded in a mass of gelatinous material (called a cupula) that detects water currents and pressure waves from nearby objects (Fig. 30.15b). Fishes also have separate **otic vesicles** that mainly function as organs of balance.

In contrast to the "ear" of fishes, the anatomy of the human ear allows sound waves to be received, amplified, and detected by sensory receptors. However, notice in Figure 30.15c that the receptors for sound are hair cells with stereocilia embedded in a tectorial membrane which is a gelatinous structure. The human ear contains sensory receptors for both hearing and balance.

Observation: Human Ear

Examine a human ear model, and identify the structures depicted in Figure 30.16 and listed in Table 30.3.

Figure 30.16 Anatomy of the human ear.

The outer ear extends from the pinna to the tympanic membrane. The middle ear extends from the tympanic membrane to the oval window. The inner ear encompasses the semicircular canals, the vestibule, and the cochlea.

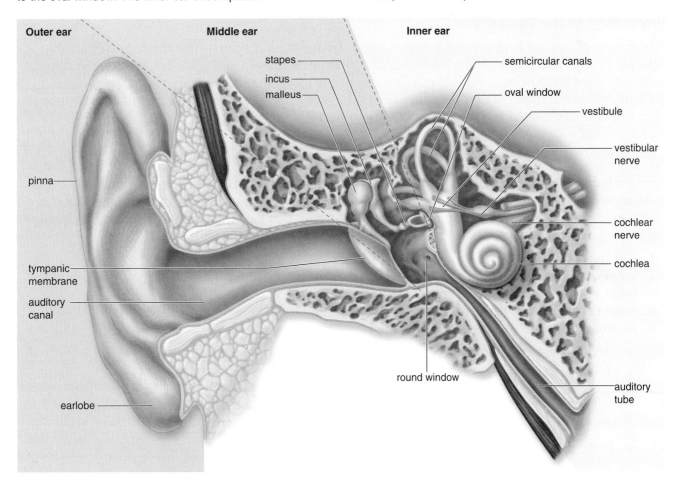

Table 30.3 Parts of the Human Ear

Part	Medium	Function	Sensory Receptors
Outer ear	Air		
Pinna		Collects sound waves	—
Auditory canal		Filters air	—
Middle ear	Air		
Tympanic membrane and ossicles		Amplify sound waves	—
Auditory tube		Equalizes air pressure	—
Inner ear	Fluid		
Semicircular canals		Rotational equilibrium	Hair cells with stereocilia embedded in cupula
Vestibule (contains utricle and saccule)		Gravitational equilibrium	Hair cells with stereocilia embedded in otolithic membrane
Cochlea (organ of Corti)		Hearing	Hair cells with stereocilia embedded in tectorial membrane

Physiology of the Human Ear

The process of hearing begins when sound waves enter the auditory canal. The sound waves are detected by the tympanic membrane and amplified by the ossicles (bones). The last ossicle, the stapes, strikes the oval window of the cochlea. As a result of the movement of the fluid within the cochlea, the hair cells in the organ of Corti are stimulated. The hair cells are sensory receptors that cause nerve impulses to be conducted in the cochlear (auditory) nerve to the brain.

Experimental Procedure: Locating Sound

Humans locate the direction of sound according to how fast it is detected by either or both ears. A difference in the hearing ability of the two ears can lead to a mistaken judgment about the direction of sound. Both you and a laboratory partner should perform this Experimental Procedure on each other. Enter the data for *your* ears, not your partner's ears, in your laboratory manual.

1. Ask the subject to be seated, with eyes closed.
2. Strike a tuning fork or rap two spoons together at the five locations listed in step 4. Use a random order.
3. Ask the subject to give the exact location of the sound in relation to his or her head.
4. Record the subject's perceptions when the sound is:

   a. directly below and behind the head _____

   b. directly behind the head _____

   c. directly above the head _____

   d. directly in front of the face _____

   e. to the side of the head _____

5. Is there an apparent difference in hearing between your two ears? _____

_____

30.4 The Senses of Human Skin

The receptors in human skin are sensitive to touch, pain, temperature, and pressure. There are individual receptors for each of these various stimuli (Fig. 30.17).

Observation: Human Skin

With the help of Figure 30.17 and a model of human skin, locate the following areas or structures, and describe the location of each:

1. Subcutaneous layer _____
2. Adipose tissue _____
3. Dermis _____
4. Epidermis _____
5. Hair follicle and hair _____
6. Oil gland _____
7. Sweat gland _____
8. Sensory receptors _____

Figure 30.17   Cutaneous receptors in human skin.
The classical view is that each cutaneous receptor has the main function shown here. However, investigators report that matters are not so clear-cut. For example, microscopic examination of the skin of the ear shows only free nerve endings (pain receptors), and yet the skin of the ear is sensitive to all sensations. Therefore, it appears that the receptors of the skin are somewhat, but not completely, specialized.

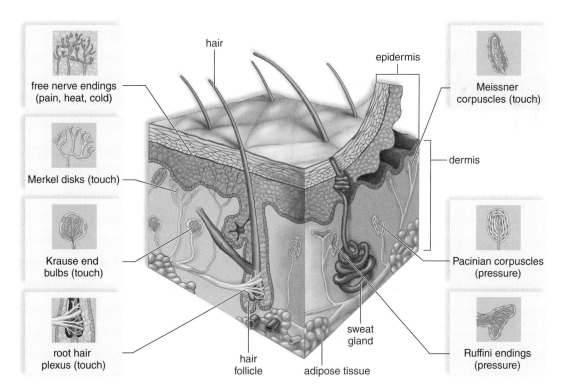

Sense of Touch

The dermis of the skin contains touch receptors whose concentration differs in various parts of the body.

Experimental Procedure: Sense of Touch

You will need a laboratory partner to perform this Experimental Procedure. Enter *your* data, not the data of your partner, in your laboratory manual.

1. Ask the subject to be seated, with eyes closed.
2. Test the subject's ability to discriminate between the two points of a hairpin or a pair of scissors at the four locations listed in step 5.
3. Hold the points of the hairpin or scissors on the given skin area, with both of the points simultaneously and gently touching the subject.
4. Ask the subject whether the experience involves one or two touch sensations.
5. Record the shortest distance between the hairpin or scissor points for a two-point discrimination in the following areas:

    a. Forearm _____ mm

    b. Back of the neck _____ mm

    c. Index finger _____ mm

    d. Back of the hand _____ mm

6. Which of these areas apparently contains the greatest density of touch receptors? _____

    Why is this useful? _____

    _____

Sense of Heat and Cold

Temperature receptors respond to a change in temperature.

Experimental Procedure: Sense of Heat and Cold

1. Obtain three 1,000 mL beakers, and fill one with *ice water,* one with *tap water* at room temperature, and one with *warm water* (45–50°C).
2. Immerse your left hand in the ice water beaker and your right hand in the warm water beaker for 30 seconds.
3. Place both hands in the beaker with room-temperature tap water.
4. Record the sensation in the right and left hands.

    a. Right hand _____

    b. Left hand _____

5. Summarize and explain your results. _____

    _____

1. Describe the cerebrum of the human brain, and state a function. _____

    _____

2. The brain stem includes the medulla oblongata, the pons, and the midbrain. Explain the expression

    brain stem as an anatomical term. _____

    _____

3. Describe the location of the gray/white matter of the spinal cord, and give a function for each.

    _____

    _____

    _____

4. State, in order from receptor to effector, the neurons associated with a spinal reflex. _____

    _____

5. Trace the path of light in the human eye—from the exterior to the retina and then from retinal nerve

    impulses to the brain. _____

    _____

6. Contrast the eye of an arthropod with the eye of a squid and human. _____

    _____

7. If you move an illustration that contains a dark circle and a dark cross toward an eye, one or the other

    may disappear. Give an explanation for this. _____

    _____

8. Trace the path of sound waves in the human ear—from the tympanic membrane to the receptors

    for hearing. _____

    _____

9. Compare the manner in which a grasshopper "hears" to the way a human hears.

    _____

    _____

10. Name four structures located in the dermis of the skin. _____

    _____

31

Musculoskeletal System

Introduction

The term **musculoskeletal** system recognizes that contraction of muscles causes the bones to move. The skeletal system consists of the bones and joints, along with the cartilage and ligaments that occur at the joints. The muscular system contains three types of muscles: smooth, cardiac, and skeletal. The skeletal muscles are most often attached to bones after crossing a joint (Fig. 31.1). In humans, the biceps brachii muscle has two **origins**, and the triceps brachii has three origins on the humerus and scapula. Find the biceps brachii **insertion** on the anterior surface of your elbow after contracting your biceps muscle. Locate the ulna at your posterior elbow. The triceps brachii tendon inserts on the ulna.

Muscles work in antagonistic pairs. For example, when the biceps brachii contracts, the bones of the forearm are pulled upward, while the triceps brachii relaxes; when the triceps brachii contracts, the bones of the forearm are pulled downward, while the biceps brachii relaxes.

🔧 **Planning Ahead** Use rotation and have some groups start with the frog exercise on page 455 to minimize the number of preserved and dissected frogs needed for this exercise.

Figure 31.1   Muscular action.
Muscles, such as these muscles of the arm (which have their origin on the scapula and their insertion on the bones of the forearm), cause bones to move.

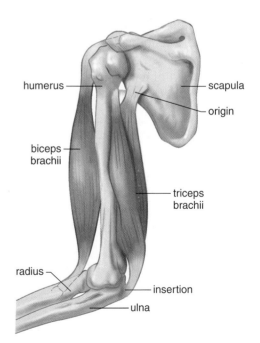

31.1 Animal Skeletons

In all animals, skeletons provide a framework and assist movement. To cause body movements, the force of muscular contractions must be specifically directed against other parts of the body. In some animals, such as segmented worms, the pressure of muscular contraction is applied to fluid-filled body compartments (Fig. 31.2); therefore, the animal is said to have a **hydrostatic skeleton.**

Figure 31.2 Earthworm locomotion.
a. Earthworms have a hydrostatic skeleton. Both circular and longitudinal muscles, which work against the fluid-filled coelom, permit them to move forward as illustrated in **b.**

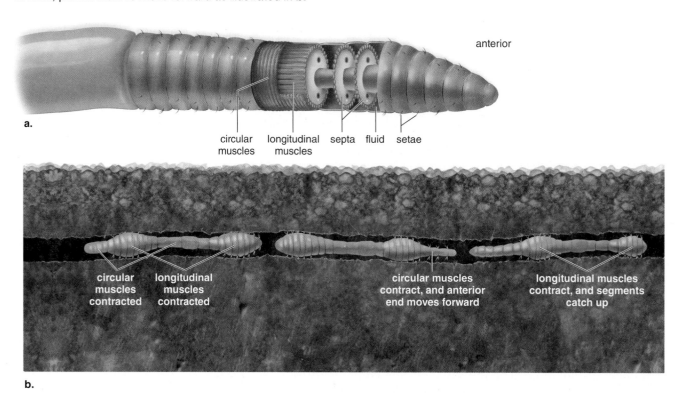

anterior

a.

circular muscles longitudinal muscles septa fluid setae

circular muscles contracted longitudinal muscles contracted

circular muscles contract, and anterior end moves forward

longitudinal muscles contract, and segments catch up

b.

Nonjointed **exoskeletons** (external skeleton) are found among such diverse animals as corals and molluscs. The jointed exoskeleton of arthropods is composed largely of chitin, which, as the animal grows, must be periodically shed by molting. The jointed **endoskeleton** (internal skeleton) of vertebrates is composed of cartilage and bone and grows with the animal. The cartilage and bone tissues are storage areas for calcium and phosphorus. The bones are also the site of blood cell production.

Tissues of the Human Skeleton

When a long bone, such as the human humerus, is split open, as in Figure 31.3*a*, it is not solid but has a cavity bounded at the sides by **compact bone** and at the ends by **spongy bone.** Beyond the spongy bone is a thin shell of compact bone and, finally, a layer of **cartilage.** While **red bone marrow,** a specialized substance that produces blood cells, is present in spongy bone, **yellow marrow,** a fat storage tissue, is present in the medullary cavity.

Observation: Tissues of the Human Skeleton

Compact Bone

Examine a prepared slide of compact bone (see page 360 and Fig. 31.3*b*). Identify:

1. **Osteons** (Haversian system): A series of concentric rings called **lamellae.**

Figure 31.3 Anatomy of a long bone.

The central shaft of **a.** long bone is composed of **b.** compact bone, but the ends are **c.** spongy bone capped by **d.** cartilage. Spongy bone can contain red bone marrow.

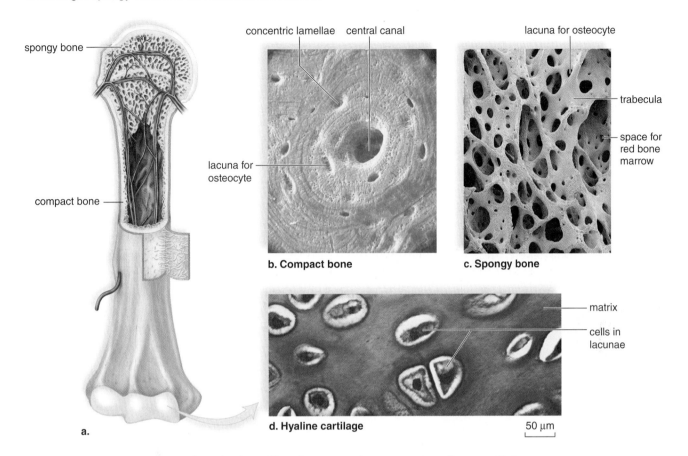

spongy bone

compact bone

a.

concentric lamellae   central canal

lacuna for osteocyte

b. Compact bone

lacuna for osteocyte

lacuna for osteocyte

trabecula

space for red bone marrow

c. Spongy bone

matrix

cells in lacunae

d. Hyaline cartilage   50 μm

2. **Lacunae:** Cavities within the lamellae that contain **osteocytes** (bone cells).
3. **Central canal:** Canal in the center of each osteon.
4. **Canaliculi:** Tiny tubules that allow nutrients to pass between the osteocytes.
5. **Matrix:** The nonliving material maintained by osteocytes. Contains mineral salts (notably calcium salts) and protein.

Spongy Bone

Examine a prepared slide of spongy bone (Fig. 31.3*a* and *c*), and identify the numerous bony bars and plates separated by irregular spaces. These spaces are often filled with red marrow.

Hyaline Cartilage

Examine a prepared slide of hyaline cartilage (Fig. 31.3*d*), and identify:

1. **Lacunae:** Cavities scattered throughout the matrix, which contain chondrocytes (cells that maintain cartilage).
2. **Matrix:** A material that is more flexible because it consists primarily of water and protein.

Experimental Procedure: Compact Bone

Compact bone is the most rigid of the connective tissues. It has an extremely hard matrix of mineral salts, primarily calcium salts, deposited around protein fibers. The minerals give bone rigidity, and the protein fibers provide elasticity and strength, much as steel rods do in reinforced concrete.

1. Examine normal chicken femurs, as well as chicken femurs that have been soaked in 5–10% acetic acid for 24 to 48 hours. Under these conditions, the calcium is removed, but the protein remains.

2. How does the shape of the treated femur compare with that of the untreated femur? _____

_____

3. Is the treated bone more flexible or less flexible than the untreated bone? _____

_____

4. Is the treated bone able to withstand more direct downward compression or less direct downward

compression than the untreated bone? _____

5. Examine other chicken bones that have been baked in an oven at a very high temperature
for several days. What remains is calcium phosphate devoid of water and without any intact

reinforcing protein fibers. What is the effect of this treatment on bone strength? _____

_____

Bones of the Human Skeleton

The human skeleton, like other vertebrate skeletons, is divided into axial and appendicular components
(Fig. 31.4). The **axial skeleton** is the main longitudinal portion. The **appendicular skeleton** includes the
bones of the appendages and their supportive pectoral and pelvic (shoulder and hip) girdles.

Figure 31.4 Human skeletal system.
a. Anterior view. **b.** Posterior view. The axial skeleton appears in blue, while the appendicular skeleton is shown in tan.

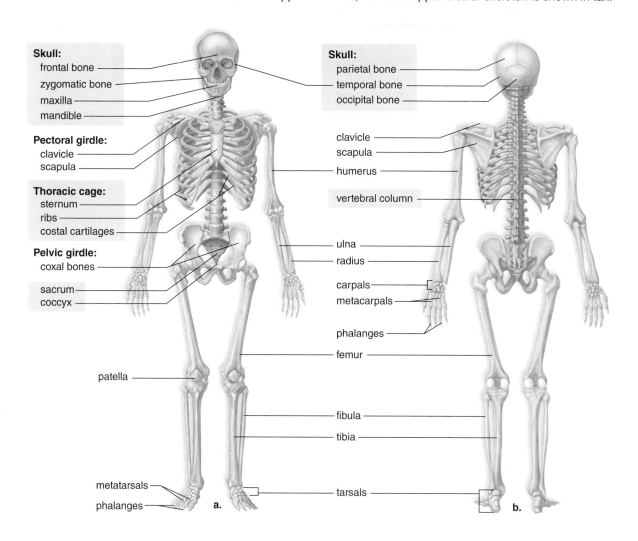

Examine a human skeleton, and with the help of Figure 31.4, identify the following bones:

1. The **skull** is composed of many small bones fused together. Particularly note the **cranium,** a bone case enclosing the brain and major sense organs; the **facial bones;** and the **mandible** (lower jaw), the only bone of the skull not immovably fused to another.
2. The **vertebral column** provides support and also houses the **spinal cord.** It is composed of many vertebrae separated from one another by intervertebral disks. The vertebral column customarily is divided into five series:
   a. Seven **cervical vertebrae** (forming the neck region).
   b. Twelve **thoracic vertebrae** (with which the ribs articulate).
   c. Five **lumbar vertebrae** (in the abdominal region).
   d. Five fused sacral vertebrae called the **sacrum.**
   e. Several caudal vertebrae (fused in humans to form the **coccyx**).
3. The **ribs** and their associated muscles form a bony case that supports the thoracic cavity wall. The ribs connect dorsally with thoracic vertebrae, and some also are attached by cartilage directly or indirectly to the sternum. Those ribs without ventral attachment are called **floating ribs.**

Observation: Appendicular Skeleton

Examine a human skeleton, and with the help of Figure 31.4, identify the following bones:

1. The **pectoral girdle,** which supports the upper limbs, is composed of the **clavicles** (collarbones) and **scapulae** (shoulder bones).
2. The upper limbs (the arm and the forearm) contain these bones:
   a. **Humerus:** The large long bone of the arm.
   b. **Radius:** The long bone of the forearm, with a pivot joint at the elbow that allows rotational motion.
   c. **Ulna:** The other long bone of the forearm, with a hinge joint at the elbow that allows motion in only one plane.
   d. **Carpals:** A group of small bones forming the wrist.
   e. **Metacarpals:** Slender bones forming the palm.
   f. **Phalanges:** The bones of the fingers.
3. The **pelvic girdle** forms the basal support for the lower limbs, and its lateral half is composed of the **coxal** (hip) **bone.** The female pelvis is much broader and more shallow, and the outlet is larger than that of the male.
4. The lower limbs (the thigh plus the leg) contain these bones:
   a. **Femur:** The long bone of the thigh.
   b. **Tibia:** The larger of the two long bones of the leg.
   c. **Fibula:** The smaller of the two long bones of the leg.
   d. **Tarsals:** A group of small bones forming the ankle.
   e. **Metatarsals:** Slender anterior bones of the foot.
   f. **Phalanges:** The bones of the toes.

Comparison of Vertebrate Skeletons

Vertebrate forelimbs (or hindlimbs) are used for various purposes. Yet they contain the same sets of bones organized in similar ways. The explanation for this unity is that the basic plan originated with a common ancestor, and then it was modified as vertebrates continued along their evolutionary pathway. Structures that are similar because they are inherited from a common ancestor are called **homologous structures.**

Forelimb Comparisons

1. The bones in the forelimbs of vertebrates are homologous but modifications make them suitable for specific functions (Fig. 31.5). In the first column of Table 31.1, enter the specific function of the forelimb for the animals listed. For example, a frog locomotes by hopping and its forelimb is used for landing. In contrast, a cat walks using only four of its phalanges. The first phalanx is much reduced.

   A bird uses its forelimb for _____.

2. Keeping function in mind, compare the forelimb bones in the animals listed in Table 31.1 to those of a human. Where *a* is noted, compare the humerus; where *b* is noted, compare the radius and ulna; and where *c* is noted, compare the carpels, metacarpals, and phalanges as a whole. Are the bones fused, longer, stronger, or more delicate than those of a human?

   Give examples to show that the forelimbs of these animals are modified to suit their means of locomotion. _____

Figure 31.5 Vertebrate forelimbs.
The forelimbs of these animals differ according to their specific functions. Homologous structures are color-coded.

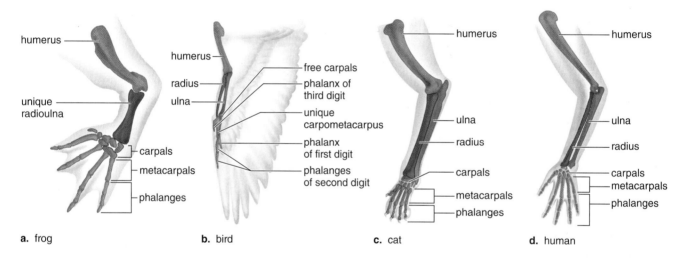

a. frog   **b.** bird   **c.** cat   **d.** human

Table 31.1	Function and Comparison of Forelimb Bones to Those of Humans*	
Animal	**Forelimb Function**	**Forelimb Bones**
Frog		a. b. c.
Bird		a. b. c.
Cat		a. b. c.

* See instructions above in text.

Hindlimbs to Forelimbs Comparison

1. Only the phalanges in the hindlimb of a frog, bird, and cat ever touch the ground (Fig. 31.6), whereas the entire foot of a human touches the ground.

2. From Figure 31.5 label the bones of the forelimbs and then compare the structure of the hindlimb bones to that of the forelimb bones in a frog, bird, cat, and human.

3. Explain any difference in the bones of each animal in Table 31.2 on the basis of the specific locomotor function of the hindlimb. For example, the hindlimbs of a frog allow it to hop while its forelimbs are used for landing. Similarly, the hindlimbs of a cat allow it to leap and the forelimbs allow it to land. However, a cat uses the phalanges of both the forelimbs and hindlimbs for walking.

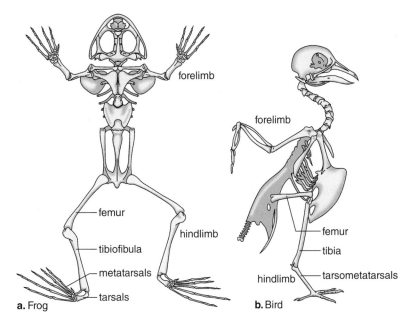

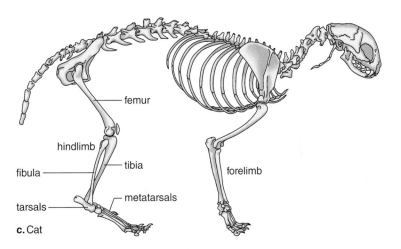

Figure 31.6   Animal skeletons.
The forelimb and hindlimb bones of these animals differ according to the locomotor functions of the limbs.

Table 31.2	Comparison of Hindlimb to Forelimb
Animal	**Differences in Hindlimb and Forelimb Bones and Explanation**
Frog	
Bird	
Cat	
Human	

31.2 Vertebrate Muscles

Muscle contraction accounts for an animal's ability to respond quickly to outside stimuli. Muscles, along with glands, are called **effectors** because they carry out the orders of the nervous system.

Muscular Tissues

The three types of vertebrate muscles are smooth, cardiac, and skeletal. **Smooth muscular tissue** is not under conscious control and is predominantly found in the walls of hollow organs, such as the bladder, blood vessels, uterus, and digestive tract. **Cardiac muscular tissue,** found in the heart, is striated, contains alternating light and dark bands, and is not under conscious control. **Skeletal muscular tissue** is under conscious control and has **striations** (alternating light and dark bands). The cells that make up muscular tissue are called muscle fibers.

Observation: Muscular Tissues

Smooth Muscle

Smooth muscle is located in the walls of hollow internal organs, and its involuntary contraction moves materials through an organ. Smooth muscle fibers are spindle-shaped cells, each with a single nucleus (uninucleated). The cells are usually arranged in parallel lines, forming sheets. Smooth muscle does not have the striations (bands of light and dark) seen in cardiac and skeletal muscle. Although smooth muscle is slower to contract than skeletal muscle, it can sustain prolonged contractions and does not fatigue easily.

Examine a prepared slide of smooth muscle (Fig. 31.7), and identify the spindle-shaped fibers and the oval-shaped nucleus found in the center of the cytoplasm. Describe what you see:

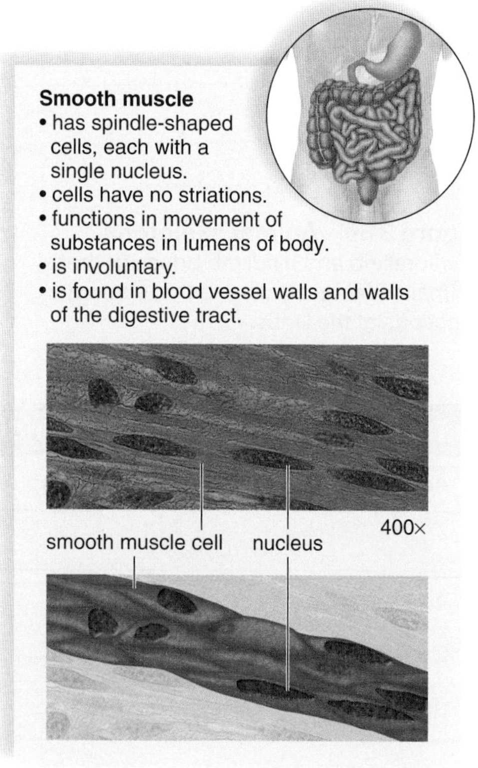

Smooth muscle
- has spindle-shaped cells, each with a single nucleus.
- cells have no striations.
- functions in movement of substances in lumens of body.
- is involuntary.
- is found in blood vessel walls and walls of the digestive tract.

smooth muscle cell   nucleus   400×

Figure 31.7 Muscular tissues.
Smooth muscle is nonstriated and involuntary. This type of muscle is composed of spindle-shaped cells. Icon gives example of smooth muscle location in human body.

Cardiac Muscle

Cardiac muscle forms the heart wall. Its fibers are uninucleated, striated, tubular, and branched, which allows the fibers to interlock at intercalated disks. Intercalated disks permit contractions to spread quickly throughout the heart. Cardiac fibers relax completely between contractions, which prevents fatigue. Contraction of cardiac muscle fibers is rhythmical; it occurs without outside nervous stimulation or control. Thus, cardiac muscle contraction is involuntary.

Examine a prepared slide of cardiac muscle (Fig. 31.8), and identify a single nucleus per cell, the branched fibers joined at intercalated disks (dark lines that pass across the fibers), and striations that show up under high power. Describe what you see:

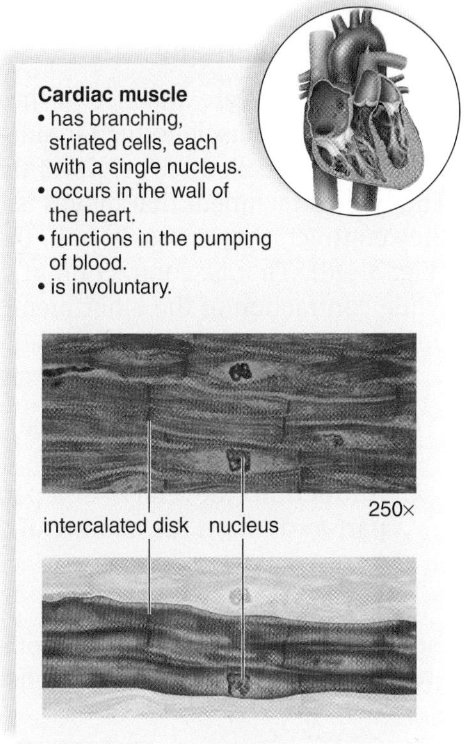

Cardiac muscle
- has branching, striated cells, each with a single nucleus.
- occurs in the wall of the heart.
- functions in the pumping of blood.
- is involuntary.

intercalated disk   nucleus   250×

Figure 31.8   Cardiac muscle.
Cardiac muscle is striated and involuntary. The branched cells join at intercalated disks. Icon shows where cardiac muscle is located in human body.

Skeletal Muscle

Skeletal muscle fibers are tubular, multinucleated, and striated. They make up the skeletal muscles attached to the skeleton. Skeletal muscle fibers can run the length of a muscle and, therefore, can be quite long. Skeletal muscle is voluntary because its contraction is always stimulated and controlled by the nervous system.

Examine a prepared slide of skeletal muscle (Fig. 31.9), and identify long fibers arranged in a parallel fashion, multinuclei per fiber, and striations. Describe what you see:

Skeletal muscle
- has striated cells with multiple nuclei.
- occurs in muscles attached to skeleton.
- functions in voluntary movement of body.

striation   nucleus

Figure 31.9   Skeletal muscle.
Skeletal muscle is striated and voluntary. The tubular cells contain many nuclei. Icon gives example of skeletal muscle location in human body.

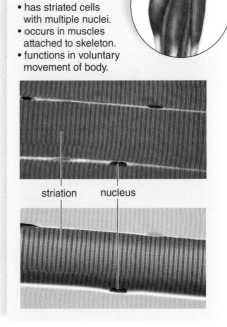

Virtual Lab   Muscle Stimulations   A virtual lab called Muscle Stimulations is available on the *Biology* website **www.mhhe.com/maderbiology11.** Use the directions provided to see the use of an oscilloscope to observe the increased amount of stimulation needed to achieve a muscle twitch when a frog muscle is subjected to different work loads. If time is limited, observe only the calf muscle.

Types of Movement

Skeletal muscles are attached to the skeleton, and when they contract, bones move. **Tendons** most often attach a muscle to the far sides of a joint (the region where two bones meet). One point of attachment is called the **origin** of the muscle; the origin stays stationary when the muscle contracts. The bone attachment that moves is called the **insertion** of the muscle. Because muscles shorten when they contract, they can only pull; they cannot push. Therefore, muscles work in **antagonistic pairs** (Fig. 31.10). That is, contraction of one member of the pair causes a bone to move in one direction, while contraction of the other member of the pair causes the same bone to move in the opposite direction.

In this section, you will be examining two sets of movements:

1. **Flexion** (movement of jointed parts toward each other) and **extension** (movement of jointed parts away from each other).
2. **Adduction** (movement of a part toward the midline of the body) and **abduction** (movement of a part away from the midline of the body).

Figure 31.10   Antagonistic pairs of muscles.
Arrows indicate direction of movement of the lower limb when upper limb muscles contract.

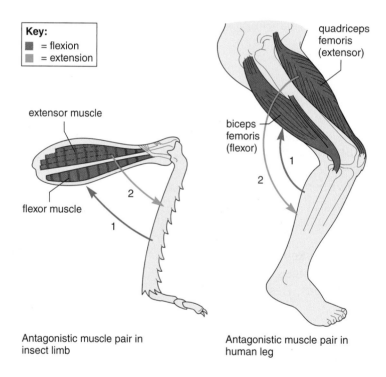

Antagonistic muscle pair in insect limb

Antagonistic muscle pair in human leg

Frog Muscles

The skin has been removed from the legs of a preserved frog so that you can study the arrangement of muscles (Fig. 31.11). The objective is to discover which of the four movements (flexion, extension, adduction, abduction) are performed by the muscles.

1. Locate the **gastrocnemius,** a large muscle on the back of the shank. Observe the white connective tissue **(fascia)** surrounding the fleshy middle, or **belly,** portion of the muscle. The fascia continues beyond the muscle at either end to form the **tendons,** which attach the muscle to bone.
2. Hold the thigh of the frog rigid with one hand, and grasp the belly of the gastrocnemius with the other. Gently pull the gastrocnemius toward its origin (lower end of the femur, the bone of the thigh). What is the action caused by the gastrocnemius when the thigh is rigid? _____

_____

3. Next, hold the shank rigid, and again pull the gastrocnemius toward its origin. When the shank is rigid what is the action caused by contraction of the gastrocnemius? _____

4. Locate the frog's **triceps femoris,** a very large ventral muscle of the thigh. It is seen best in the dorsal view. This muscle has two origins on the pelvic girdle and inserts into the femur by way of the broad, white tendon that passes over the anterior surface of the knee. Hold the thigh firmly, and pull on the triceps. Is the triceps femoris a shank extensor or flexor? _____

Figure 31.11   Dorsal view of hindleg muscles in frogs.
The contraction of muscles makes the bones move in particular directions. In this illustration note that the left shows surface muscles and the right shows deep muscles of the frog.

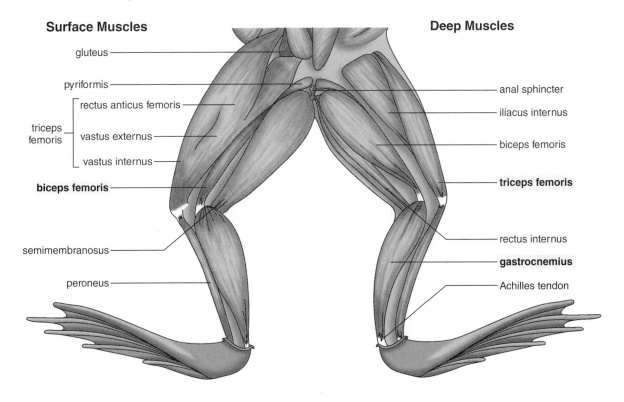

5. Now find your specimen's **biceps femoris,** a narrow muscle seen best in dorsal view just posterior to the triceps. Its origin is on the pelvic girdle. Where does the biceps femoris insert? _____

   _____

6. What action does the biceps femoris cause? _____

   _____

Human Muscles

Locate the following antagonistic pairs (muscles that act in opposition to each other) in Figure 31.12. In each case, state their contrary actions by inserting one of these functions—*flexes, extends, adducts,* or *abducts*—in the following:

1. The biceps brachii _____ the forearm.

   The triceps brachii _____ the forearm.

2. The sartorius _____ the thigh.

   The adductor longus _____ the thigh.

3. The quadriceps femoris _____ the leg.

   The biceps femoris _____ the leg.

Muscular Contractions

Isometric and Isotonic Contractions

During an **isometric contraction,** the length of the muscle does *not* change. During an **isotonic contraction,** the length of the muscle *does* change.

Experimental Procedure: Types of Contraction

Isometric Contraction

1. Place the palm of your left hand underneath a tabletop. Push up against the table while you have your right hand cupped over the anterior surface of your left upper arm so that you can feel the muscle there undergo an isometric contraction.
2. Is the biceps brachii or the triceps brachii located on the ventral surface of the arm?

   _____

3. What change did you notice in the firmness of this muscle as it contracted? _____

   _____

4. Did your hand or forearm move as you pushed up against the table? _____

5. Given your answer to question 4, did this muscle's fibers shorten as you pushed up against the tabletop? _____

Isotonic Contraction

1. Start with your left forearm resting on a table. Watch the anterior surface of your left upper arm while you slowly bend your elbow and bring your left forearm toward the upper arm. An isotonic contraction of the biceps brachii produces this movement.
2. What makes this contraction isotonic rather than isometric?

   _____

Figure 31.12 Human superficial skeletal muscles.

a. Anterior view. **b.** Posterior view. The muscles highlighted here are those noted in the exercise entitled "Human Muscles."

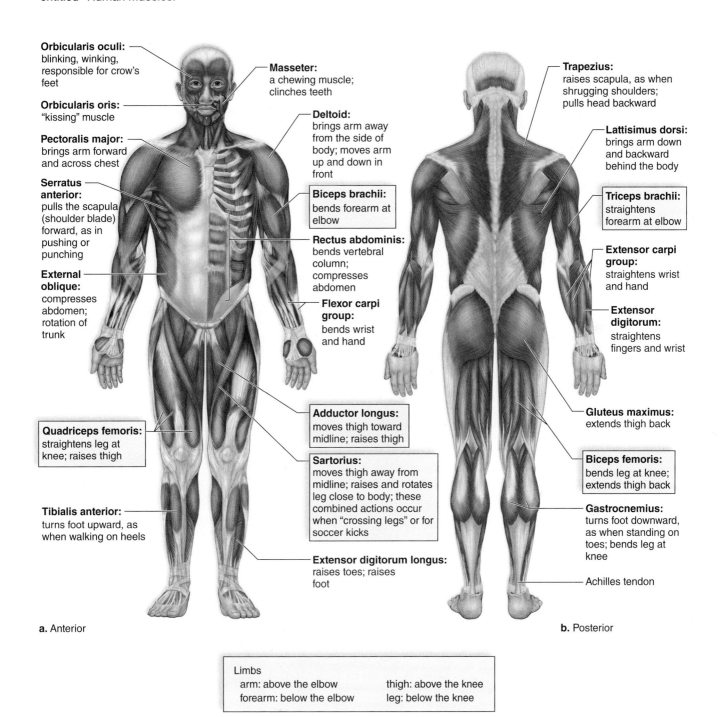

Orbicularis oculi:
blinking, winking, responsible for crow's feet

Orbicularis oris:
"kissing" muscle

Pectoralis major:
brings arm forward and across chest

Serratus anterior:
pulls the scapula (shoulder blade) forward, as in pushing or punching

External oblique:
compresses abdomen; rotation of trunk

Quadriceps femoris:
straightens leg at knee; raises thigh

Tibialis anterior:
turns foot upward, as when walking on heels

Masseter:
a chewing muscle; clinches teeth

Deltoid:
brings arm away from the side of body; moves arm up and down in front

Biceps brachii:
bends forearm at elbow

Rectus abdominis:
bends vertebral column; compresses abdomen

Flexor carpi group:
bends wrist and hand

Adductor longus:
moves thigh toward midline; raises thigh

Sartorius:
moves thigh away from midline; raises and rotates leg close to body; these combined actions occur when "crossing legs" or for soccer kicks

Extensor digitorum longus:
raises toes; raises foot

Trapezius:
raises scapula, as when shrugging shoulders; pulls head backward

Lattisimus dorsi:
brings arm down and backward behind the body

Triceps brachii:
straightens forearm at elbow

Extensor carpi group:
straightens wrist and hand

Extensor digitorum:
straightens fingers and wrist

Gluteus maximus:
extends thigh back

Biceps femoris:
bends leg at knee; extends thigh back

Gastrocnemius:
turns foot downward, as when standing on toes; bends leg at knee

Achilles tendon

a. Anterior

b. Posterior

Limbs
arm: above the elbow
forearm: below the elbow
thigh: above the knee
leg: below the knee

Contraction of Muscle Fibers

Note in Figure 31.13 that electron microscopy has shown that striations in skeletal muscle are due to the placement of **myosin** and **actin filaments.** During contraction, actin filaments slide past myosin filaments, and units of the muscle called **sarcomeres** shorten. ATP serves as the immediate energy source for sarcomere contraction. Potassium (K^+) and magnesium ions (Mg^{2+}) are cofactors for the breakdown of ATP by myosin.

The current model for muscle contraction states that as muscle contraction occurs, actin filaments _____ past myosin filaments. This causes _____ to shorten.

bundle of muscle fibers

myofibrils

A muscle contains bundles of muscle fibers, and a muscle fiber has many myofibrils.

skeletal muscle fiber

sarcolemma

mitochondrion

calcium storage sites

sarcoplasm

one myofibril

sarcoplasmic reticulum

T tubule

nucleus

Z line ← ——— one sarcomere ——— → Z line

A myofibril has many sarcomeres.

6000×

Figure 31.13 Microscopic structure of a skeletal muscle fiber.

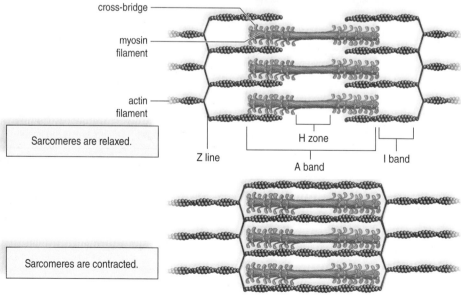

cross-bridge

myosin filament

actin filament

Sarcomeres are relaxed.

Z line

A band

H zone

I band

Sarcomeres are contracted.

Experimental Procedure: Contraction of Muscle Fibers

1. Label two slides 1 and 2. Mount a strand of glycerinated fibers in a drop of glycerol on each slide. Place each slide on a millimeter ruler, and measure the length of the strand. Record these lengths in the first row in Table 31.3.
2. If there is more than a small drop of glycerol on the slides, soak up the excess on a piece of lens paper held at the edge of the glycerol farthest from the fiber strand.
3. To slide 1, add a few drops of salt solution containing potassium (K^+) and magnesium (Mg^{2+}) ions, and measure any change in strand length. Record your results in Table 31.3.
4. To slide 2, add a few drops of ATP solution, and measure any change in strand length. Record your results in Table 31.3.
5. Now add ATP solution to slide 1. Measure any change in strand length, and record your results in Table 31.3.
6. To slide 2, add a few drops of K^+/Mg^{2+} salt solution, and measure any change in strand length. Record your results in Table 31.3.
7. You probably observed muscle contraction (as demonstrated by shortening of the fiber strand) only after both the K^+/Mg^{2+} salt solution and the ATP solution were applied to the muscle. To demonstrate that you understand the requirements for contraction, state the function of each of the substances listed in Table 31.4.

Table 31.3 Glycerinated Muscle Contraction

Solution	Length (mm)	
	Slide 1	Slide 2
Glycerol alone		
K^+/Mg^{2+} salt solution alone		
ATP alone		
Both salt solution and ATP		

Table 31.4 Summary of Muscle Fiber Contraction

Substance	Function
Myosin filament	
Actin filament	
K^+/Mg^{2+} salt solution	
ATP	

Laboratory Review 31

1. An earthworm has no hard parts; why is it said to have a skeleton? _____
_____

2. Name and describe three features that would allow you to identify a slide of compact bone.
_____

_____

_____

3. What bones make up the pectoral girdle of humans?
_____

4. What bones protect the thoracic cavity? _____
_____

5. Diversity of function explains why the bones in a bird's forelimb differ in appearance from those in its
hindlimb. Explain. _____
_____

6. Name the features that would allow you to identify the following muscular tissues:
   a. Smooth muscle _____
   b. Cardiac muscle _____
   c. Skeletal muscle _____

7. In frogs, the triceps femoris and the biceps femoris are an antagonistic pair. The triceps femoris extends
the hindlimb. What does the biceps femoris do?
_____

8. Sequence the components of a skeletal muscle fiber from the largest to the smallest components.
_____

_____

9. When you observe glycerinated muscle shorten, what is happening to actin filaments microscopically?
_____

10. When skeletal muscle fibers contract, what happens to sarcomeres? _____
_____

***Biology* Website**

Enhance your study of the text and laboratory manual with study
tools, practice tests, and virtual labs. Also ask your instructor
about the resources available through ConnectPlus, including
the media-rich eBook, interactive learning tools, and animations.

www.mhhe.com/maderbiology11

McGraw-Hill Access Science Website

An Encyclopedia of Science and Technology Online
which provides more information including videos
that can enhance the laboratory experience.

www.accessscience.com

32

Animal Development

Learning Outcomes

Introduction

The early development of animals is quite similar, regardless of the species. The fertilized egg, or zygote, undergoes successive divisions by cleavage, forming a mulberry-shaped ball of cells called a morula and then a hollow ball of cells called a blastula. The fluid-filled cavity of the blastula is the blastocoel. Later, some of the surface cells fold inward, or invaginate, eventually forming a double-walled structure. The outer layer is called the ectoderm, and the inner layer is the endoderm. Between these layers, a middle layer, or mesoderm, arises. The embryo is now called a gastrula. In particular, the presence of yolk (nutrient material) influences how the gastrula comes about.

All later development can be associated with the three **germ layers** that give rise to different tissues and systems: (1) The **ectoderm** forms the nervous system and the skin plus its accessory structures (hair, nails, scales, feathers, etc.); (2) the **endoderm** forms the lining of the digestive system and respiratory system; and (3) the **mesoderm** gives rise to the cardiovascular, muscular, reproductive, and skeletal systems and to connective tissue.

Human development involves embryonic development and fetal development. **Embryonic development** comes to a close when all the basic organs have formed. Refinements and an increase in size occur during **fetal development**.

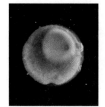

zygote

embryo at one week; implants in uterine wall

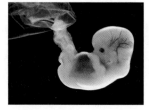

embryo at eight weeks

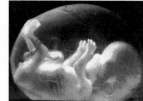

fetus at three months

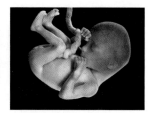

fetus at five months

32.1 Early Embryonic Stages

Successive division of the **zygote** (fertilized egg) results in two-cell, four-cell, eight-cell, and sixteen-cell stages, and finally, in a many-celled stage called a **morula.** The morula becomes a **blastula,** and the blastula becomes a **gastrula.**

Sea Star Development

Echinoderms (e.g., sea stars, sea urchins) are useful for illustrating the stages of early development for multicellular animals. Sea stars develop in an aquatic environment and develop quickly into a larva capable of feeding itself. Explain why you would not expect a sea star's egg to be heavily laden with yolk.

_____

_____

Observation: Sea Star Embryos

Examine whole-mount microscope slides of stained sea star embryos at the stages of development shown in Figure 32.1. Try to identify the following:

1. **Unfertilized egg:** Observe the large nucleus and the darkly staining nucleolus. The plasma membrane surrounds the cytoplasm, which contains a small amount of yolk. After fertilization, a fertilization membrane, which prevents the entrance of other sperm, can be seen outside the plasma membrane, and the distinct nucleus disappears.
2. **Cleavage:** Successive division of the embryo results in a two-cell, four-cell, eight-cell, sixteen-cell, and finally, a many-celled stage. How does the size of the two- to eight-cell stage of a sea star

   compare to the size of the unfertilized egg? _____

   Does growth occur during cleavage? _____
3. **Morula:** The morula is a ball of cells about the same size as the original zygote.

   Explain why the morula and zygote are about the same size. _____

   _____
4. **Blastula:** The large number of embryonic cells of the morula arrange themselves into a blastula, a single-layered ball with a fluid-filled cavity, called the **blastocoel,** in the middle. _Label the blastocoel in Figure 32.1f._
5. **Early gastrula:** The cells of the blastula fold inward to form a two-layered gastrula. The cavity produced by the infolded layer of cells is the **archenteron,** or primitive gut, which has an opening to the outside called the **blastopore.** The outer layer of cells is the **ectoderm,** and the inner layer is

   the **endoderm.** Why are the ectoderm and endoderm called germ layers? _____

   Recall that the sea star is a deuterostome. In the sea star, does the blastopore become the mouth or

   the anus? _____
6. **Late gastrula:** As development continues, two pouches (the coelomic sacs) form by outpocketing from the endoderm surrounding the gut. These pouches become part of the **coelom** (body cavity), and the walls of the lateral pouches become the third germ layer, the **mesoderm.** Recall that

   a sea star has a true coelom. Do you expect that mesoderm will line the coelom of a sea star? _____

   Explain. _____

Figure 32.1 Photographs of sea star developmental stages.

All animal embryos go through these same early developmental stages. (**a–f:** Magnification ×75.)
Courtesy of Carolina Biological Supply Company, Burlington, N.C.

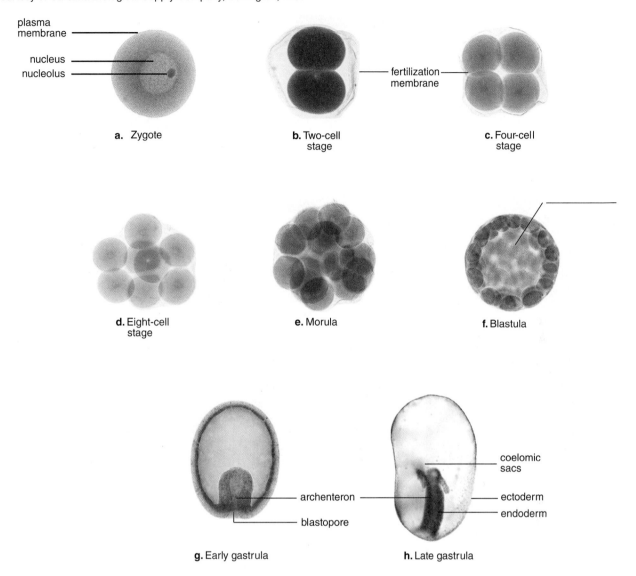

plasma
membrane

nucleus
nucleolus

a. Zygote

fertilization
membrane

b. Two-cell
stage

c. Four-cell
stage

d. Eight-cell
stage

e. Morula

f. Blastula

archenteron

blastopore

g. Early gastrula

coelomic
sacs

ectoderm

endoderm

h. Late gastrula

Frog Development

Like sea star embryos, frog embryos develop in water, but the process takes a bit longer. Frog eggs contain more yolk than sea star eggs.

Observation: Preserved Frog Embryos

Examine preserved frog embryos at various stages of development (Fig. 32.2) in a petri dish, or use models, and identify the following stages:

1. **Fertilized egg:** The eggs are partially pigmented. The black side contains very little yolk (nutrient material) and is called the **animal pole.** The unpigmented, yolky side is called the **vegetal pole.**

   Did the sea star unfertilized egg have an animal pole and a vegetal pole? _____ Explain. _____

   _____

ⅰⅱe presence of yolk alters the appearance of the usual stages somewhat. (**a:** Magnification ×25; **i:** Magnification ×22.5.)
Courtesy of Carolina Biological Supply, Burlington, N.C.

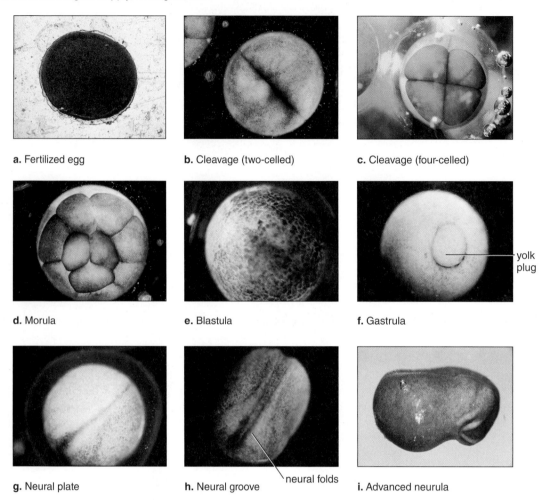

a. Fertilized egg   **b.** Cleavage (two-celled)   **c.** Cleavage (four-celled)

d. Morula   **e.** Blastula   **f.** Gastrula

g. Neural plate   **h.** Neural groove   neural folds   **i.** Advanced neurula

2. **Morula:** Cleavage begins at the animal pole (Fig. 32.3*a*). The first two cell divisions are polar. The third division is horizontal or equatorial between the poles. Cleavage continues until there is a morula. Which pole, animal or vegetal, contains the largest cells?_____ Explain. _____

_____

The yolk-laden cells of the vegetal pole are slower to divide than those of the animal pole.

3. **Blastula:** The blastula is a hollow ball of cells. However, the blastocoel (fluid-filled cavity) is found only at the animal pole. The yolk-laden cells of the vegetal pole do not divide rapidly and, therefore, do not help in the formation of the blastocoel.

Compare the frog blastula to that of the sea star. _____

_____

_____

4. **Early gastrula:** Gastrulation is recognized by the presence of a crescentic slit. This is the location of the blastopore, where cells are invaginating (Fig. 32.3*b*). The blastopore later takes on a circular shape, but is plugged by yolk cells that do not invaginate rapidly. In the frog, invagination is accomplished primarily by yolkless cells from the animal pole.

5. **Late gastrula:** The moderate amount of yolk also influences the formation of the mesoderm. This germ layer develops by invagination of cells at the lateral and ventral lips of the blastopore.

Compare formation of the mesoderm in the frog to that in the sea star. _____

_____

6. **Neurula:** During neurulation in the frog, two folds of ectoderm grow upward as the neural folds with a groove between them (Fig. 32.3c). The flat layer of ectoderm between them is the **neural plate.** The tube resulting from closure of the folds is the **neural tube,** which will become the spinal (nerve) cord and brain. An examination of the neurula in cross section shows that the nervous system develops directly above the **notochord,** a structure that arises from invaginated cells in the middorsal region. Draw a series of sketches that shows how the neural tube develops.

What has to happen to ectoderm in order for a neural tube to form? _____

_____

Figure 32.3   Drawings of frog developmental stages.
Compare these series of drawings with the photographs in Figure 32.2.

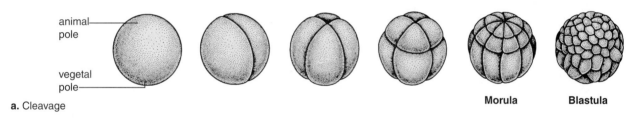

a. Cleavage

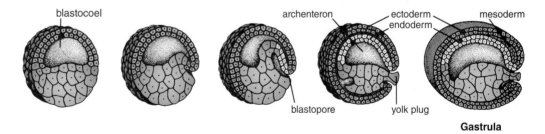

b. Gastrulation

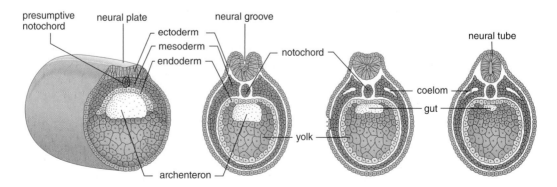

c. Neurulation

2 Germ Layers

As mentioned previously, early development proceeds to the formation of three germ layers. In Table 32.1, list the three germ layers and the major organs that develop from each in the frog.

Table 32.1 Germ Layer Organization	
Germ Layers	**Organs/Systems Associated with Germ Layer**
1.	
2.	
3.	

Induction

The notochord is said to induce the formation of the nervous system.

1. A chick is a chordate. What is the function of the notochord in a chick? _____
_____

Experiments have shown that, if contact with notochord tissue is prevented, no neural plate is formed. Even more dramatic are experiments in which presumptive (soon-to-be) notochord is transplanted under an area of ectoderm not in the dorsal midline. This ectoderm then is induced to differentiate into neural plate tissue, something it would not normally do.

2. For each of these scenarios, tell whether a neural tube will develop:

   a. Notochord is removed. Does ectoderm in this location become a neural tube? _____

   b. Ectoderm above notochord is replaced with belly ectoderm. Does belly ectoderm become a neutral tube? _____

   c. Develop a hypothesis to explain these suggested outcomes. _____
   _____

Induction is believed to be one means by which development is usually orderly. The part of the embryo that induces the formation of an adjacent organ is said to be an **organizer** and is believed to carry out its function by releasing one or more chemical substances.

32.3 Chick Development

Unlike sea stars and frogs, chicks develop on land, and there is no larval stage. Therefore, chick development is markedly different from that of sea stars and frogs because of three features: (1) The embryo is enclosed by extraembryonic membranes, (2) there is a large amount of yolk to sustain development, and (3) there is a hard outer shell.

Extraembryonic Membranes

The embryos of land vertebrates (reptiles, birds, and mammals) are surrounded and covered by membranes that do not become a part of the animal. These **extraembryonic membranes** are the **chorion, amnion, yolk sac,** and **allantois** (Fig. 32.4).

The chorion is the outermost membrane, and in chicks, it lies just below the porous shell, where it functions in gas exchange. In mammals, such as humans, the chorion forms the fetal portion of the placenta through which nutrients, gases, and wastes are exchanged with the blood of the mother.

The amnion is a delicate membrane containing amniotic fluid that bathes the embryo. The embryo, suspended in this fluid medium, is thus protected from drying out and from mechanical shocks.

The yolk sac surrounds the yolk in bird and reptile eggs. In mammals, it is the first site of blood cell formation.

The allantois serves as a storage area for metabolic waste products in reptiles, including birds. In mammals, the allantoic blood vessels are a part of the umbilical cord.

Complete Table 32.2.

Table 32.2 Functions of Extraembryonic Membranes in Chick and Human

Membrane	Chick	Human
Amnion		
Allantois		
Chorion		
Yolk sac		

Figure 32.4 Extraembryonic membranes.

The chick and a human have the same extraembryonic membranes, but except for the amnion, they have different functions.

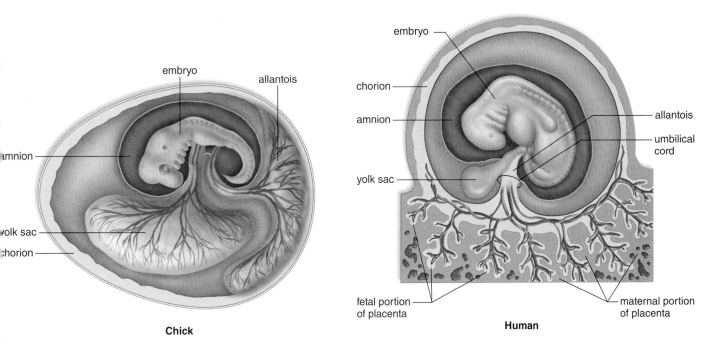

Chick

Human

1. Carefully crack an unfertilized chick egg into a 5-in finger bowl.
2. Examine the contents of the raw egg (Fig. 32.5). Some of the egg white **(albumen)** is denser and thicker than the remainder. These thicker masses are the **chalazae,** two spirally wound strands of albumen that result from the passage of the egg through the oviduct. Drag a probe through the albumen. The presence of albumen, a protein, accounts for the viscous nature of uncooked egg white. Egg white does not participate in the development of a chick. What could be its function?

_____

3. On the top of the yolk mass, note the germinal vesicle, a small, white spot containing the clear cytoplasm of the egg cell and the egg nucleus. Only the yolk with its germinal vesicle is the ovum. The various layers of albumen, including the chalazae, the shell membranes, and the shell, are all secreted by the oviduct as the ovum travels through it. Most of a chick's egg is yolk. Why does a

chick have need of a large amount of yolk? _____

_____

4. Why would you expect a chick's egg to be porous (permeable) instead of impervious

(impermeable)? _____

_____

Figure 32.5 Unfertilized chick egg.
Chalazae are two spirally wound strands of albumen.

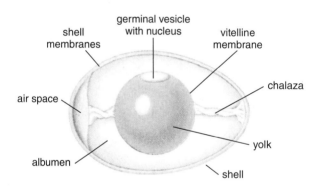

Observation: Twenty-Four-Hour Chick Embryo

1. Follow the standard procedure (see page 470) for selecting and opening an egg containing a 24-hour chick embryo.
2. Neurulation is occurring anteriorly, and invagination is occurring posteriorly. Refer to Figure 32.6, and identify the following:
   a. **Embryo,** which lies atop the yolk.
   b. **Head fold,** which is the beginning of the embryo.

    c. **Nervous system,** which is in the process of developing and has a **head fold** at the anterior end and **neural folds** toward the posterior end, because the neural tube is still in the process of forming.

    d. **Primitive streak,** an elongated mass of cells. In the chick, the germ layers develop as flat sheets of cells. The mesodermal cells invaginate as the germ layers form and give rise to the early organs.

    e. **Somites,** blocks of developing muscle tissue that differentiate from mesoderm.

3. What two organ systems develop first in a chick? _____

_____

Would you expect a human to also have a stage that looks like Figure 32.6? _____

Why or why not? _____

Figure 32.6   Dorsal view of 24-hour chick embryo.
The head fold is the amnion that will eventually envelop the entire embryo.
Courtesy of Carolina Biological Supply, Burlington, N.C.

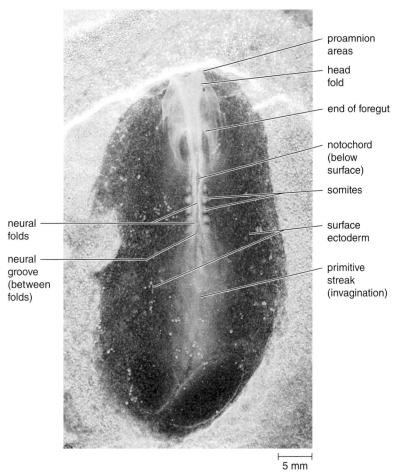

5 mm

a. Photograph

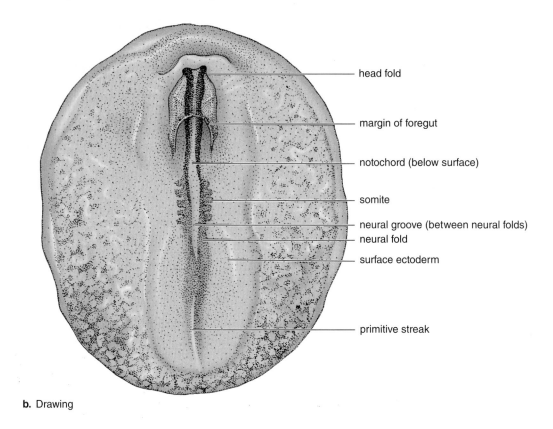

head fold

margin of foregut

notochord (below surface)

somite

neural groove (between neural folds)
neural fold

surface ectoderm

primitive streak

b. Drawing

Observing Live Chick Embryos

Use the following procedure for selecting and opening the eggs of live chick embryos:

1. Choose an egg of the proper age to remove from the incubator, and put a penciled X on the uppermost side. The embryo is just below the shell.
2. Add warmed chicken Ringer solution to a finger bowl until the bowl is about half full. (Chicken Ringer solution is an isotonic salt solution for chick tissue that maintains the living state.) The chicken Ringer solution should not cover the yolk of the egg.
3. On the edge of the dish, gently crack the egg on the side opposite the X.
4. With your thumbs placed over the X, hold the egg in the chicken Ringer solution while you pry it open from below and allow its contents to enter the solution. If you open the egg too slowly or too quickly, the shell may damage the delicate membranes surrounding the embryo.

Observation: Forty-Eight-Hour Chick Embryo

1. Follow the standard procedure (see page 470) for selecting and opening an egg containing a 48-hour chick embryo.
2. The embryo has started to twist so that the head region is lying on its side. Refer to Figure 32.7, and identify the following:
   a. **Shape of the embryo,** which has started to bend. The head is now almost touching the heart.
   b. **Heart,** contracting and circulating blood. Can you make out a ventricle, an atrium, and the aortic arches in the region below the head? Later, only one aortic arch will remain.
   c. **Vitelline arteries** and **veins,** which extend over the yolk. The vitelline veins carry nutrients from the yolk to the embryo.

Figure 32.7   Forty-eight-hour chick embryo.
The most prominent organs are labeled.
Courtesy of Carolina Biological Supply, Burlington, N.C.

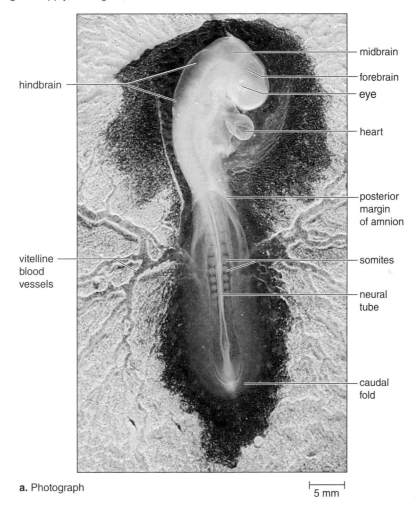

a. Photograph

5 mm

d. **Brain** divided into three regions: forebrain, midbrain, and hindbrain.
e. **Eye,** which has a developing lens.
f. **Margin (edge) of the amnion,** which can be seen above the vitelline arteries.
g. **Somites,** which now number 24 pairs.
h. **Caudal fold** of the amnion. The embryo will be completely enveloped when the head fold and caudal fold meet the margin of the amnion.

Figure 32.7 Forty-eight-hour chick embryo—*continued.*
The most prominent organs are labeled.

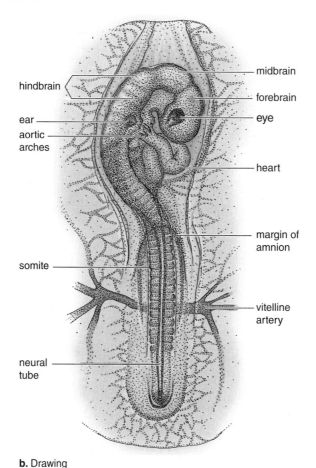

b. Drawing

Observation: Seventy-Two-Hour Chick Embryo

1. Follow the standard procedure (see page 470) for selecting and opening an egg. For this Observation, the egg should contain a 72-hour chick embryo.
2. The chick is now almost completely on its side, and the brain is even more flexed than before. Refer to Figure 32.8, and identify the following:
   a. **Brain,** which has a Ↄ shape and is very prominent.
   b. **Eye,** which has a distinctive lens.

Figure 32.8 Seventy-two-hour chick embryo.
Sense organs and limb buds have appeared.
Courtesy of Carolina Biological Supply, Burlington, N.C.

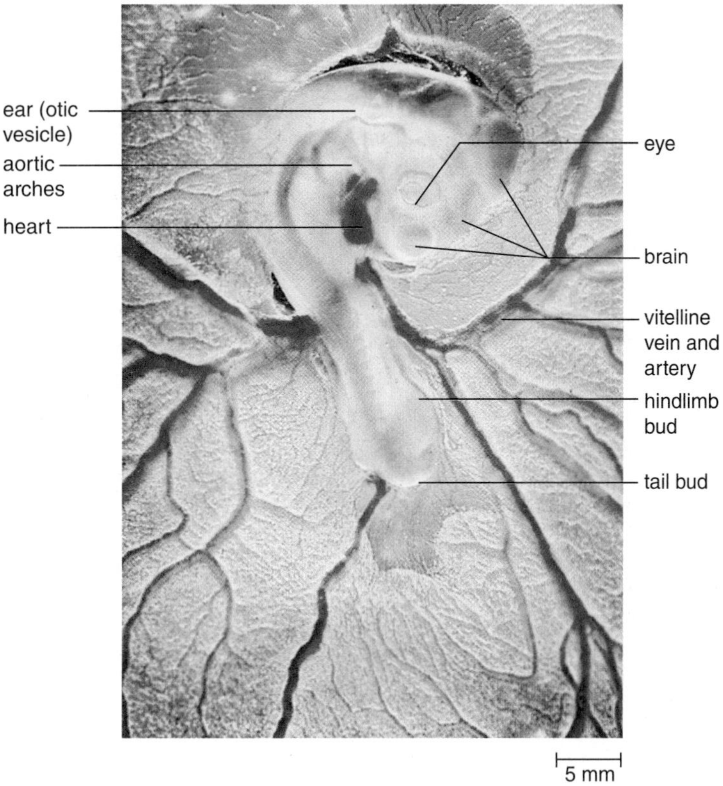

a. Photograph

c. **Ear (otic vesicle),** enlarged and more noticeable than before.
d. **Heart,** which has a distinctive ventricle and atrium and is actively pumping blood. There are pharyngeal pouches between the aortic arches.
e. **Vitelline blood vessels,** which extend over the yolk.
f. **Limb buds,** which become the wings and hindlimbs. The posterior limb buds are easier to see at this point.
g. **Somites,** which now number 36 pairs.
h. **Tail bud,** which marks the end of the embryo.

Figure 32.8   Seventy-two-hour chick embryo—*continued.*
Sense organs and limb buds have appeared.

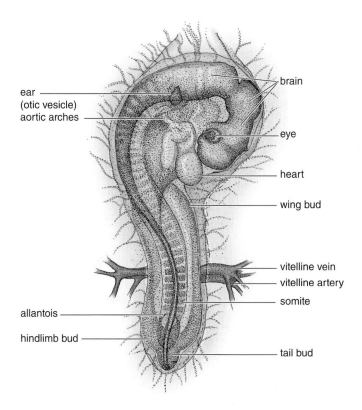

b. Drawing

As a chick embryo continues to grow, various organs differentiate further (Fig. 32.9). The neural tube closes along the entire length of the body and is now called the spinal cord. The allantois, an extraembryonic membrane, is seen as a sac extending from the ventral surface of the hindgut near the tail bud. The digestive system forms specialized regions, and there is both a mouth and an anus. The yolk sac, the extraembryonic membrane that encloses the yolk, is attached to the ventral wall, but when the yolk is used up, the ventral wall closes.

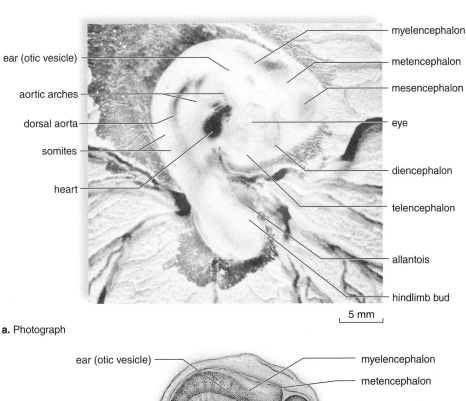

5 mm

a. Photograph

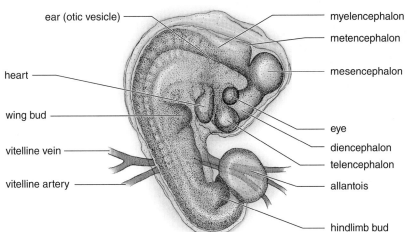

b. Drawing

Key:	
Brain region	**Becomes:**
telencephalon	cerebrum
diencephalon	thalamus, hypothalamus, posterior pituitary gland, pineal gland
mesencephalon	midbrain
metencephalon	cerebellum, pons
myelencephalon	medulla oblongata

Figure 32.9 Ninety-six-hour chick embryo.
Brain regions listed in the key can now be seen.
Courtesy of Carolina Biological Supply, Burlington, N.C.

4 Human Development

As illustrated in Figure 32.10, the early stages of human development are quite similar to those of the chick. Differences become marked only as development proceeds.

Figure 32.10 Comparison of mammalian embryos.
Successive stages in the embryonic development of chick and human. Early stages are similar; differences become apparent only during later stages.

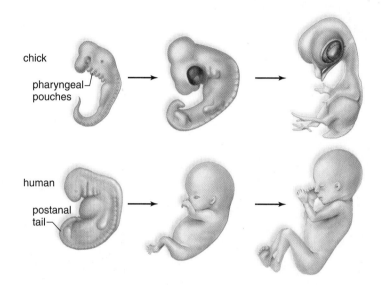

Stages of Human Development

Human development is divided into embryonic development and fetal development. During **embryonic development** (first two months), the germ layers and extraembryonic membranes develop; then the internal organs form (Fig. 32.11). At the end of this period, the embryo has a human appearance.

1. Review frog development on pages 463–65. What are the stages you would expect to see during early embryonic development in humans? _____

_____

2. Review chick development and decide what two organs will most likely be first to make their appearance during human development.

_____

3. Figure 32.11 shows in particular the development of the extraembryonic membranes in humans. Circle the labels for the membranes in Figure 32.11. Review the function of these membranes in humans (page 469).

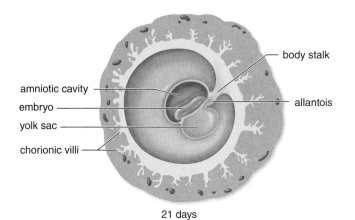

21 days

Figure 32.11 Human development.
Changes occurring during the third to the fifth week include the development of the extraembryonic membranes and the umbilical cord.

- body stalk
- amniotic cavity
- embryo
- yolk sac
- chorionic villi
- allantois

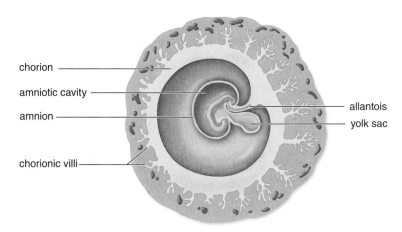

25 days

- chorion
- amniotic cavity
- amnion
- chorionic villi
- allantois
- yolk sac

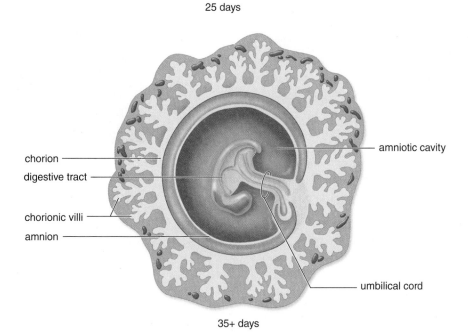

35+ days

- chorion
- digestive tract
- chorionic villi
- amnion
- amniotic cavity
- umbilical cord

During **fetal development** (last seven months), the skeleton becomes ossified (bony), reproductive organs form, arms and legs fully develop, and the fetus enlarges in size and gains weight (Fig. 32.12).

Figure 32.12 Human development.

Changes occurring from the fifth week to the eighth month.

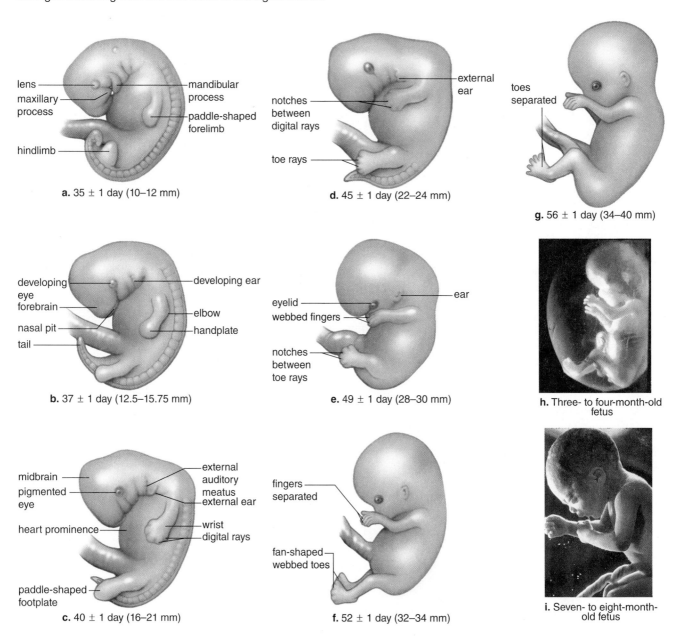

a. 35 ± 1 day (10–12 mm)

lens
maxillary process
hindlimb
mandibular process
paddle-shaped forelimb

b. 37 ± 1 day (12.5–15.75 mm)

developing eye
forebrain
nasal pit
tail
developing ear
elbow
handplate

c. 40 ± 1 day (16–21 mm)

midbrain
pigmented eye
heart prominence
paddle-shaped footplate
external auditory meatus
external ear
wrist
digital rays

d. 45 ± 1 day (22–24 mm)

notches between digital rays
toe rays
external ear

e. 49 ± 1 day (28–30 mm)

eyelid
webbed fingers
notches between toe rays
ear

f. 52 ± 1 day (32–34 mm)

fingers separated
fan-shaped webbed toes

g. 56 ± 1 day (34–40 mm)

toes separated

h. Three- to four-month-old fetus

i. Seven- to eight-month-old fetus

Laboratory Review 32

Complete this table by placing a ✓ in the appropriate square if the feature pertains to the organism.

Comparison of Embryonic Features of a Developing Sea Star, Frog, and Chick			
Feature	Sea Star	Frog	Chick
1.a. Has the most yolk.			
b. Blastula is a circular cavity.			
2.a. Germ layers are present.			
b. Primitive streak is present.			
3.a. Notochord is present.			
b. Waste is deposited in water.			

4. Describe how an embryo becomes a morula, blastula, and gastrula. _____

_____

5. What factor causes a frog's morula, blastula, and gastrula to appear differently from that of a sea star?

_____

6. Describe how induction may control development.

_____

Name the four extraembryonic membranes, and state the function of each in birds and mammals.

| **Extraembryonic Membrane** | **Function in Birds** | **Function in Mammals** |

7. a. _____

b. _____

8. a. _____

b. _____

9. Name three features that are quite noticeable in a 48-hour chick embryo. _____

_____

10. List the two stages of human development, and state a reason for dividing human development into

these two stages. _____

_____

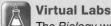

Virtual Labs

The *Biology* website **www.mhhe.com/maderbiology11** has three virtual laboratories of particular interest that pertain to the study of ecology:

In **Model Ecosystems**, place organisms at the correct trophic level in five ecological pyramids and then click on the number of organisms and amount of energy per level. It is not necessary to calculate energy conversion efficiency ratios; instead note the 10% rule. Each higher trophic level has 10% of the energy of the lower level.

In **Ecosystems, Organisms, and Trophic Levels,** view organisms found in five ecosystems. Click on any organism and answer questions about its place in the food web, how it acquires energy, and its trophic level in that ecosystem.

In **Population Biology,** conduct a laboratory experiment with paramecia in which the results support the competitive exclusion principle—no two species can indefinitely occupy the same niche.

LABORATORY

33

Sampling Ecosystems

Learning Outcomes

Introduction
- Define ecology and an ecosystem. Identify the abiotic and biotic components of an ecosystem.   481–82

33.1 Terrestrial Ecosystems
- Define and give examples of producers in terrestrial ecosystems.   482–85
- Define and give examples of consumers in terrestrial ecosystems.   482–85
- Define and give examples of decomposers in terrestrial ecosystems.   482–85

33.2 Aquatic Ecosystems
- Define aquatic ecosystem.   486
- Give examples of producers, consumers, and decomposers in aquatic ecosystems.   486–89

Introduction

Ecology is the study of interactions between organisms and their physical environment within an **ecosystem.** The **abiotic** (nonliving) components of an ecosystem include soil, water, light, inorganic nutrients, and weather variables. The **biotic** (living) components can be organized according to the **trophic** (feeding) level, in which each organism belongs. This includes producers, consumers, and decomposers.

Producers are autotrophic organisms with the ability to carry on photosynthesis and to make food for themselves (and indirectly for the other populations as well). In terrestrial ecosystems, the predominant producers are green plants, while in freshwater and saltwater ecosystems, the dominant producers are various species of algae.

Consumers are heterotrophic organisms that eat available food. Three types of consumers can be identified, according to their food source:

1. **Herbivores** feed directly on green plants and are termed *primary consumers.* A caterpillar feeding on a leaf is a herbivore.

2. **Carnivores** feed on other animals and are therefore *secondary* or *tertiary consumers.* A blue heron feeding on a fish is a carnivore.

3. **Omnivores** feed on both plants and animals. A human who eats both leafy green vegetables and beef is an omnivore.

Decomposers and **detritivores** are organisms of decomposition, such as bacteria, fungi, and millipedes, that break down **detritus** (nonliving organic matter) to inorganic matter, which can be used again by producers. In this way, the same chemical elements are constantly recycled in an ecosystem.

The trophic structure (feeding relationships) of an ecosystem is represented in the form of a pyramid, such as the one shown in Figure 33.1. **Biomass** is the weight of all the organisms at each trophic level.

Figure 33.1 Ecological pyramid.

Organisms at lower trophic levels are higher in number and have greater biomass than organisms at higher trophic levels. The examples of organisms on the left form an aquatic food chain in which herons eat fish, which eat zooplankton, which eat phytoplankton. The examples of organisms on the right form a terrestrial food chain in which owls eat shrews, which eat beetles, which eat plants.

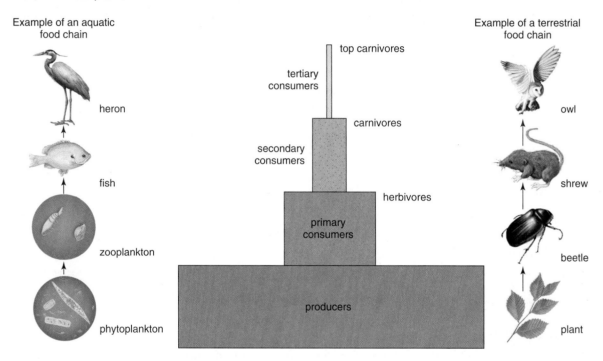

33.1   Terrestrial Ecosystems

Examining an ecosystem in a scientific manner requires concentrating on a representative portion of that ecosystem and recording as much information about it as possible. Representative areas or plots should be selected randomly. For example, random sampling of a terrestrial ecosystem often involves tossing a meterstick gently into the air in the general area to be sampled and then sampling the square meter where the stick lands. A **terrestrial ecosystem** sampling should include samples from the air above and from the various levels of plant materials growing on and in the ground beneath the selected plot.

Study of Terrestrial Sampling Site

The objective of the next Experimental Procedure is to characterize the abiotic and biotic components of a terrestrial ecosystem and to determine how those factors affect the trophic structure of an ecosystem. Although terrestrial ecosystems include deciduous forest, prairie, scrubland, and desert, a weedy field, if dominated by annual and perennial herbaceous plants up to a meter or so in height, is also a good site choice.

1. Gather all necessary equipment, such as metersticks, jars and bags (brown paper and plastic) with labels for collecting specimens, nets, pH paper, thermometers, and other testing equipment, to take with you to the site. Do not forget data-recording materials. Number your collection jars and bags so that you can use these numbers when recording data later (see Table 33.2).
2. When you arrive at the site, take several minutes to observe the general area. Describe what you observe, including weather conditions.

_____

_____

3. Choose two sampling locations at the site. The two locations should differ in significant features (e.g., northern or southern exposure, high- or low-slope position, time since last disturbance, native versus exotic species composition). Formulate hypotheses about differences between the two sampling locations in regard to the following variables:

   a. Air temperature, humidity, and light intensity _____

   _____

   b. Soil temperature, moisture, and pH _____

   _____

   c. Producer biomass and diversity (variety of producers and how many of each type) _____

   _____

   d. Consumer biomass and diversity _____

   _____

Experimental Procedure: Terrestrial Ecosystems

Your instructor will organize class members into teams, and each team will be assigned specific tasks at each of the two chosen sampling locations. At each sampling location (e.g., shaded versus not shaded), you will randomly choose three 1 m square plots. Within each of these plots, randomly choose one 0.1 m square area (subplot) for all of the following abiotic variables and record your measurements in Table 33.1. For each of the biotic variables, randomly choose separate one 0.1 m square subplots to sample and complete Table 33.2. Keep outside the 1 m square plots when sampling to minimize disturbance.

Abiotic Components
1. Measure air temperature, relative humidity, and light intensity at 0.5 m above the surface of the ground. Calculate the average for all three subplots (replicates).
2. Measure soil temperature, soil moisture, and soil pH (and any standing water) at 0.2 m below the surface using a soil corer. Calculate the average for all three replicates.

Biotic Components—Plants
In each subplot:
1. Count the total number of live plants.
2. Harvest each entire living plant and wash any soil from the roots. Place plant material in labeled brown paper bags.

Biotic Components—Animals
In each subplot:
1. Sweep with a net as thoroughly as possible about three to five times to capture different organisms on the vegetation in each plot. Empty the contents of the net into a labeled jar of alcohol for later sorting and identification.
2. Collect leaf litter samples in labeled plastic bags for Berlese (or Tullgren) funnel analysis.

Table 33.1 Abiotic Components of a Terrestrial Ecosystem

	Abiotic Factor	Location 1				Location 2			
		a	b	c	avg	a	b	c	avg
Air	Temperature								
	Humidity								
	Light intensity								
Soil	Temperature								
	Moisture								
	pH								

Experimental Procedure: Terrestrial Ecosystem

Laboratory Work

1. Examine collected plants and animals using a stereomicroscope, or a compound microscope, if appropriate. Group organisms into different types based on morphological features (morphotypes) and classify them as producers, consumers, or decomposers. Further classify invertebrates into herbivores, detritivores, and carnivores. Complete Table 33.2.
2. Determine the dry biomass (weight) of the plant material, or wet biomass if a drying oven is not available. Dry biomass, although more time consuming to measure, is preferable when comparing biomasses among sites and between trophic levels.
3. Determine the dry biomass of the animal material (or wet weight if plant wet weight was used).
4. Construct graphs comparing the abiotic conditions of each terrestrial sampling location.
5. Construct a pyramid for biomass, morphotype, and the total number of producers, herbivores (including detritivores), and carnivores.
6. Select any three producers and any three consumers from the organisms collected, and explain how each has adapted to its terrestrial environment.

   Producer 1 _____

   Producer 2 _____

   Producer 3 _____

   Consumer 1 _____

   Consumer 2 _____

   Consumer 3 _____

7. Return all living creatures and litter samples to their respective collection sites, as explained by your instructor. If any organisms were preserved, ask your instructor what to do with them.

Table 33.2 Biotic Components of a Terrestrial Ecosystem

	Biotic Factor		Location 1				Location 2			
			a	b	c	avg	a	b	c	avg
Plants	Producers	Total number								
		Number of morphotypes								
		Biomass								
Animals on vegetation	Herbivores	Total number								
		Number of morphotypes								
		Biomass								
	Carnivores	Total number								
		Number of morphotypes								
		Biomass								
Animals in litter	Herbivores	Total number								
		Number of morphotypes								
		Biomass								
	Carnivores	Total number								
		Number of morphotypes								
		Biomass								
	Detritivores	Total number								
		Number of morphotypes								
		Biomass								

Conclusions: Terrestrial Ecosystems

Compare the results of this terrestrial ecosystem analysis with the hypotheses you formulated (see page 483).

- Air temperature, humidity, and light intensity _____

- Soil temperature, moisture, and pH

- Producer biomass and diversity

- Consumer biomass and diversity

33.2 Aquatic Ecosystems

An **aquatic ecosystem** sampling should include samples from the air above the water column, the column of water itself, and the soil beneath the water column.

Study of Aquatic Sampling Site

The objective of the next Experimental Procedure is to characterize the abiotic and biotic components of an aquatic ecosystem and to determine how those factors affect the trophic structure of an ecosystem. Aquatic ecosystems consist of freshwater ecosystems (e.g., lakes, ponds, rivers, and streams) and marine ecosystems (e.g., oceans). A good site for this study is a large pond, small lake, or reservoir (with a shallow margin having rooted aquatic plants and a deeper zone with water 1–2 m deep).

1. Gather all necessary equipment, such as metersticks, collection jars and bags with labels, nets, pH paper, thermometers, and other testing equipment, to take with you to the site. Do not forget data-recording materials. Number your collection jars and bags so that you can use these numbers when recording data in Table 33.3.

2. When you arrive at the site, take several minutes to observe the general area. Describe what you observe, including weather conditions.

   _____

   _____

3. Choose two sampling locations that differ in significant features (e.g., sheltered by trees versus unsheltered, near stream inflow versus far from stream inflow). Plan to sample conditions near shore (shallow-water zone) and in deeper water away from shore (or only one or the other, if logistics are limiting).

   Formulate hypotheses about differences between the two sampling locations in:

   a. Temperature, humidity, and light intensity above the surface _____

   _____

   b. Temperature, dissolved oxygen, pH, and visibility below the surface _____

   _____

   c. Producer biomass and diversity (variety of producers and how many of each type) _____

   _____

   d. Consumer biomass and diversity _____

   _____

Table 33.3 Abiotic Components of an Aquatic Ecosystem

	Abiotic Factor	Location 1				Location 2			
		a	b	c	avg	a	b	c	avg
Air	Temperature								
	Humidity								
	Light intensity								
Water	Temperature								
	Dissolved oxygen								
	pH								
	Visibility								

Experimental Procedure: Aquatic Ecosystem

Your instructor will organize class members into teams, and each team will be assigned specific tasks at each of the two sampling locations. At each sampling location (e.g., shaded versus not shaded), you will randomly choose three 1 m square plots. Within each of these plots, randomly choose one 0.1 m square area (subplot) for all of the following abiotic variables and record your measurements in Table 33.3. For each of the biotic variables, randomly choose separate one 0.1 m square subplots to sample and complete Table 33.4. Keep outside the 1 m square plots when sampling to minimize disturbance.

Abiotic Components

1. Measure air temperature, relative humidity, and light intensity at 0.5 m above the surface of the water.
2. Measure water temperature, dissolved oxygen, pH, and visibility at 0.5 m below the surface.

Biotic Components: Plants

In each subplot:

1. Count the total number of live plants in each subplot.
2. Harvest each entire plant and wash any sediment from the roots. Place plant material in labeled brown paper bags.

Biotic Component: Plankton

Lower a plankton net with attached collecting bottle into the water to a depth of 0.5 m. Slowly raise the net vertically two to three times to collect plankton in the water column. Pour the sample into a labeled collecting jar and preserve with Lugol's solution for later sorting and identification.

Biotic Component: Animals

Choose an area of the plot not disturbed by other sampling but near emergent vegetation. Sweep through the water inside the plot three to five times. Be sure to sample around the vegetation and bump the substrate several times to dislodge benthos from the sediment. Empty the contents of the net into a sieve placed over a bucket of water. Examine, wash, and collect any macroinvertebrates and place them in the bucket. Transfer collected organisms into a labeled jar of alcohol for later sorting and identification.

Laboratory Work

1. Examine collected plants, plankton, and animals using a stereomicroscope, or a compound microscope, if appropriate. Group organisms into different types based on morphological features (morphotypes) and classify them as producers, consumers, or decomposers. Further classify plankton and invertebrates into herbivores, detritivores, and carnivores. Complete Table 33.4.
2. Determine the dry biomass (weight) of the plant material, or wet biomass if a drying oven is not available.

Table 33.4	Biotic Components of an Aquatic Ecosystem		Location 1				Location 2			
	Biotic Factor									
			a	b	c	avg	a	b	c	avg
Plants	Producers	Total number								
		Number of morphotypes								
		Biomass								
Plankton	Producers	Total number								
		Number of morphotypes								
		Biomass								
	Herbivores	Total number								
		Number of morphotypes								
		Biomass								
	Carnivores	Total number								
		Number of morphotypes								
		Biomass								
Animals	Herbivores	Total number								
		Number of morphotypes								
		Biomass								
	Carnivores	Total number								
		Number of morphotypes								
		Biomass								
	Detritivores	Total number								
		Number of morphotypes								
		Biomass								

3. Determine the dry biomass of the plankton and animal material (or wet weights if plant wet weight was used).
4. Construct graphs comparing the abiotic conditions of each aquatic sampling location.
5. Construct a pyramid for biomass, morphotype, and the total number of individuals of producers, herbivores (including detritivores), and carnivores.
6. Select any three producers and any three consumers from the organisms collected, and explain how each has adapted to its aquatic environment.

Producer 1 _____

Producer 2 _____

Producer 3 _____

Consumer 1 _____

Consumer 2 _____

Consumer 3 _____

7. Return all living organisms and water samples to their respective collection sites, as explained by your instructor. If any organisms were preserved, ask your instructor what to do with them. Place any exposed petri dishes in a designated area for incubation until the next laboratory.

Conclusions: Aquatic Ecosystems

Compare the results of this aquatic ecosystem analysis with the hypotheses you formulated (see page 486).

* Temperature, humidity, and light intensity above the surface _____

_____

* Temperature, dissolved oxygen, pH, and visibility below the surface _____

_____

* Producer biomass and diversity _____

_____

* Consumer biomass and diversity _____

_____

1. What is an ecosystem? _____

_____

2. Define and give several examples of abiotic components of an ecosystem.

_____

_____

3. Define and give several examples of biotic components of an ecosystem.

_____

_____

4. In a forest ecosystem, what type of organisms are the predominant producers?

_____

5. In an aquatic ecosystem, what type of organisms are the predominant producers?

_____

6. What is the role of producers in an ecosystem? _____

7. Define a consumer, and give examples of consumers in terrestrial and aquatic ecosystems.

_____

_____

8. In an ecological pyramid, describe the final consumers in terms of their diet.

_____

9. In an ecological pyramid, where would you place parasites of humans such as a tapeworm?

_____

10. Regardless of the ecosystem, what types of organisms would be the predominant decomposers?

_____

Biology **Website**

Enhance your study of the text and laboratory manual with study tools, practice tests, and virtual labs. Also ask your instructor about the resources available through ConnectPlus, including the media-rich eBook, interactive learning tools, and animations.

www.mhhe.com/maderbiology11

McGraw-Hill Access Science Website

An Encyclopedia of Science and Technology Online which provides more information including videos that can enhance the laboratory experience.

www.accessscience.com

34

Effects of Pollution on Ecosystems

Learning Outcomes

34.1 Studying the Effects of Pollutants
- Predict the effect of oxygen deprivation on species composition and diversity of ecosystems.   491–95
- Predict the effect of acid deposition on the species composition and diversity of ecosystems.   491–95
- Predict the effect of enrichment on species composition and diversity of ecosysytems.   491–95

34.2 Studying the Effects of Cultural Eutrophication
- Predict the effect of cultural eutrophication on food chains so that pollution results.   495–96

Introduction

This laboratory will consider three causes of aquatic pollution: thermal pollution, acid pollution, and cultural eutrophication. **Thermal pollution** occurs when water temperature rises above normal. The enzymes of cells work best at a temperature of 37°C, and most cells require a substantial supply of oxygen. As water temperature rises, the amount of oxygen dissolved in water decreases. Deforestation, soil erosion, and the burning of fossil fuels contribute to thermal pollution but the chief cause is use of water from a lake or the ocean as a coolant for the waste heat of a power plant.

When sulfur dioxide and nitrogen oxides enter the atmosphere, usually from the burning of fossil fuels, they are converted to acids, which return to Earth as **acid deposition** (acid rain or snow). Acid deposition kills aquatic invertebrates and also decomposers, threatening the entire ecosystem.

Figure 34.1   Cultural eutrophication.
Eutrophic lakes tend to have large populations of algae and rooted plants.

Cultural eutrophication, or overenrichment, is due to runoff from agricultural fields, wastewater from sewage treatment plants, or even excess detergents. These sources of excess nutrients cause an algal bloom seen as a green scum on a lake (Fig. 34.1). When algae overgrow and die, decomposition robs the lake of oxygen, causing a fish die-off.

34.1 Studying the Effects of Pollutants

We are going to study the effects of pollution by observing its effects on hay infusion organisms, on seed germination, and on an animal called *Gammarus*.

Study of Hay Infusion Culture

A hay infusion culture (hay soaked in water) contains various microscopic organisms, but we will be concentrating on how the pollutants in our study affect the protozoan populations in the culture. We will consider both of these aspects:

species composition: number of different types of protozoans, and
species diversity: relative abundance of each type protozoan.

> #### *Experimental Procedure: Effect of Pollutants on a Hay Infusion Culture*

During this experimental procedure you will examine, by preparing a wet mount, hay infusion cultures that have been treated in the following manner.

1. Control culture. This culture simulates the species composition and diversity of an untreated culture. Prepare a wet mount and answer the following questions:
    With the assistance of Figure 34.2 and any guides available in the laboratory, identify as many different types of protozoans as possible in the hay infusion culture. State whether species composition is high, medium, or low? _____ Record your estimation in the second column of Table 34.1. Are certain species more abundant than other species? _____ Do you judge species diversity to be high, medium, or low?_____ Record your estimation in the third column of Table 34.1.

2. Oxygen-deprived culture. Thermal pollution causes water to be oxygen deprived; therefore, when we study the effects of low oxygen on a hay infusion culture, we are studying an effect of thermal pollution. Prepare a wet mount of this culture and determine if there is a change in species composition and diversity. Again record the species composition and species diversity as high, medium, or low in Table 34.1.

Figure 34.2 Microorganisms in hay infusion cultures.

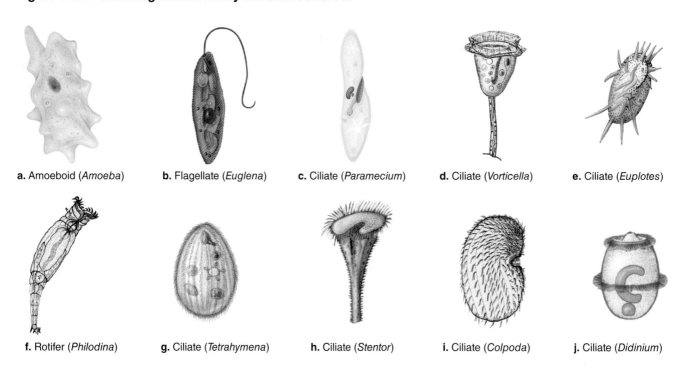

a. Amoeboid (*Amoeba*) **b.** Flagellate (*Euglena*) **c.** Ciliate (*Paramecium*) **d.** Ciliate (*Vorticella*) **e.** Ciliate (*Euplotes*)

f. Rotifer (*Philodina*) **g.** Ciliate (*Tetrahymena*) **h.** Ciliate (*Stentor*) **i.** Ciliate (*Colpoda*) **j.** Ciliate (*Didinium*)

Table 34.1 Effect of Pollution on a Hay Infusion Culture

Type of Culture	Species Composition (High, Medium, or Low)	Species Diversity (High, Medium, or Low)	Explanation
Control			
Oxygen-deprived			
Acidic			
Enriched			

3. Acidic culture. In this culture, the pH has been adjusted to 4 with sulfuric acid (H_2SO_4). This simulates the effect of acid rain on a hay infusion culture. Prepare a wet mount of this culture and determine if there is a change in species composition and diversity. Again record the species composition and species diversity as high, medium, or low in Table 34.1.
4. Enriched culture. More organic nutrients have been added to this culture. These nutrients will cause the algae population, which is food for most protozoans, to increase. In the short-term, their species composition should increase. Eventually, as the algae die off decomposition will rob the water of oxygen and the protozoans may start to die off. Prepare a wet mount of this culture and determine if there is a change in species composition and diversity. Again record the species composition and species diversity as high, medium, or low in Table 34.1.

Conclusions

- What could be a physiological reason for the adverse effects of oxygen deprivation on a hay infusion culture? If consistent with your results enter this explanation in the last column of Table 34.1.
- What could be a physiological reason for the adverse effects of a low pH on a hay infusion culture? If consistent with your results enter this explanation in the last column of Table 34.1.
- What could be an environmental reason for the adverse affects of an enriched culture? If consistent with your results enter this explanation in the last column of Table 34.1.

Effect of Acid Rain on Seed Germination

Seeds depend on favorable environmental conditions of temperature, light, and moisture to germinate, grow, and reproduce. Like any other biological process, germination requires enzymatic reactions that can be adversely affected by an unfavorable pH.

Experimental Procedure: Effect of Acid Rain on Seed Germination

In this experimental procedure we will test whether there is a negative correlation between acid concentration and germination. In other words, it is hypothesized that as acidity increases, the more likely seeds will _____.

Your instructor has placed 20 sunflower seeds in each of five containers with water of increasing acidity: 0% vinegar (tap water), 1% vinegar, 5% vinegar, 20% vinegar, and 100% vinegar.
1. Test and record the pH of solutions having the vinegar concentrations noted above. Record the pH of each solution in Table 34.2.
2. Count the number of germinated sunflower seeds in each container, and complete Table 34.2.

Table 34.2 Effect of Increasing Acidity on Germination of Sunflower Seeds

Concentration of Vinegar	pH	Number of Seeds that Germinated	Percent Germination
0%			
1%			
5%			
20%			
100%			

Conclusions: Effect of Increasing Acidity on Germination of Sunflower Seeds

- As you know, each enzyme has an optimum pH. Explain why acid rain is expected to inhibit metabolism, and therefore, seedling development. _____
- Do the data support or falsify your hypothesis? _____

Study of *Gammarus*

A small crustacean called *Gammarus* lives in ponds and streams (Fig. 34.3) where it feeds on debris, algae, or anything smaller than itself, such as some of the protozoans in Figure 34.1. In turn, fish like to feed on *Gammarus*.

Experimental Procedure: Gammarus

Control Culture

1. Add 25 mL of spring water to a container.
2. Measure the pH and record it here: _____.
   Add four *Gammarus* to the container.
3. Observe the behavior of *Gammarus* for 10–15 minutes, and then answer the following questions. Retain these *Gammarus* for subsequent exercises.

   a. Where do the *Gammarus* spend their time in the container? _____

   b. How do they spend their time? _____

   c. What percentage of their time is spent moving? _____

   d. Do they use all their legs in swimming? _____ _____

Figure 34.3   *Gammarus*.
Gammarus is a type of crustacean classified in a subphylum that also includes shrimp.

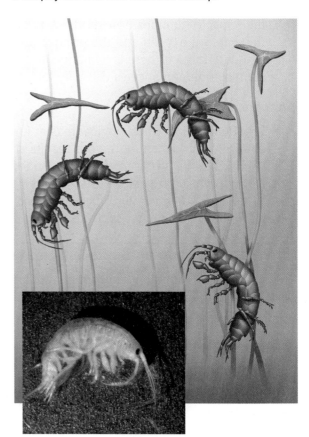

e. Which legs are used in jumping and climbing? _____

f. Do *Gammarus* avoid each other? _____

g. What do *Gammarus* do when they "bump" into each other? _____

_____

4. Create a food chain that shows who eats whom and includes algae, protozoans, *Gammarus*, fish and humans. _____

   a. Predict what would happen to this food chain if the water was

   oxygen deprived _____

   acidic _____

   enriched with inorganic nutrients (short term and long term) _____

   _____

   b. If so directed by your instructor, put a *Gammarus* in a beaker of spring water adjusted to pH 4. Or examine a sample on display. How does *Gammarus* react to acid conditions? _____

   _____

Conclusions: Studying the Effects of Pollutants

- Give an example to show that the hay infusion study pertains to real ecosystems. _____

  _____

- What are the potential consequences of acid rain on crops that reproduce by seeds? _____

  _____

  On the food chains of the ocean? _____

- How does the addition of nutrients affect species composition and species diversity of an ecosystem over time? _____

34.2 Studying the Effects of Cultural Eutrophication

Chlorella, the green alga used in this study, is considered to be representative of algae in bodies of fresh water. The protozoan *Daphnia* feeds on green algae such as *Chlorella* (Fig. 34.4). First, you will observe how *Daphnia* feeds, and then you will determine the extent to which *Daphnia* could keep cultural eutrophication from occurring in a hypothetical example. Keep in mind that this case study is an oversimplification of a generally complex problem.

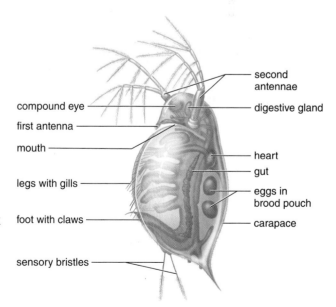

Figure 34.4   Anatomy of *Daphnia*.

Observation: Daphnia Feeding

1. Place a small pool of petroleum jelly in the center of a small petri dish.
2. Use a dropper to take a *Daphnia* from the stock culture, place it on its back (covered by water) in the petroleum jelly, and observe it under the stereomicroscope (see Fig. 34.4).
3. Note the clamlike carapace and the legs waving rapidly as the *Daphnia* filters the water.
4. Add a drop of *carmine solution,* and observe how the *Daphnia* filters the "food" from the water and passes it through the gut. The gut is more visible if you push the animal onto its side. In this position, you may also observe the heart beating in the region above the gut and just behind the head.
5. Allow the *Daphnia* to filter-feed for up to 30 minutes, and observe the progress of the carmine particles through the gut. Does the carmine travel completely through the gut in 30 minutes?

Experimental Procedure: Daphnia Feeding on Chlorella

This exercise requires the use of a spectrophotometer. Absorbance will be a measure of the algal population level; the greater the number of algal cells, the greater the absorbance. The higher the absorbance, the greater the amount of light absorbed and *not* passed through the solution.

1. Obtain two spectrophotometer tubes (cuvettes) and a Pasteur pipette.
2. Fill one of the cuvettes with distilled water, and use it to zero the spectrophotometer. Save this tube for step 6.
3. Use the Pasteur pipette to fill the second cuvette with *Chlorella.* Gently aspirate and expel the sample several times (without creating bubbles) to give a uniform dispersion of the algae.
4. Add ten hungry *Daphnia,* and following your instructor's directions, immediately measure the absorbance with the spectrophotometer. If a *Daphnia* swims through the beam of light, a strong deflection should occur; do not use any such higher readings—instead, use the lower figure for the absorbance. Record your reading in the first column of Table 34.3.
5. Remove the cuvette with the *Daphnia* to a safe place in a test tube rack. Allow the *Daphnia* to feed for 30 minutes.
6. Rezero the spectrophotometer with the distilled water cuvette.
7. Measure the absorbance of the experimental cuvette again. Record your data in the second column of Table 34.3, and explain your results in the third column.

Table 34.3 Spectrophotometer Data/*Daphnia* Feeding on *Chlorella*

Absorbance Before Feeding	Absorbance After Feeding	Explanation

The following problem will test your understanding of the value of a single species—in this case, *Daphnia.* Please realize that this is an oversimplification of a generally complex problem.

1. Assume that developers want to build condominium units on the shores of Silver Lake. Homeowners in the area have asked the regional council to determine how many units can be built without altering the nature of the lake. As a member of the council, you have been given the following information:

   The current population of *Daphnia,* 10 animals/liter, presently filters 24% of the lake per day, meaning that it removes this percentage of the algal population per day. This is sufficient to keep the lake essentially clear. Predation—the eating of the algae—will allow the *Daphnia* population to increase to no more than 50 animals/liter. Therefore, 50 *Daphnia*/liter will be available for feeding on the increased number of algae that would result from building the condominiums.

   Using this information, complete Table 34.4.

Table 34.4 *Daphnia* Filtering

Number of *Daphnia*/liter	Percent of Lake Filtered
10	24%
50	

2. The sewage system of the condominiums will add nutrients to the lake. Phosphorus output will be 1 kg per day for every 10 condominiums. This will cause a 30% increase in the algal population. Using this information, complete Table 34.5.

Table 34.5 Cultural Eutrophication

Number of Condominiums	Phosphorus Added	Increase in Algal Population
10	1kg	30%
20		
30		
40		
50		

Conclusion: Cultural Eutrophication

• Assume that phosphorus is the only nutrient that will cause an increase in the algal population and that *Daphnia* is the only type of zooplankton available to feed on the algae. How many condominiums would you allow the developer to build? _____

• What other possible impacts could condominium construction have on the condition of the lake?

_____

1. What type of population would you expect to be the largest in most ecosystems? _____
   Explain. _____
   _____

2. What causes acid rain? _____

3. Why is acid deposition harmful to organisms? _____
   _____

4. Name the type of pollution that results when water from rivers and ponds is used for cooling power
   plants, and explain why it has detrimental effects. _____
   _____

5. Give an example to show that the pollutants studied in this laboratory can have an affect on the human
   population. _____
   _____

6. When excess nutrients enter an aquatic ecosystem, long-term effects can result. Why?
   _____
   _____

7. Describe how the cultural eutrophication study supports the hypothesis that a balance of population
   sizes in ecosystems is beneficial. _____
   _____

8. When pollutants enter an ecosystem, they have far-ranging effects. Use acid rain and a food chain to
   support this statement. _____
   _____
   _____

9. Contrast species composition with species diversity of ecosystems. _____
   _____

10. Suppose among sunflower seeds, a particular variety can germinate despite acidic conditions. What do
    you predict about the survival of that sunflower variety in today's acidic environment compared to the
    rest of the population? _____
    _____

    What do we call a change in a population's phenotype composition due to the presence of an
    environmental agent? _____

Preparing a Laboratory Report

A laboratory report has the sections noted in the outline that follows. Use this outline and a copy of the Laboratory Report Form on page A–3 to help you write a report assigned by your instructor. In general, do not use the words *we, my, our, your, us,* or *I* in the report. Use scientific measurements and their proper abbreviations. (For example, cm is the proper notation for centimeter and sec is correct for seconds.)

1. **Introduction:** Tell the reader what the experiment was about.
   a. **Background information:** Begin by giving an overview of the topic. Look at the Introduction to the Laboratory (and/or at the introduction to the section for which you are writing the report). Do not copy the information, but use it to get an idea about what background information to include.

   For example, suppose you are doing a laboratory report on "Solar Energy" in Laboratory 6 (Photosynthesis). You might give a definition of photosynthesis and explain the composition of white light.

   b. **Purpose:** Think about the steps of the experiment and state what the experiment was about. Tell the independent and dependent variable.

   For example, you might state that the purpose of the photosynthesis experiment was to determine the effect of white light versus green light on the photosynthetic rate. The independent variable was the color of light and the dependent variable was the rate of photosynthesis.

   c. **Hypothesis:** Consider the expected results of the experiment in order to state the hypothesis. It's possible that the Introduction to the Laboratory might hint at the expected results. State this in the form of a hypothesis.

   For example, you might state: It was hypothesized that white light would be more effective than green light for photosynthesis.

2. **Method:** Tell the reader how you did the experiment.
   a. **Equipment and sample used:** Use any illustrations in the laboratory manual that show the experimental setup to describe the equipment and the sample (subject) used.

   For example, for the photosynthesis experiment look at Figure 6.4 and describe what you see. You might state that a 150-watt lamp was the source of white light directed at *Elodea,* an aquatic plant, placed in a test tube filled with a solution of sodium bicarbonate ($NaHCO_3$). A beaker of water placed between the lamp and test tube was a heat absorber.

   b. **Collection of data:** Think about what you did during the experiment such as what you observed or what you measured. Look at any tables you filled out in order to recall how the data were collected and what control(s) were.

   For example, for the photosynthesis experiment, you might state that the rate of photosynthesis was determined by the amount of oxygen released and was measured by how far water moved in a side arm placed in a stopper of the test that held *Elodea.* A control was the same experimental setup except the test tube lacked *Elodea.*

3. **Results:**   Present the data in a clear manner.
   a. **Graph or table:**   If at all possible, show your data in table or graph form. You could reproduce a table you filled in or a graph you drew to show the results of the experiment. Be sure to include the title of the table; do not include any interpretation of the data column in the table.

   For example, for the photosynthesis experiment you might reproduce Table 6.3.

Table 6.3 Rate of Photosynthesis (Green Light)	
	Data
Gross Photosynthesis (mm/10 min)	
White (from Table 6.2)	33.5 mm/10 min
Green	12.5 mm/10 min
Rate of Photosynthesis (mm/hr)	
White (from Table 6.2)	201 mm/hr
Green	75 mm/hr

   Or for 5.2 Effect of Temperature on Enzyme Activity you might show this graph as your results.

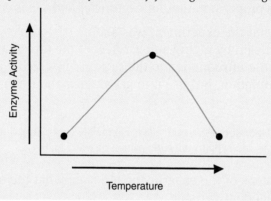

   b. **Description of data:**   Examine your data, and decide what they tell you. Then, below any table or graph add a description to help the reader understand what the table or graph is showing. Define any terms in the table that are not readily understandable.

   For example, below Table 6.3 you might state that these data indicate that the rate of photosynthesis with white light is faster than with green light. Also, you should define gross photosynthesis. Or below the graph that shows the effect of temperature on enzyme activity, you might state that these data show that the rate of enzymatic activity speeds up until boiling occurs and then it drops off.

4. **Conclusion:**   Tell if the data support or do not support the hypothesis.
   a. **Compare the hypothesis with the data:**   Do your data agree or disagree with the hypothesis?

   For example, for the photosynthesis experiment you might state: These results support the hypothesis that white light is more effective for photosynthesis than green light.

   b. **Explanation:**   Explain why you think you obtained these results. Look at any questions you answered while in the laboratory, and use them to help you decide on an appropriate explanation.

   For example, the answers to the questions in 6.2 Solar Energy might help you state that white light gives a higher rate of photosynthesis because it contains all the visible light rays. Green light gives a lower rate because green plants such as *Elodea* do not absorb green light.

   If your results do not support the hypothesis, explain why you think this occurred.

   In this instance you might state that while white light contains all visible light rays and green light is not absorbed by a green plant, the experiment did not support the hypothesis because of failure to use a heat absorber when doing the green light experiment.

Laboratory Report for _____

1. **Introduction**
   a. Background information

   b. Purpose

   c. Hypothesis

2. **Methods**
   a. Equipment and sample used

   b. Collection of data

3. **Results**
   a. Graph or table
   (Place these on attached sheets.)

   b. Description of data

4. **Conclusion**
   a. Compare the hypothesis with the data

   b. Explanation

   c. Conclusion

Metric System

Unit and Abbreviation	Metric Equivalent	Approximate English-to-Metric Equivalents	Units of Temperature

Length

nanometer (nm) $= 10^{-9}$ m $(10^{-3}$ μm)
micrometer (μm) $= 10^{-6}$ m $(10^{-3}$mm)
millimeter (mm) $= 0.001$ (10^{-3}) m
centimeter (cm) $= 0.01$ (10^{-2}) m   1 inch $= 2.54$ cm
   1 foot $= 30.5$ cm
meter (m) $= 100$ (10^2) cm   1 foot $= 0.30$ m
   $= 1{,}000$ mm   1 yard $= 0.91$ m
kilometer (km) $= 1{,}000$ (10^3) m   1 mi $= 1.6$ km

Weight (mass)

nanogram (ng) $= 10^{-9}$ g
microgram (μg) $= 10^{-6}$ g
milligram (mg) $= 10^{-3}$ g
gram (g) $= 1{,}000$ mg   1 ounce $= 28.3$ g
   1 pound $= 454$ g
kilogram (kg) $= 1{,}000$ (10^3) g $= 0.45$ kg
metric ton (t) $= 1{,}000$ kg   1 ton $= 0.91$ t

Volume

microliter (μL) $= 10^{-6}$ liter $(10^{-3}$ mL)
milliliter (mL) $= 10^{-3}$ liter
   $= 1$ cm^3 (cc)   1 tsp $= 5$ mL
   $= 1{,}000$ mm^3 1 fl oz $= 30$ mL
liter (L) $= 1{,}000$ mL   1 pint $= 0.47$ liter
   1 quart $= 0.95$ liter
   1 gallon $= 3.79$ liter
kiloliter (kL) $= 1{,}000$ liter

Units of Temperature

°F   °C

212° — 210 ... 100 — 100°

160° — 160 ... 70 — 71°

134°
131° — 130 ... 57°

105.8° — 110 ... 41°
98.6° — 100 ... 37°

56.66° — 60 ... 13.7°

32° — 30 ... 0 — 0°

-40 ... -40 ... 105.8°

To convert temperature scales:

$$°C = \frac{(°F - 32)}{1.8}$$

$$°F = 1.8°C + 32$$

Common Temperatures

°C	°F	
100	212	Water boils at standard temperature and pressure.
71	160	Flash pasteurization of milk
57	134	Highest recorded temperature in the United States, Death Valley, July 10, 1913
41	105.8	Average body temperature of a marathon runner in hot weather
37	98.6	Human body temperature
13.7	56.66	Human survival is still possible at this temperature.
0	32.0	Water freezes at standard temperature and pressure.

C

Tree of Life

The tree of life depicted in this appendix shows how the three domains of life—Bacteria, Archaea, and Eukarya—are related, and indeed, how all organisms may be related to one another through the evolutionary process.

PROKARYOTES

Laboratories 4 and 14 review the prokaryotes. Prokaryotic organisms are characterized by their simple structure but a complex metabolism. The chromosome of a prokaryote is not bounded by a nuclear envelope, and therefore, these organisms do not have a nucleus. Prokaryotes carry out all the metabolic processes performed by eukaryotes and many others besides. However, they do not have organelles, except for plentiful ribosomes.

DOMAIN BACTERIA

Laboratory 14 covers the diversity of bacteria, which are the most plentiful of all organisms. Bacteria are capable of living in most habitats, and carry out many different metabolic processes. Most bacteria are aerobic heterotrophs, but some are photosynthetic, and some are chemosynthetic. Motile forms move by flagella consisting of a single filament. Their cell wall contains peptidoglycan, and they have distinctive RNA sequences.

DOMAIN ARCHAEA

Domain Archaea is briefly mentioned in Laboratory 14. Archaean cell walls lack peptidoglycan, their lipids have a unique branched structure, and their ribosomal RNA sequences are distinctive. Examples are methanogens and the extremophiles.

DOMAIN EUKARYA

Laboratories 14–17 and 22–24 explore the eukaryotes. Eukaryotes have a complex cell structure with a nucleus and several types of organelles that compartmentalize the cell. Mitochondria that produce ATP and chloroplasts that produce carbohydrate are derived from prokaryotes that took up residence in a larger nucleated cell. They may be unicellular (the majority of the Protists), or multicellular (the plants, fungi, animals). Each multicellular group is characterized by a particular mode of nutrition. Flagella, if present, have a 9 + 2 organization.

PROTISTS

The protists (Laboratory 14) are a catchall group for any eukaryote that is not a plant, fungus, or animal. The protists are currently divided among six supergroups whose evolutionary relationship are being actively investigated. A supergroup is a major eukaryotic group, and six supergroups encompass all members of the domain Eukarya, including all protists, plants, fungi, and animals. Examples of protists include amoebas, the green algae, and the paramecium.

PLANTS

Plants (Laboratories 16 and 17) are photosynthetic eukaryotes that became adapted to living on land. This group includes aquatic green algae called charophytes, which have a haploid life cycle and share certain traits with the land plants.

The land plants exhibit an alternation-of-generations life cycle; protect a multicellular sporophyte embryo; produce gametes in gametangia; possess apical tissue that produces complex tissues; and possess a waxy cuticle that prevents water loss. Examples include mosses, ferns, the conifers, and the flowering plants.

FUNGI

Fungi (Laboratory 15) have multicellular bodies composed of hyphae; usually absorb food and lack flagella; and produce nonmotile spores during both asexual and sexual reproduction. Chytrids (Chytridiomycota) are aquatic fungi with flagellated spores and gametes. Examples include the mushrooms, cup fungim and molds.

ANIMALS

Animals (Laboratories 22–24) are multicellular, usually with specialized tissues and a digestive cavity; they ingest or absorb food; and have a diploid life cycle. This diverse group includes invertebrate organisms such as sponges, worms, mollusks, and insects, and vertebrates such as fishes, reptiles, and humans.

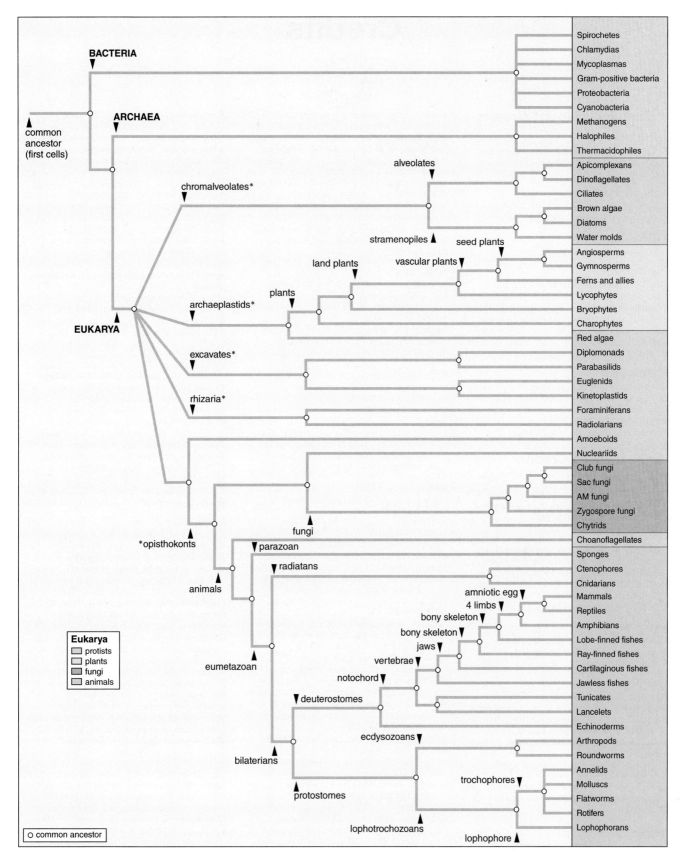

* See the textbook for a description of these supergroups. See the lab manual for a description of the specific organisms studied in the manual.

Credits

Text and Line Art Credits

Laboratory 1
Laboratory adapted from Kathy Liu, "Eye to Eye with Garden Snails." Reprinted by permission of Kathy Liu, Port Townsend WA.

Laboratory 20
Courtesy of Wisconsin FAST PLANTS™ Program. University of Wisconsin—Madison.

Figure 23.12b
After Carolina Biological Supply.

Figure 30.16
From The Complementarity of Structure and Function: A Laboratory Block by A. Glenn Richards. Copyright © 1965 by D.C. Heath, a division of Houghton Mifflin Harcourt Publishing Company. All rights reserved. Reprinted by permission of Holt McDougal, a division of Houghton Mifflin Harcourt Publishing Company.

Figure 34.3
Drawing by Kristine A. Kohn. From Harriett Stubbs' "Acid Precipitation Awareness Curriculum Materials in the Life Sciences: The American Biology Teacher (1963), 45(4), 221. With permission from the National Association of Biology Teachers.

Photo Credits

Laboratory 1
Page 1 (pillbugs): © James Robinson/Animals Animals; 1.1: © Arnulf Husmo/Getty Images; 1.2: Courtesy Leica Microsystems, Inc.

Laboratory 2
Figure 2.5a: © Michael Ross/Photo Researchers, Inc.; 2.5b: © CNRI/SPL/Photo Researchers, Inc.; 2.5c: © Steve Gschmeissner/Photo Researchers, Inc.; 2.6–2.7: Courtesy Leica Microsystems, Inc.; 2.10: © Kevin and Betty Collins/Visuals Unlimited; 2.11: © Dr. Gopal Murti/Photo Researchers, Inc.

Laboratory 3
Figure 3.2a: © Jeremy Burgess/SPL/Photo Researchers, Inc.; 3.6: © Ed Reschke; 3.7–3.9: © Dwight Kuhn.

Laboratory 4
Figure 4.1: © Howard Sochurek/The Medical File/Peter Arnold/Photolibrary; 4.4a, b: Courtesy Ray F. Evert, University of Wisconsin; 4.8 (all): © David M. Phillips/Visuals Unlimited; 4.9a: © Dwight Kuhn; 4.9b: © Alfred Owczarzak/Biological Photo Service.

Laboratory 5
Figure 5.2: © The McGraw Hill Companies, Inc. Jill Braaten, photographer and Anthony Arena, Chemical Consultant.

Laboratory 7
Figure 7.6: © The McGraw Hill Companies, Inc. John Thoeming, photographer.

Laboratory 8
Figure 8.2 (left): © Andrew Syred/Photo Researchers, Inc.; 8.3 (animal-early prophase, prophase, metaphase, anaphase, telophase): © Ed Reschke; 8.3 (prometaphase): © Michael Abbey/Photo Researchers, Inc.; 8.3 (plant-early prophase, prometaphase): © Ed Reschke; 8.3 (prophase, metaphase, anaphase, telophase): © R. Calentine/Visuals Unlimited; 8.4 (top): © Thomas Deerinck/Visuals Unlimited; 8.4 (bottom): © Getty RF; 8.5: © B.A. Palevitz & E.H. Newcomb/BPS/Tom Stack & Assoc.; 8.7 (all): © Ed Reschke; 8.9: © CNRI/SPL/Photo Researchers, Inc.; 8.11: © Ed Reschke/Peter Arnold/Photolibrary; 8.12b (top): © Secchi-Leaque/CNR/SPL/Photo Researchers, Inc.; 8.12b (bottom): © Ed Reschke.

Laboratory 9
Page 110 (anesthetizing flies), 9.5: © Carolina Biological Supply/Phototake.

Laboratory 10
Figure 10.1: © James King-Holmes/SPL/Photo Researchers, Inc.; p. 122 (Down syndrome): © Stockbyte/Veer; 10.3: © CNRI/SPL/Photo Researchers, Inc.; p. 123 (Patau, Edwards): © Custom Medical Stock Photo; p. 123 (Turner): Courtesy UNC Medical Illustration and Photography; p. 123 (Poly X): Courtesy The McElligott Family; p. 124 (Klinefelter): Courtesy Stefan D. Schwarz, http://www.klinefeltersyndrome.org; p. 124 (Jacob): Courtesy The Giles Family; 10.4 (all): © CNRI/SPL/Photo Researchers, Inc.; 10.5 (all): © The McGraw-Hill Companies, Inc.Bob Coyle, photographer.

Laboratory 11
Figure 11.9 (both): © Bill Longcore/Photo Researchers, Inc.

Laboratory 13
Figure 13.1 (left): © Mark Chappell/Animals Animals; 13.1 (right): © OSF/Andrewartha-Sur/Animals Animals; 13.3a: © Michael Tweedie/Photo Researchers, Inc.; 13.3b: © John T. Rotenberry; 13.5: © Bruce Watkins/Animals Animals.

Laboratory 14
Figure 14.1 (bacteria): © Dr. Dennis Kunkel/Phototake; 14.1 (paramecium): © M. Abbey/Visuals Unlimited; 14.1 (morel): © Corbis RF; 14.1 (sunflower): © Photodisc Green/Getty RF; 14.1 (snow goose): © Charles Bush Photography; 14.2: © Science Source/Photo Researchers, Inc.; 14.3 (both): © Kathy Park Talaro; 14.4a: © Dr. Richard Kessel & Dr. Gene Shih/Visuals Unlimited; 14.4b: © Gary Gaugler/Visuals Unlimited; 14.4c: © SciMAT/Photo Researchers, Inc.; 14.5: © Science Source/Photo Researchers, Inc.; 14.6a: Courtesy Steven R. Spilatro, Marietta College, Marietta, OH; 14.6b: © Sherman Thomas/Visuals Unlimited; 14.7 (right): © M.I. Walker/Photo Researchers, Inc.; 14.8: © R. Knauft/Photo Researchers, Inc.; p. 183 (red alga): © Steven P. Lynch; p. 183 (Chlamydomonas): © Dr. Richard Kessel & Dr. Gene Shih/Visuals Unlimited; p. 183 (green alga): © Melba Photo Agency/PunchStock RF; p. 183 (Blepharisma): © Eric Grave/Photo Researchers, Inc.; p. 183 (diatoms): © M.I. Walker/Photo Researchers, Inc.; p. 183 (Ceratium): © D.P. Wilson/Photo Researchers, Inc.; p. 183 (Giardia): © CDC/Dr. Stan Erlandsen and Dr. Dennis Feely; p. 183 (Euglena): © Stephen Durr; p. 183 (Dictyostelium): © Dr. Owen Gilbert, Rice University; p. 183 (amoeba): © Melba Photo Agency/PunchStock; p. 183 (Nonionina): © Astrid & Hanns-Frieder Michler/Photo Researchers, Inc.; p. 183 (radiolarians): © Dennis Kunkel Microscopy/Visuals Unlimited; p. 183 (choanoflagellate): Image by D. J. Patterson, provided courtesy microscope. mbl.edu; 14.9b: © M.I. Walker/Photo Researchers, Inc.; 14.10 (main): © John D. Cunningham/Visuals Unlimited; 14.10 (inset): © Carolina Biological Supply/Visuals Unlimited; 14.11: © D.P. Wilson/Eric & David Hosking/Photo Researchers, Inc.; 14.12a: © Steven P. Lynch; 14.12b: © Gary R. Robinson/Visuals Unlimited; 14.13: © Andrew Syred/Photo Researchers, Inc.; 14.14: © Biophoto Associates/Photo Researchers, Inc.; 14.15c: © Eye of Science/Photo Researchers, Inc.; 14.15d: © Dr. Gopal Murti/Visuals Unlimited; 14.16 (left): © Carolina Biological Supply/Visuals Unlimited; 14.16 (right): © V. Duran/Visuals Unlimited.

Laboratory 15
Figure 15.2 (top): © James Richardson/Visuals Unlimited; 15.2 (bottom): © David M. Phillips/Visuals Unlimited; p. 196 (both): © Carolina Biological Supply/Phototake; p. 197: © Dr. John D. Cunningham/Visuals Unlimited; 15.3a (cup): © Walter H. Hodge/Peter Arnold/Photolibrary; 15.3b (morel): © Corbis RF; 15.3b (ascospores): © James Richardson/Visuals Unlimited; 15.4: © Biophoto Associates/Photo Researchers, Inc.; 15.5b: © Dr. Jeremy Burgess/SPL/Photo Researchers, Inc.; 15.5c: © Stephen Sharnoff/Visuals Unlimited; 15.5d: © Kerry T. Givens; 15.6: © Runk/Schoenberger/Grant Heilman Photography.

Laboratory 16
Figure 16.2 (left): © Dr. John D. Cunningham/Visuals Unlimited; 16.2 (right): © Kingsley Stern; 16.4 (sporophyte): © Heather Angel/Natural Visions; 16.4 (gametophyte): © Steven P. Lynch; 16.5–16.6: © J.R. Waaland/Biological Photo Service; 16.7b: © Ed Reschke; 16.8a: © Ed Reschke/Peter Arnold/Photolibrary; 16.8b: © J.M. Conrarder/Nat'l Audubon Society/Photo Researchers, Inc.; 16.8c: © R. Calentine/Visuals Unlimited; 16.9: © Steve Solum/Bruce Coleman/Photoshot; 16.10 (both): © Steven P. Lynch; 16.11: © Carolina Biological Supply/Phototake; 16.12: © Robert P. Carr/Bruce Coleman/Photoshot; 16.13: © Matt Meadows/Peter Arnold/Photolibrary; 16.14: © The McGraw-Hill Companies, Inc. Carlyn Iverson, photographer; 16.15: © Ken Wagner/Visuals unlimited; 16.16 (top): © R. Knauft/Biology Media/Photo Researchers, Inc.; 16.16 (bottom): © Carolina Biological Supply/Phototake.

Laboratory 17
Page 221 (ovule): © W. P. Armstrong 2004; p. 223 (cycad): © D. Cavagnaro/Visuals Unlimited; p. 224 (all Ginkgo): © Bert Wiklund; p. 225 (top left, Ephedra main photo): Courtesy Dan Nickrent/Southern Illinois Botanical Image Fund; p. 225 (top left, strobili): Courtesy K.J. Niklas; p. 225 (right center): © Steve Robinson/NHPA/Photoshot; p. 225 (conifers, a): © Creatas/Jupiterimages; p. 225 (conifers, b): © Walt Anderson/Visuals Unlimited; p. 225 (conifers, c): © Evelyn Jo Johnson; 17.3 (bottom): © Carolina Biological Supply/Phototake; 17.6a: © J.R. Waaland/Biological Photo Service;

17.6b: © Ed Reschke; 17.7: © Carolina Biological Supply/Phototake; 17.10: © W. P. Armstrong 2004; p. 235 (both): © Evelyn Jo Johnson.

Laboratory 18
Figure 18.2 (shoot tip): © Steven P. Lynch; 18.2 (root tip), 18.3: Courtesy Ray F. Evert, University of Wisconsin; 18.6b: © Carolina Biological Supply/Phototake; 18.7a: © John D. Cunningham/Visuals Unlimited; 18.7b: Courtesy George Ellmore, Tufts University; 18.8a: © Dr. Robert Calentine/Visuals Unlimited; 18.8b: © Evelyn Jo Johnson; 18.8c: © David Newman/Visuals Unlimited; 18.8d: © Terry Whittaker/Photo Researchers, Inc.; 18.9a: © Ed Reschke; 18.9b: Courtesy Ray F. Evert, University of Wisconsin; 18.10 (left): © Carolina Biological Supply/Phototake; 18.10 (right): © Kingsley Stern; 18.11a: Evelyn Jo Johnson; 18.11b: © Science Pictures Limited/Photo Researchers, Inc.; 18.11c, d: © The McGraw Hill Companies, Inc. Carlyn Iverson, photographer; 18.13b: © Carolina Biological Supply/Phototake.

Laboratory 19
Figure 19.2: © Evelyn Jo Johnson; 19.3 (left): © J.R. Waaland/Biological Photo Service; 19.5b: © Andrew Syred/SPL/Photo Researchers, Inc.

Laboratory 20
Figure 20.2: © Kingsley Stern; 20.5b: © Runk/Schoenberger/Grant Heilman Photography; 20.7a: © Robert E. Lyons/Visuals Unlimited; 20.7b: © Sylvan Whittwer/Visuals Unlimited; 20.9: © Nigel Cattlin/Photo Researchers, Inc.

Laboratory 21
Figure 21.1: © Corbis RF; 21.4: © Barry Blackburn/Shutterstock images; 21.5a: © Robert Maier/Animals Animals; 21.5b: © Andrew Darrington/Alamy; 21.5c: © Anthony Mercieca/Photo Researchers, Inc.; 21.6 (proembryo, globular, heart): Courtesy Dr. Chun-Ming Liu; 21.6 (torpedo): © Biology Media/Photo Researchers, Inc.; 21.6 (mature embryo): © Jack M. Bostrack/Visuals Unlimited; 21.7a (blossoms): © Joe Munroe/Photo Researchers, Inc.; 21.7a (nuts): © Inga Spence/Photo Researchers, Inc.; 21.7b (tomatoes): © Werner Layer/Animals Animals; 21.7b (slice): © Josh Westrich/zefa/Corbis; 21.7c (both): © Dwight Kuhn; 21.7d (blossoms): © William Weber/Visuals Unlimited; 21.7 (berry): Courtesy Robert A. Schlising; 21.8 (germinating bean plant): © Ed Reschke; 21.9 (left): © James Mauseth; 21.9 (right): © Scott Sinklier/AgStock Images/Corbis.

Laboratory 22
Figure 22.3a: © Amar and Isabelle Guillen, Guillen Photography/Alamy; 22.3b: © Andrew J. Martinez/Photo Researchers, Inc.; 22.3c: © OSF/Bernard/Animals Animals; 22.5: © Carolina Biological Supply/Visuals Unlimited; 22.6 (both): © Kim Taylor/npl/Minden Pictures; 22.7a: © Azure Computer & Photo Services/Animals Animals; 22.7b: © Ron Taylor/Bruce Coleman/Photoshot; 22.7c: © Therisa Stack; 22.7d: © Amos Nachoum/Corbis; 22.9: © Carolina Biological Supply/Phototake; 22.10 (planarian eating, top): Photography by Marc C. Perkins, Orange Coast College, Costa Mesa, CA; image blending by Heather Bartell, Huntington Beach, CA; 22.10 (planarian, bottom): © Tom E. Adams/Peter Arnold/Photolibrary; 22.12a: © C. James Webb/Phototake; 22.12b: © Carolina Biological Supply/Visuals Unlimited; 22.13: © Biophoto Associates/Getty Images; 22.14: © SPL/Photo Researchers, Inc.; 22.15a, c: © Larry Jenson/Visuals Unlimited; 22.16: © Carolina Biological Supply/Phototake; 22.17: © Vanessa Vick/The New York Times/Redux; 22.18: © Wim van Egmond/Visuals Unlimited.

Laboratory 23
Figure 23.1a: © Fred Bavendam/Minden; 23.1b: © NHPA/Photoshot; 23.1c: © Rosemary Calvert/Getty Images; 23.1d: © Douglas Faulkner/Photo Researchers, Inc.; 23.3b, 23.4b: © Ken Taylor/Wildlife Images; 23.6a: © Roger K. Burnard/Biological Photo Service; 23.6b: © R. DeGoursey/Visuals Unlimited; 23.6c: © Brian Parker/Tom Stack & Assoc.; 23.6d: © C.P. Hickman/Visuals Unlimited; 23.8b: © Ken Taylor/Wildlife Images; 23.9: © John Cunningham/Visuals Unlimited; 23.10a (honeybee): © R. Williamson/Visuals Unlimited; 23.10a (millipede): © Bill Beatty/Visuals Unlimited; 23.10a (centipede): © Adrian Wenner/Visuals Unlimited; 23.10b (spider): © W.J. Weber/Visuals Unlimited; 23.10b (scorpion): © David M. Dennis; 23.10b (horseshoe crab): © E.R. Degginger/Photo Researchers, Inc.; 23.10c (crab): © Tom McHugh/Photo Researchers, Inc.; 23.10c (shrimp): © Alex Kerstitch/Visuals Unlimited; 23.10c (barnacles): © Kjell Sandved/Visuals Unlimited; 23.11b: © Ken Taylor/Wildlife Images; 23.15a: © Daniel Gotshall/Visuals Unlimited; 23.15b: © Hal Beral/Visuals Unlimited; 23.15c: © Robert Dunne/Photo Researchers, Inc.; 23.15d: © Neil McDan/Photo Researchers, Inc.; 23.15e: © Robert Clay/Visuals Unlimited; 23.15f: © Alex Kerstitch/Visuals Unlimited; 23.16 (both): © BiologyImaging.com.

Laboratory 24
Figure 24.2: © Rick Harbo; 24.3: © Heather Angel/Natural Visions; 24.4: © Stan Sims/Visuals Unlimited; 24.5 (shark): © Hal Beral/Visuals Unlimited; 24.5 (fish): © Patrice/Visuals Unlimited; 24.5 (frog): © Rod Planck/Photo Researchers, Inc.; 24.5 (turtle): © Suzanne and Joseph Collins/Photo Researchers, Inc.; 24.5 (bird): © Robert and Linda Mitchell Photography; 24.5 (fox): © Craig Lorenz/Photo Researchers, Inc.; 24.6: © Rod Planck/Photo Researchers, Inc.; 24.7b: © Carolina Biological Supply/Phototake; 24.8, 24.9b, 24.10b, 24.11b: © Ken Taylor/Wildlife Images.

Laboratory 25
Figure 25.1 (simple squamous, pseudostratified ciliated, simple cuboidal, simple columnar, cardiac, skeletal, neuron, adipose, bone, cartilage: © Ed Reschke; 25.1 (smooth, dense): © The McGraw Hill Companies, Inc. Dennis Strete, photographer; 25.1 (blood): © National Cancer Institute/Photo Researchers, Inc.; pp. 356–358 (all), and p. 359 (loose): © Ed Reschke; p. 359 (dense): © The McGraw Hill Companies, Inc./Dennis Strete, photographer; p. 360 (both), p. 361 (hyaline), 25.2a–d: © Ed Reschke; 25.2e: © R. Kessel/Visuals Unlimited; p. 363 (skeletal, cardiac): © Ed Reschke; p. 364 (smooth): © The McGraw Hill Companies, Inc. Dennis Strete, photographer; 25.3a (neuron); 25.4: © Ed Reschke; p. 367 (skin): © John Cunningham/Visuals Unlimited; p. 368 (a): © The McGraw Hill Companies, Inc. Dennis Strete, photographer; p. 368 (b): © National Cancer Institute/Photo Researchers, Inc.; p. 368 (c, d): © Ed Reschke.

Laboratory 26
Figure 26.3b, 26.6: © Ken Taylor/Wildlife Images.

Laboratory 27
Page 383: © SIU/Visuals Unlimited; 27.6: © Ralph T. Hutchings/Visuals Unlimited; 27.8, 27.11: © The McGraw Hill Companies, Inc. Carlyn Iverson, photographer.

Laboratory 28
Figure 28.2 (gastric gland): © Ed Reschke/Peter Arnold/Photolibrary.

Laboratory 29
Figure 29.4: © Ilene MacDonald/Alamy; 29.6 (emphysema): © CMSP/Getty Images; 29.6 (normal lung): © Dr. Keith Wheeler/Photo Researchers, Inc.; 29.8: Photo courtesy of Phipps & Bird, Inc., Richmond, VA.

Laboratory 30
Figure 30.2 (all): Courtesy Dr. J. Timothy Cannon; 30.5: © Kage-mikrofotografie; 30.7a: Copyright © 2005 The Regents of the University of California. All Rights Reserved. Used by permission; 30.7b: © P.H. Gerbier/SPL/Photo Researchers, Inc.; 30.8 (ommatidia): © S.L. Flegler/Visuals Unlimited.

Laboratory 31
Figure 31.3b: © Dr. Richard Kessel & Dr. Randy Kardon/Visuals Unlimited, 31.3d: © Ed Reschke; 31.3c: © Susumu Nishinaga/Photo Researchers, Inc.; 31.7: © The McGraw Hill Companies, Inc./Dennis Strete, photographer; 31.8–31.9: © Ed Reschke; 31.13: © Biology Media/Photo Researchers, Inc.

Laboratory 32
Page 461 (zygote): © Anatomical Travelogue/Photo Researchers, Inc.; p. 461 (embryo, 1 week): © Bettman/Corbis; p. 461 (embryo, 8 weeks): © Neil Harding/Getty Images; p. 461 (fetus, 3 months): © Petit Format/Photo Researchers, Inc.; p. 461 (fetus, 5 months): © John Watney/Photo Researchers, Inc.; 32.1a–g: © Carolina Biological Supply/Phototake; 32.1h: © Ed Reschke/Peter Arnold/Photolibrary; 32.2a: © Martin Rotker/Phototake; 32.2b–h: © Carolina Biological Supply/Phototake; 32.2i: © Alfred Owczarzak/Biological Photo Service; 32.6a–32.9a: © Carolina Biological Supply/Phototake; 32.12h–i: © Petit format/Photo Researchers, Inc.

Laboratory 34
Figure 34.1: © Michael Gadomski/Animals Animals; 34.3: © NOAA/Visuals Unlimited.

Index

Note: page references followed by *f* and *t* refer to figures and tables, respectively.